Mine Drainage

By H.C. Behrs

with an introduction by Kerby Jackson

This work contains material that was originally published in 1896.

This publication was created and published for the public benefit, utilizing public funding and is within the Public Domain.

This edition is reprinted for educational purposes and in accordance with all applicable Federal Laws.

Introduction Copyright 2014 by Kerby Jackson

Introduction

It has been over one hundred and ten years since the State of California released it's important publication "Mine Drainage, Pumps, Etc". First released in 1896 this important volume has now been out of print for over a century and has been unavailable to the mining community since those days, with the exception of expensive original collector's copies and poorly produced digital editions.

It has often been said that "*gold is where you find it*", but even beginning prospectors understand that their chances for finding something of value in the earth or in the streams of the Golden West are dramatically increased by going back to those places where gold and other minerals were once mined by our forerunners. Despite this, much of the contemporary information on local mining history that is currently available is mostly a result of mere local folklore and persistent rumors of major strikes, the details and facts of which, have long been distorted. Long gone are the old timers and with them, the days of first hand knowledge of the mines of the area and how they operated. Also long gone are most of their notes, their assay reports, their mine maps and personal scrapbooks, along with most of the surveys and reports that were performed for them by private and government geologists. Even published books such as this one are often retired to the local landfill or backyard burn pile by the descendents of those old timers and disappear at an alarming rate. Despite the fact that we live in the so-called "Information Age" where information is supposedly only the push of a button on a keyboard away, true insight into mining properties remains illusive and hard to come by, even to those of us who seek out this sort of information as if our lives depend upon it. Without this type of information readily available to the average independent miner, there is little hope that our metal mining industry will ever recover.

This important volume and others like it, are being presented in their entirety again, in the hope that the average prospector will no longer stumble through the overgrown hills and the tailing strewn creeks without being well informed enough to have a chance to succeed at his ventures.

Kerby Jackson
Josephine County, Oregon
July 2014

MINE DRAINAGE, PUMPS, ETC.

By HANS C. BEHR, Mechanical Engineer.

INTRODUCTORY.

CONTROLLING THE WATER IN MINES.

Mines worked through shafts are subject to flooding by penetrating water-bearing ground. Even if not encountered at first, water is liable to be struck at any time, and appliances should therefore always be in readiness to handle it. For moderate inflow the water can generally be hoisted in bailing-tanks without encroaching too much on the time required for other hoisting operations. When, however, a large permanent flow is struck, the entire hoisting capacity may be required for bailing until a suitable pumping-plant can be installed.

In deciding upon the capacity of a proposed pumping-plant, it is necessary to ascertain as nearly as possible the maximum quantities of water that may be encountered at different levels. In well-opened mines this can generally be done without difficulty; not so in sinking a shaft in new ground. But, if other mines are adjacent, a record of their water production is a very good guide.

The pumping-plant should be able to handle a much larger quantity of water than any recorded maximum, so that a considerable increase can be taken care of without resorting to bailing. Bailing arrangements should, however, also be in readiness to meet at once any extraordinary increase that may occur at any time.

The water is generally found in a mine at various levels, and, where economy rather than simplicity is the object, any considerable quantity of water should be collected and led to pumps at the levels where it issues, and not be permitted to first find its way to the bottom, from where it would have to be raised the entire height to the surface, thereby increasing the cost of pumping in proportion to the increased lift for that part of the water.

In many mines, the quantity of water varies, not only as new bodies are tapped or opened ones drained, but also with the seasons of the year; and observations extending over at least a year should therefore be available for fixing on the capacity of a pumping-plant. In the Kennedy Mine, Amador County, California, the water production varies from 75,000 gals. per day during the dry season to 150,000 gals. during the wet season, and is handled by bailing-tanks.

The generally variable nature of water inflow necessitates a corresponding variation in the work of the water-raising apparatus. Bailing adapts itself most readily to such variation, as it gives equal though low mechanical efficiency for a very wide range of capacity. With pumps the case is different, since the number or length of strokes can only be varied economically within certain limits.

Mine pumping-plants should be designed and constructed with the aim of obtaining the greatest possible security against breakdowns, and at the same time admitting of rapidly making repairs and replacing worn parts. If possible, the pumping-plant should also be so designed that it will give the highest mechanical efficiency for that rate of flow which prevails most of the time and furnishes the largest proportion of all the water. Large excess over this, if known to be of short duration, can be taken care of by bailing-tanks or cheaper and less efficient emergency-pumps. A sudden influx of large quantities of water can be handled by bailing with powerful direct-acting hoisting engines, which bring the tanks to the surface rapidly. Often a mechanically less efficient plant may, owing to other conditions, prove to be commercially the most efficient.

Timbered shafts are universally used in the West. They are generally arranged with three compartments—two for hoisting, and one for the pumps. The latter should be partitioned off from the hoisting-compartment, so that it can be made to serve as upcast to ventilate the bottom of the shaft, because the pump-shaft is usually warmer than the hoisting-compartments, due either to steam-pipes for operating direct-acting pumps, or to the warm water in the column-pipes.

Where the mine has two separate shafts connected below, so that one serves as upcast shaft, the pumps should, if possible, be placed in the latter.

The kind of pumps, source of power, and the means of transmitting this to the pumps underground, depend on surrounding conditions, and only a careful study of these can decide the proper kind of plant to be adopted.

SECTION I.

GENERAL FEATURES OF MINE PUMPING-PLANTS.

CHAPTER I.

Preliminary Remarks on Mining Pumps.

1.1.01.* *Water-Raising Machines Used in Mining.* The pumps used in pumping out mines are chiefly reciprocating. Centrifugal pumps find some application for low lifts, and generally in open workings. Of other water-raising appliances used, the bailing-tank is the principal one, and finds a wide range of application. Pulsometers are used as a low-lift auxiliary to pumps, etc. The same is true of ejectors. It is also occasionally possible to employ siphons for raising water over an eminence.

1.1.02. Reciprocating pumps may be divided into plunger, piston, and bucket (or lift) pumps.

1.1.03. The oldest pump used in mines is the draw-lift pump, with a valved bucket working in the barrel. The modern forms of this type of pump are much used for sinking where the pumps are operated by rods. They are not suitable for working against heads of over 200'. The pump-barrels and bucket-packing also are exposed to great wear, particularly when the water carries sand. The bucket cannot be packed while the pump is running. Nevertheless, their use in mining is very extensive. In the Cornish system they are generally arranged so that the bucket can be hauled up through the column-pipe for repairs.

1.1.04. Plunger, or force, pumps are suitable for much higher lifts. Vertical, single-acting plungers are the typical form of the modern pumprod system. In these the plunger-packing can be taken up while the pump is running, and, as the packing is located at the highest part of the pump-barrel, away from the course of the water, little sand or grit is liable to reach it. The pump can, therefore, run quite a long time before repairs are required at that point.

1.1.05. Horizontal, double-acting plungers are generally used for high-pressure, direct-driven pumps. These are arranged either with or without cranks and flywheels. In the former case they are called direct-acting pumps; in the latter, rotative pumps. Flywheel, or double-crank pumps of this class, with mechanically actuated valves designed by Riedler, have been used continuously for single lifts of 1,300'.

1.1.06. Piston pumps are suitable only for lower pressures. The piston-packing and cylinder are subject to wear, while the pump must be stopped and the piston taken out to pack it.

1.1.07. Centrifugal pumps are, as generally constructed, only suitable for low lifts, but are capable of handling large volumes of water.

---

*The numbers at the beginning of the paragraphs are so arranged that the first figure denotes the *Section*, the next two figures the *Chapter*, and the last two the *Paragraph*. Thus, 1.5.17, means the 17th Paragraph in Chapter V of Section I.

As they have no valves, the water may contain large quantities of sand and gravel without impairing the efficiency of the pumps while they last. The capacity of centrifugal pumps can only be varied economically within very narrow limits, as they require to be run at a certain speed to pump against a given head.

1.1.08. Injectors, pulsometers, etc., are not economical water-raising machines, and can only be considered as temporary appliances or as substitutes for better apparatus during its repair. The steam used to operate them acts so that a large proportion of its energy is wasted by being applied to heat the water which they deliver. For admissible application, see 2.3.33.

1.1.09. *Conditions Affecting the Working of Pumps.* The operation of pumps is influenced by many conditions: the height above sea-level; the barometric pressure; the temperature of the water pumped; the size, length, and course of the suction- and delivery-pipes; the area, weight, and lift of valves; etc. The height above sea-level, and therefore the existing atmospheric pressure, limits the height to which water may be drawn by suction or the velocity with which it will follow the piston or plunger, thereby limiting the speed of the pump for a given suction lift. The higher the temperature of the water the less will be the admissible suction lift, because if the reduction of pressure at the upper end of the suction-pipe be sufficient, the water will begin to boil at a temperature much below that at which it would boil under atmospheric pressure, and give off steam, which will fill the pump-barrel, instead of the water doing so. The suction lift must therefore be kept so low that the pressure will be sufficient to prevent steam from forming. The suction height is the vertical distance from the level of the suction water to the highest point of the piston displacement and spaces connected with it. The greater also the head pumped against the less is the admissible speed, because with the longer column shocks are more severe.

1.1.10. The influence of the pipes connected with pumps on their action is treated of in the succeeding chapter.

1.1.11. The effect of different constructions of valves is also relegated to another chapter, and is further considered in connection with the various pump constructions described in other parts of this paper.

1.1.12. *Starting Pumps.* In starting a reciprocating pump it is necessary to remove the air from the pump-barrel and the spaces communicating with it. Where these waste spaces are large compared with the piston or plunger displacement, and the head pumped against is high, the air, particularly in high altitudes, will not be sufficiently compressed on the working-stroke to lift the discharge-valve and escape into the discharge-pipe in case it is full of water. Again, if atmospheric pressure exist at the beginning of the suction-stroke, the air in the pump may not be sufficiently expanded and lowered in pressure on completion of the suction-stroke so that the outer air can lift the water in the suction-pipe, cause it to force open the suction-valve, and enter the pump.

1.1.13. *Priming and Draining.* The operation of expelling the air from a pump and filling it with water is called *priming*. Means are generally provided in a by-pass pipe with a cock for priming the pump from the discharge-pipe in case the latter already contains water, the escape of air being then generally effected through a cock near the highest part of the space communicating with the working-barrel. When no air-escape is provided, the air will be forced out through the discharge-valve into the discharge-pipe, as soon as the pump is put in motion. When there is no water in the discharge-pipe, pumps with large waste spaces generally require independent means for priming them, such as an opening with a funnel, through which water may be poured. Pumps placed below the supply from which they draw do not require priming. Pumps and pipes should be fitted with means for draining them to prevent freezing and to draw off sediment.

1.1.14. *Methods of Driving Pumps.* Main pumps for shafts are either operated through rods from a motor or engine at the surface, like in the familiar Cornish system of pumping, or, as in more modern methods, by transmitting power to motors directly coupled to the pumps, either through pipes, in the form of steam, compressed air, or pressure water, or as electricity through wires. Some one of these modes of transmission is required, where, as is usually the case, pumps or other machines are used to raise water from winzes or low places, and force it up to the nearest station-tank at the pump-shaft. Hand pumps are also similarly used to raise small quantities of water from low places into launders in the drifts. Pumps should be started in motion gradually, and not in such a manner as results from throwing them suddenly into gear with driving machinery already in motion.

1.1.15. *Distribution of Pumps.* The distribution of pumps along the line of the shaft depends, first of all, upon the lift allowable for the individual pumps. This condition determines the spacing of pumps in the Cornish system, in which they are generally 200' to 250' apart. Where, however, the pumps are capable of working against a very high head, as in some of the modern direct-acting types, they should, for economical reasons, be spaced according to the levels at which water issues.

1.1.16. Though the water which is generally encountered in sinking a shaft does not always issue at the lowest point, it is nevertheless usually necessary, if pumps are put in, to have the lowest pump so arranged that it can follow close to the shaft bottom as it goes down, in order to be prepared to handle any water that may be struck there, or which may flow down from upper levels. Pumps used for this purpose are called sinking-pumps.

1.1.17. When the sinking-pump has been lowered so far that the limit of its admissible lift is reached in raising water to the next higher pump, another permanent pump is put in near the bottom of the shaft. The sinking-pump then delivers its water to this lowest fixed pump, and is made ready to proceed with further sinking.

1.1.18. *Desirable Features of Mining Pumps.* The welfare of a mine, if subject to influx of water, depends largely upon the reliability of the pumps. These should therefore be so constructed and arranged that there may be the least possible chance of their failure. The following are some of the main desirable features: (1) They should be capable of running a long time without requiring packing, repairs, or adjustment; (2) They should, if possible, be capable of being operated and repaired under water. This is particularly desirable in the lowest, or sinking, pump; (3) They should be able to handle sandy and sometimes acid water, without too rapid wear or deterioration.

1.1.19. In addition, they should be so arranged with reference to the driving power that they can be operated for a wide range of capacities to adapt them to the varying conditions of the water production of the mine.

CHAPTER II.

Pipes.

1.2.01. Pipes used in connection with mining pumps are, firstly, those for conveying the water handled by the pumps, constituting in reality a part of the pumps; and, secondly, those used for conveying power to the motors operating the pumps, in the form of pressure water, steam, or compressed air. While the main object of this chapter is to treat more at length of the former, it is proper, though perhaps to a more limited extent, to consider also the latter, as they are intimately connected with the operation and care of pumps in mines.

1.2.02. The suction- or inlet-pipes and the discharge-pipes of a pump or hydraulic pumping-engine affect the working of these to a great extent, and it is necessary to consider them in a different manner from ordinary continuous-flow water-pipes, in order to fix upon the most advantageous arrangement, size of pipes and pumps, and admissible speed of the latter.

1.2.03. *Material of Pipes.* Cast-iron, formerly used exclusively for larger pipes subjected to pressure underground, is now rarely employed in American mines for this purpose. While this material is less subject to corrosion than either wrought-iron or steel, the pipes made from it have to be very heavy with a proper factor of safety to withstand the pressure, and the sections are therefore more difficult to handle.

1.2.04. The cheapness of wrought-iron pipes, their greater security under water-hammer, and the facility with which sections of any length can be cut off and fitted to place at the mine, have led to their almost universal use in general practice.

1.2.05. In cases where the corrosive action of the mine-water on the iron pipes is very strong, and their destruction rapid, pipes of other materials have been used.

1.2.06. At the Barranca Mine, Mexico, drawn copper tubes were put in at great cost. Wooden pipes, where the pressure is not great, or, for higher pressure, iron pipes lined with wood, are sometimes used.

1.2.07. *Wrought-Iron Pipe.* Formerly, column-pipes larger than 14" in diameter for mine use were made of boiler plate, riveted hot, often with butt-joints and lap-strips; the rivets being countersunk on the inside. Now, iron and steel lap-welded tubes up to 24" diameter can be obtained, and manufacturers are preparing machinery for sizes up to 30" in diameter.

1.2.08. Welded pipes are either lap-welded or butt-welded. The latter should be used only for smaller sizes, and for moderate pressure, as they are liable to split open at the weld. Lap-welded tubes or hot-riveted pipes of boiler plate are the only wrought pipes suitable for pump-columns in shafts, and for all purposes where heavy pressures and water-hammer are encountered. Lap-welded tubes are also used for steam- and compressed-air-pipes. Iron boiler plates, including those of which welded tubes are made, have less strength in the direction of their width than their length, which latter is the direction of strain when manufactured into a welded pipe. Sheets of mild steel are homogeneous in this respect, besides possessing greater strength; therefore, for larger sizes steel pipes are nearly always used. Welded pipes may be obtained in lengths up to 20'. For the sake of facility in handling, however, the sections composing a line of pipe in a mine are usually not over 16' in length.

1.2.09. Ordinary pipes, either lap- or butt-welded, having screwed ends for connection by threaded flanges or couplings, are classified by manufacturers according to nominal inside diameter. The actual diameter is generally in excess of nominal diameter. Lap-welded tubes connected by other means than the regular coarser pipe threads, that is, by flanges shrunk or riveted on, or by leaded joints, or finely threaded sleeves or flanges, are known according to their exact outside diameter. Such pipe is generally called tubing; when connected by fine thread, it is known as casing.

1.2.10. The different sizes of lap-welded tubing can each be obtained of different thickness of material to suit different pressures. The following table of standard sizes and thickness may prove useful for reference:

TABLE I.

Dimensions, etc., of Lap-Welded Tubes of a Prominent Manufacturer.

Outside Diameter of Pipe, in inches.	Inside Diameter of Pipe, in inches.	Thickness of Metal.		Weight per Foot, in pounds.	Bursting Pressure, lbs. per sq. inch.
		Birmingham Wire Gauge.	Inches.		
3	2.73	10	.135	4.05	5,900
4	3.70	9	.150	6.00	4,800
5	4.67	8	.165	8.40	4,200
6	5.64	7	.180	11.00	3,800
7	6.64	7	.180	13.00	3,200
8	$7\frac{5}{8}$	--	$\frac{3}{16}$	15.65	2,900
9	$8\frac{1}{2}$	--	$\frac{1}{4}$	23.10	3,500
10	$9\frac{1}{2}$	--	$\frac{1}{4}$	25.75	3,100
12	$11\frac{1}{2}$	--	$\frac{1}{4}$	31.00	2,600
13	$12\frac{1}{2}$	--	$\frac{1}{4}$	33.40	2,400
14	$13\frac{1}{2}$	--	$\frac{1}{4}$	36.35	2,220
15	$14\frac{1}{2}$	--	$\frac{1}{4}$	39.00	2,070
16	$15\frac{1}{2}$	--	$\frac{1}{4}$	42.00	1,930
18	$17\frac{3}{8}$	--	$\frac{5}{16}$	58.40	2,150
20	$19\frac{3}{8}$	--	$\frac{5}{16}$	65.15	1,970
22	$21\frac{3}{8}$	--	$\frac{5}{16}$	85.00	1,750
24	$23\frac{1}{4}$	--	$\frac{3}{8}$	93.50	1,930

1.2.11. The thickness given in the table is known as standard. Pipe can be made one or two gauges lighter, but would not come any cheaper per foot. On special orders, the pipe can be made thicker to almost any extent. Numerous experiments have demonstrated that in properly welded pipes the weld is practically as strong as the rest of the metal.

1.2.12. *Heavy Riveted Pipes* used for pump-columns should, if possible, be made of mild steel, because then they can usually be made from a single sheet, requiring only one longitudinal joint. Steel admits of this method of construction, because, as stated in 1.2.08, it has about the same strength across the sheet as lengthwise. Iron, being fibrous in its nature, and having less strength across the sheet, should therefore be bent so that the fiber runs around the pipe, in order to secure the greatest strength. As the sheets are limited in width, this necessitates making a wrought-iron riveted pipe section of several sheets riveted together by circular seams. Longitudinal seams should be double-riveted.

1.2.13. Heavy riveted column-pipe sections are usually connected by cast- or wrought-iron flanges riveted to the sections. Where laid on the ground and not liable to be disturbed, they are often connected by lead-caulked joints, with cast or wrought-iron rings to hold the lead.

1.2.14. Riveted sinking-columns, inside of which a pumprod works as in the Cornish system, should have the rivets countersunk on the inside, and the circular seams made as butt-joints with outside lap-strips, so that the lift-pump-bucket can be drawn up and lowered through the column-pipe wihout catching on obstructions.

1.2.15. *Light Riveted Pipes** are used principally for water supply for power or for hydraulic mining where the pressure is constant and where the pipe is not subject to being crowded out of line, as in a shaft. The sheets, rarely thicker than $\frac{1}{8}''$, are riveted up cold, often, on account of transportation, at the point where put in use. They are now almost universally made of steel. If made of iron, the sheets must, for reasons previously stated, be bent in the direction of the fiber. The longitudinal seams should be double-riveted. The lengths of pipe, except for very heavy pressure (when both internal and external sleeves caulked with lead are used), are generally joined by simply slipping the ends into each other like the sections of a stovepipe. The sections are made larger at one end for this purpose. These pipes will stand considerable pressure when it is constant, but they are not suitable for withstanding any water-ram. Iron pipes of this kind have been subjected continuously for many years to a constant fiber-stress of 17,000 lbs. per square inch on the section of the sheet. At the line of the rivets, where their insertion reduces the iron section of the sheets, the stress would in that case be about 22,000 lbs.

1.2.16. *Wooden Pipes*, made of staves like a continuous barrel, hooped with steel bands, as in Fig. 1, have been in use for a number of years in connection with irrigation and gravity water supplies for cities. They are economical, especially for light pressures, and may be used for pressures of 200', if steady, the spacing of the bands varying with the pressure. They are very smooth on the inside, and offer little resistance to the flow of water. They are not suitable for pump-columns, but there are cases in mining where this class of pipe can be used to advantage. The water does not come in contact with the steel bands, and cannot corrode them; and if the pipe is continuously filled with water, the wood will at all times be saturated and cannot decay. Where

---

*This class of pipe has been ably discussed by Hamilton Smith in his "Hydraulics," and also by Aug. J. Bowie in his "Practical Treatise on Hydraulic Mining." Numerous examples of (completed) pipe-lines, with experiments on flow, leakage, and stress on material, are given in those two works.

a pipe-line is required in a mountainous country, difficult of access, it is an advantage that the parts of which this pipe is composed are all light, can be closely packed, and easily transported. The entire pipe-line can be taken down without any injury to its parts, and be re-erected

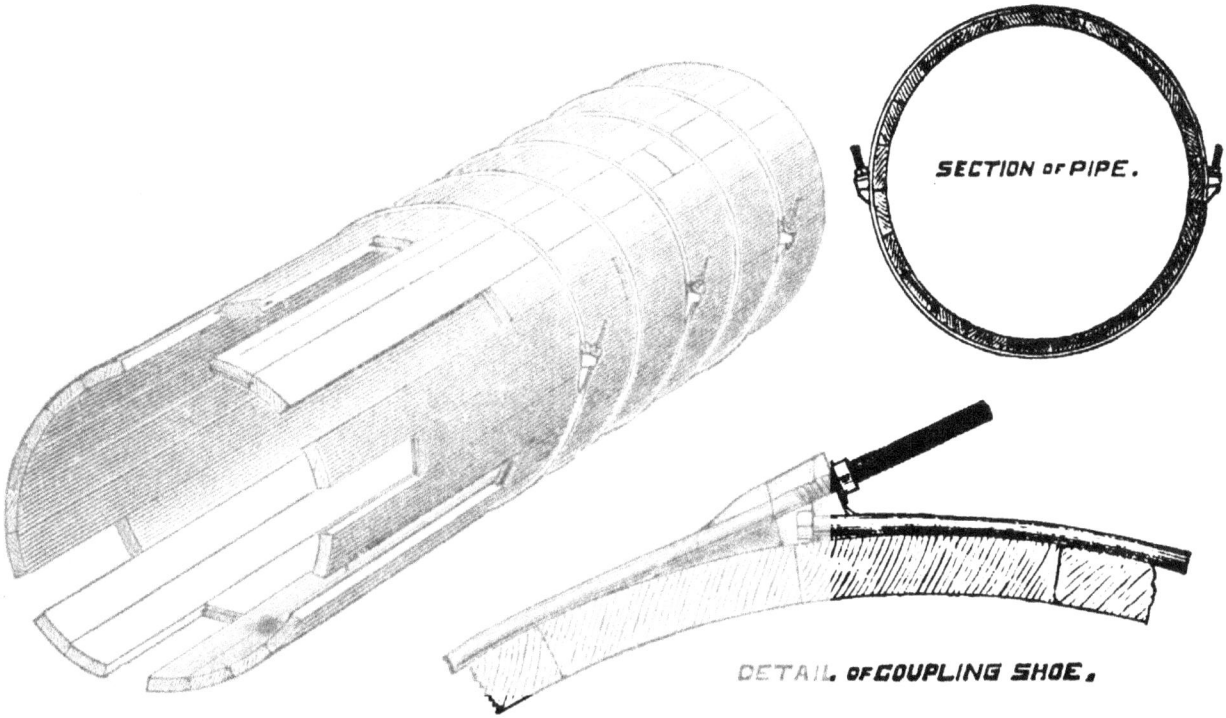

Fig. 1.

Fig. 2.

elsewhere. These pipes do not contract and expand with heat, and can, if necessary, be left on the surface. The pipe is very rigid and not readily flattened by snow or landslides. Fig. 2 is a view of a completed pipe-line of this kind.

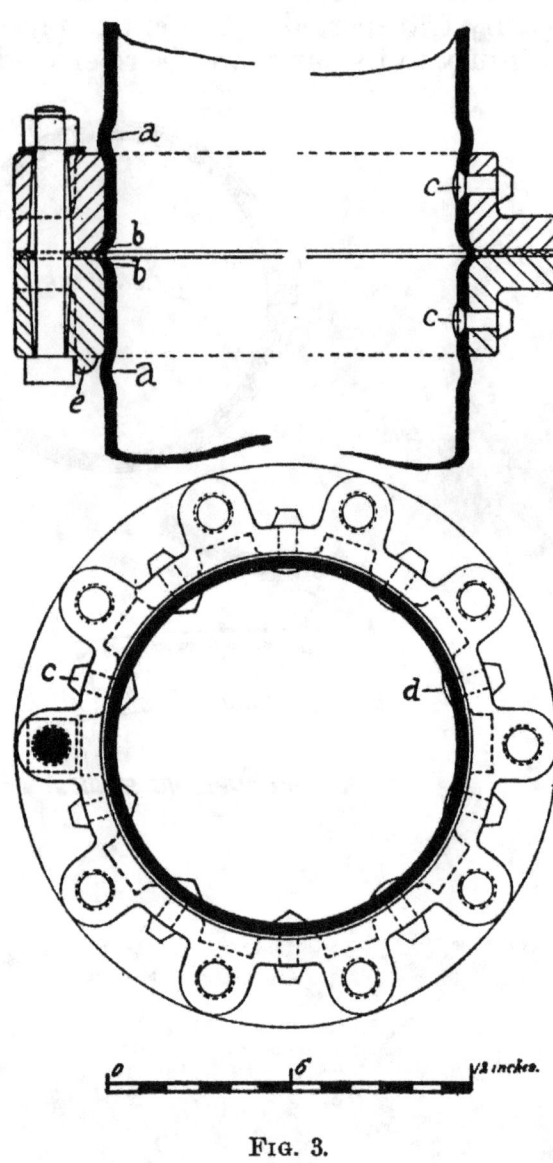

Fig. 3.

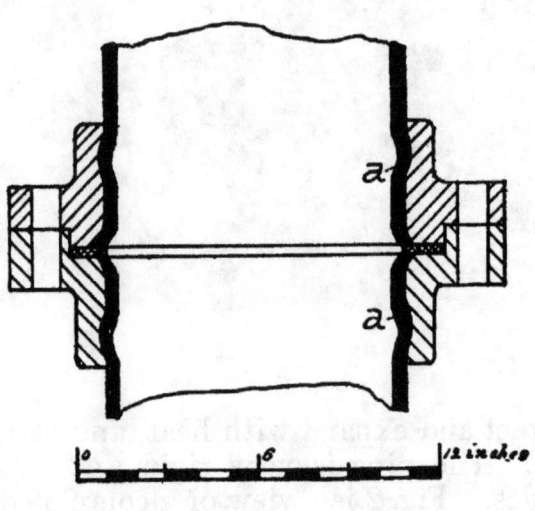

Fig. 4.

1.2.17. Pipe Connections. Wrought-iron pipes are connected principally by flanges, screwed ends and couplings, leaded or cemented sleeves, or by simply slipping the smaller end of one length into the larger end of the next. In underground work, shafts, etc., welded tubes with flanges or screwed connections are used for water-, steam-, and air-pipes. Leaded joints are only used where a water-pipe is permanently located and not liable to be disturbed, such as pipes for water distribution. They are not suitable for pump-columns in shafts or inclines.

1.2.18. Flanges are the usual means of connecting pipes underground. They are commonly made of cast-iron, and, in case of welded pipe, either screwed or shrunk on the ends of the tube, which, in the latter case, is expanded behind the flange, as at a, Fig. 3, and then beaded over in front, as at b. Instead of expanding the pipe behind the flanges, it is preferable to have the bore of the flange recessed, and to hammer the pipe into the recess, as shown at a, Fig. 4. This gives a firmer hold on the flange. It is sometimes necessary to put in rivets, as at c, Fig. 3, in case of riveted pipe or where the pipe-line is subject to lateral disturbance. Flanges for sinking-pumps, where the pumprod works inside of the pipe, should have these rivets countersunk on the inside, as at d, Fig. 3. In putting flanges on pipes care must be taken, in the first place, to have their faces come square with the pipe, and also to have the bolt-holes of the two flanges in line, so that the lengths of a column or pipe-line are interchangeable. In order to allow for inaccuracies in this respect, and also to provide for possibly required variations in position of elbows or other connections, the bolt-holes are sometimes made oblong, as in Fig. 5, so that one flange can be slightly rotated upon its mate. In this case a wrought-iron washer must be placed below the nut to give it an even bearing. Such a washer is an

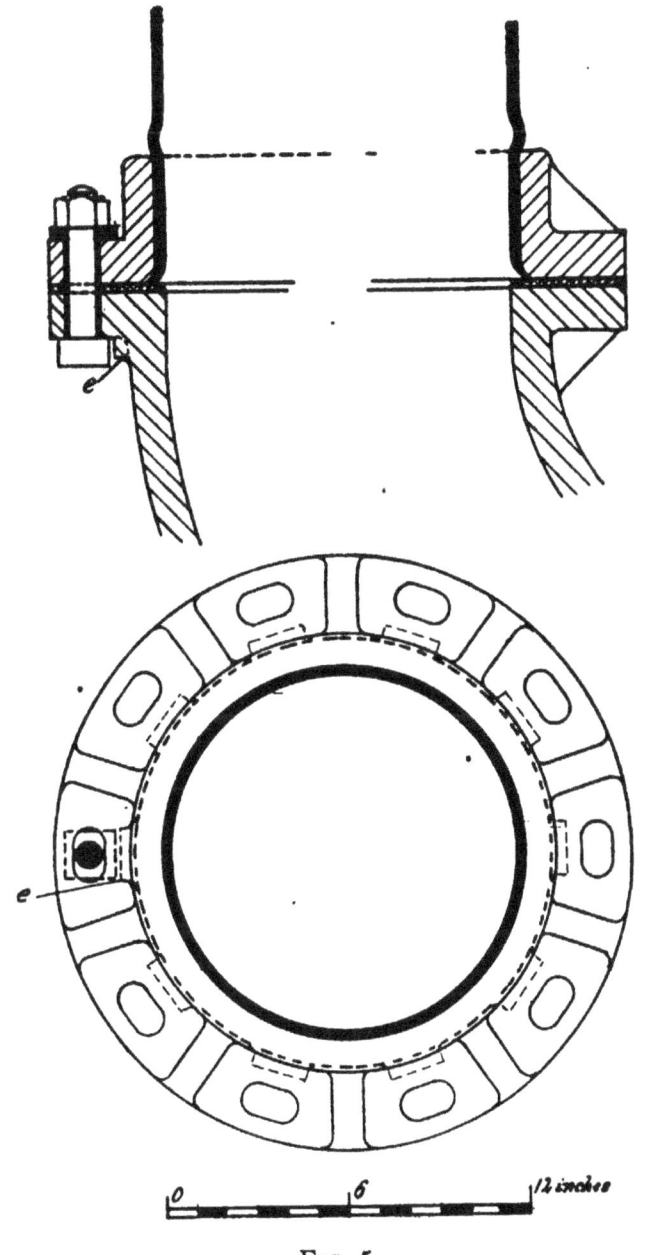

Fig. 5.

advantage also for ordinary round holes, as it provides a better bearing for the nut than the rough casting. A projection e, Fig. 3 and Fig. 5, should be cast on one flange of each pair, to absolutely prevent the bolt from turning when the nut is screwed up. Where it is desirable to get the flanges of as small diameter as possible, bosses are carried up around the bolt-holes to the full depth of the flange, Fig. 3. In this way the bolts can be brought closer to the body of the pipe than in the form shown in Fig. 5, while the thinner metal between the bosses affords facility for riveting to the pipe. Flanges of larger diameter are, however, always required where the pipes connect to cast elbows or nozzles, so as to allow room for the bolt-heads or nuts on the back of the flange of the casting. (See Fig. 5.) Where a greater number of such connections are required, it is sometimes preferable to make all the flanges of the larger size. Such flanges should be ribbed between the bolt-holes. Nearly all flanges obtained from dealers in pipe are much too light, and have too few bolts to be suitable for pipes subjected to heavy pressure and deflecting strains, like pump-columns or high-pressure steam-mains.

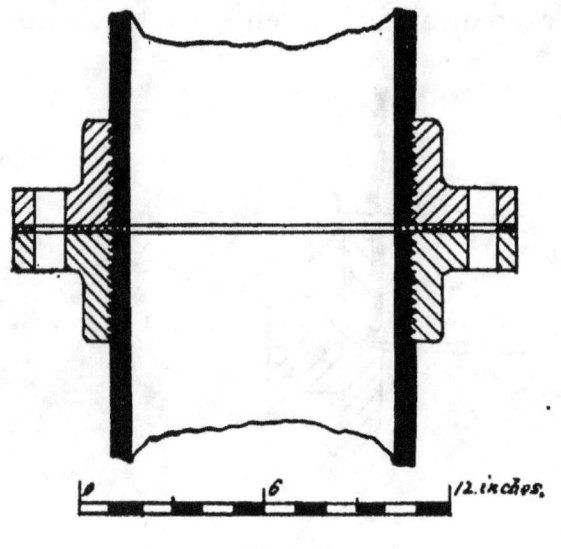

FIG. 6.

1.2.19. The smaller sizes of pipes, and often larger ones also, have their ends threaded, and are connected together by threaded sleeve-couplings or by flanges screwed on. The flanges of large steam-pipes are often put on in this manner. This method of securing flanges is generally also necessary for pipe of extra thickness. Where a tight pipe is required under high pressure, the ends are sometimes screwed through the flange, so as to project beyond its face, and then faced off level with the flange. This is a good plan for high-pressure steam-pipes where leaks are liable to occur at the threads. By the construction described, and illustrated in Fig. 6, it will be seen that the packing entirely prevents leakage at the thread by covering the joint between pipe and flange. Ordinarily a putty of red lead is used with threaded joints. For air-pipes, shellac varnish makes a very tight joint.

1.2.20. A water-tight threaded joint may also be secured by cutting away a portion of the thread, as at x, Fig. 7, and wrapping hemp or wicking into the groove before screwing into place. This joint is stated to be water tight under heavy pressure, even when the thread is so loose that the pipe can be rotated by hand. For column-pipes of Cornish pumps screwed flanges are not generally used; nor are they used in such cases where it is occasionally necessary to cut pipes to lengths, and where screw-cutting machinery of large size is not at hand. Where flanges are used on riveted pipe they are not shrunk on, but simply riveted to the pipe, and the latter caulked, if the metal be sufficiently heavy.

FIG. 7.

1.2.21. A flange connection which has been used with success in English collieries, is shown in Fig. 8. The ends of the tubes are expanded after the flanges a a are slipped on. When put together the double cone-ring b, with packing c c encircling each end, is inserted, and the bolts in the flanges drawn up. This joint has been used for water and steam; for the former, under pressures up to 4,000 lbs. per square inch. The inside ring and the outer flanges are not machined.

1.2.22. *Leaded Joints.* Leaded joints are usually adopted on pipes which are not liable to be disturbed in position, such as those for water supply on the surface. For such cases they make the most suitable

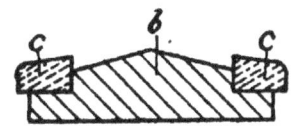

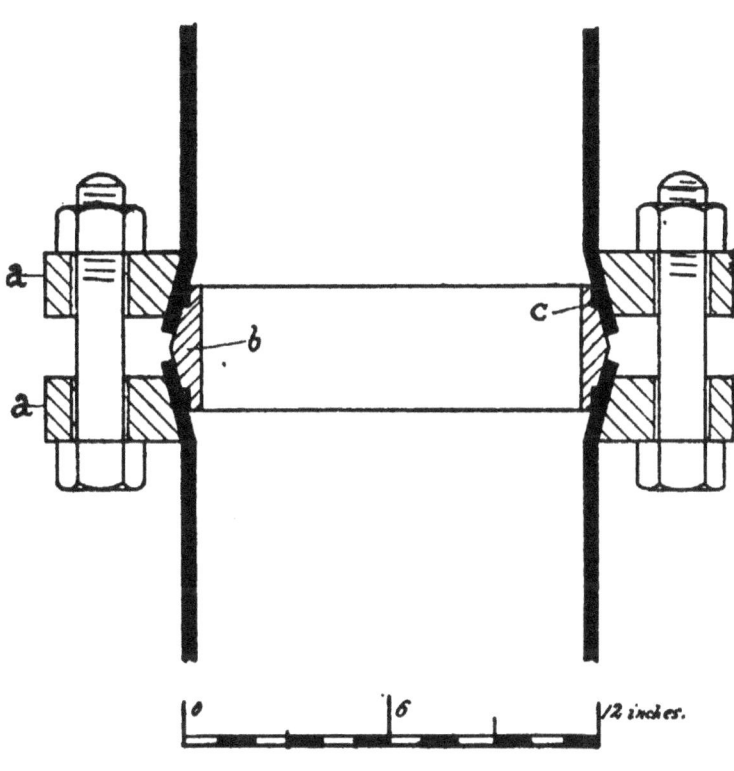

Fig. 8.

joint. The lead serves both for securing the connection and as packing. A lead joint much used for riveted pipe is shown in Fig. 9. The ends of the pipes abut on each other; and an internal sleeve prevents the lead from flowing into the pipe. An outer sleeve, usually welded, holds the lead, and must be sufficiently strong to resist pressure and caulking. Fig. 10 illustrates the Converse patent leaded sleeve-joint for wrought-iron pipe. The rivets serve to lock the pipe into the sleeve by their entering the recesses shown in the cut. Fig. 11 illustrates the pouring clamp, which fits the pipe and sleeve, and does away with the necessity of clay to form a mold for the lead when poured. After pouring, the lead is caulked firmly into place.

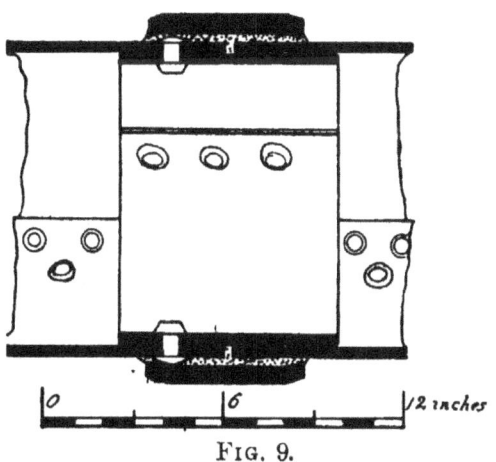

Fig. 9.

1.2.23. *Packing.* The material commonly used for securing tightness of flanged steam- and air-, as well as water-pipes, in mines, is the so-

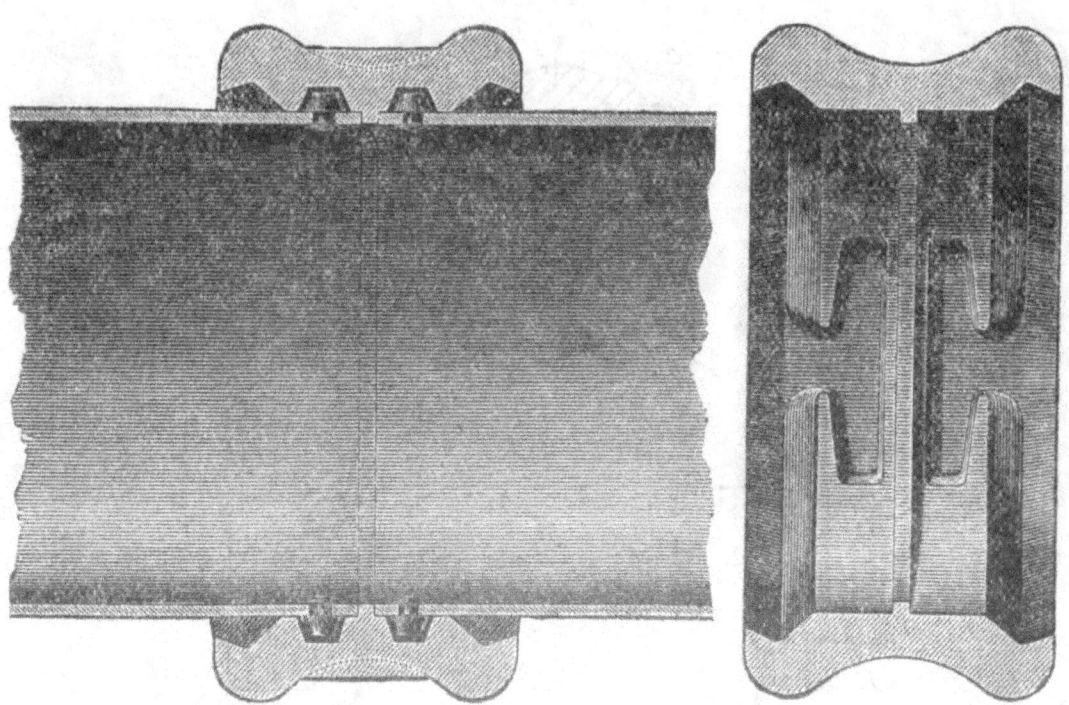

Fig. 10.

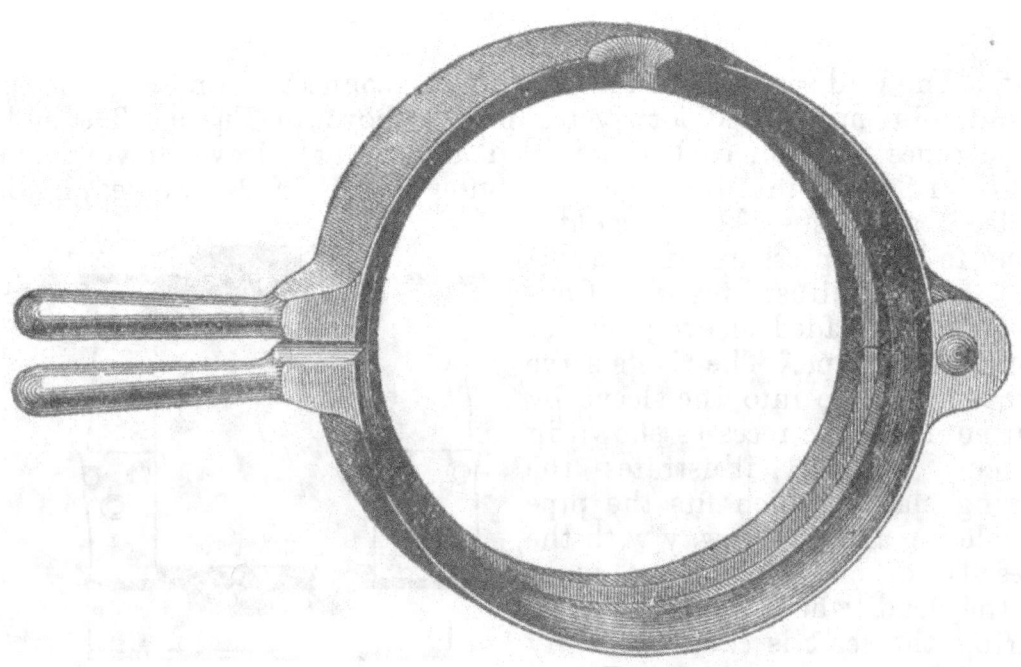

Fig. 11.

called sheet-rubber packing, composed of alternate layers of rubber and canvas. For water-pipes the gaskets are made of a thickness, ranging from $\frac{1}{8}''$ for small pipes, to $\frac{3}{16}''$ or $\frac{1}{4}''$ in larger pipes. Where the flanges are rough, thicker rubber must be used than where faced. In steam-pipes the rubber is usually not over $\frac{1}{16}''$ thick, in order to present less surface for the deteriorating action of the steam and hot water. It is always economy to use the best grades of sheet rubber. Rubber gaskets, if they have been in place for some time, particularly where subjected to heat, adhere very firmly to the flanges, and usually tear on being removed, thus necessitating new ones. Adhesion may be prevented by rubbing graphite on the surface of the gasket before putting in place. For heavier pressure the flanges are sometimes made in pairs, "male and female," as at a, Fig. 4, the recess being somewhat deeper than the thickness of the rubber which is laid in it, and which is prevented from

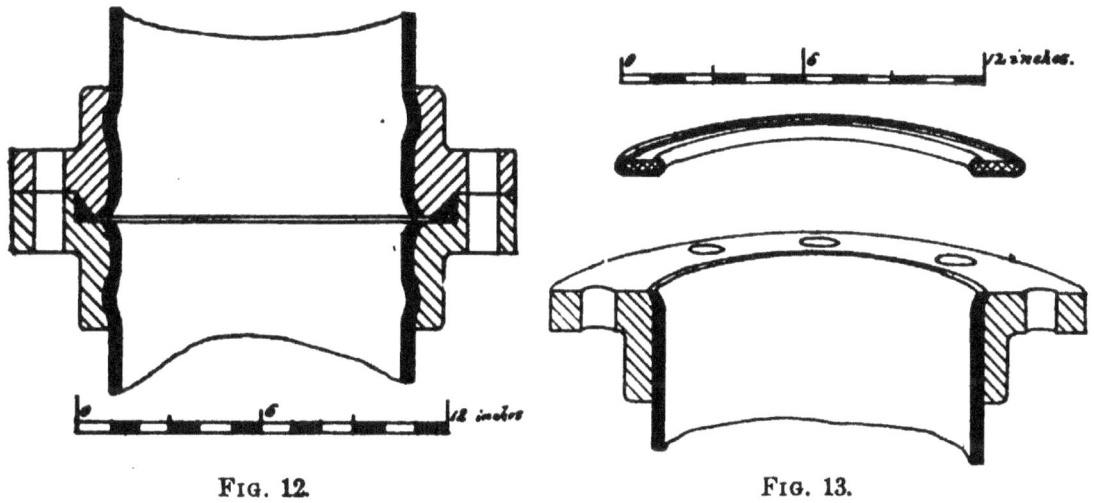

Fig. 12.   Fig. 13.

being blown out by the inclosing ring of metal. The packing shown in Fig. 12 is particularly adapted for heavy pressures, but requires continuous rubber rings, of circular or square cross-section, to obtain the best results.

1.2.24. *Lead Gaskets.* For heavy pressures, where rubber is liable to be forced out of the joint, sheet-lead gaskets are sometimes used between water-pipe flanges, these being usually machined in such a manner that their faces present a close succession of annular ridges, which sink into the lead and grip it tightly. Lead gaskets are, however, not sufficiently elastic for most purposes, and are liable to leak upon the least crowding out of line of the pipe. These gaskets are also sometimes used with male and female flanges, as shown in Fig. 4.

1.2.25. *Elastic Copper Gaskets.* A very efficient and durable gasket for steam-pipes is shown in Fig. 13. It is made of a ring of thin copper, the inner and outer edges being turned over the corresponding edges of a rubber gasket. The copper is about $\frac{1}{32}''$ thick, and the rubber $\frac{1}{16}''$. These gaskets are best made small enough to go inside the circle of bolts in the flange.

1.2.26. A flange-packing used for a head of 1,700' at the Mayrau shaft, Kladno, Bohemia, and which has been very satisfactory, is shown in Fig. 14. Here one of the flanges is recessed at *a* to admit a ring *b* of leather, rubber, or metal, of L-shaped section. This elastic ring is held in place by a rigid metal ring *c*, the whole forming a packing similar to that used for hydraulic-press plungers. (Modifications of this form

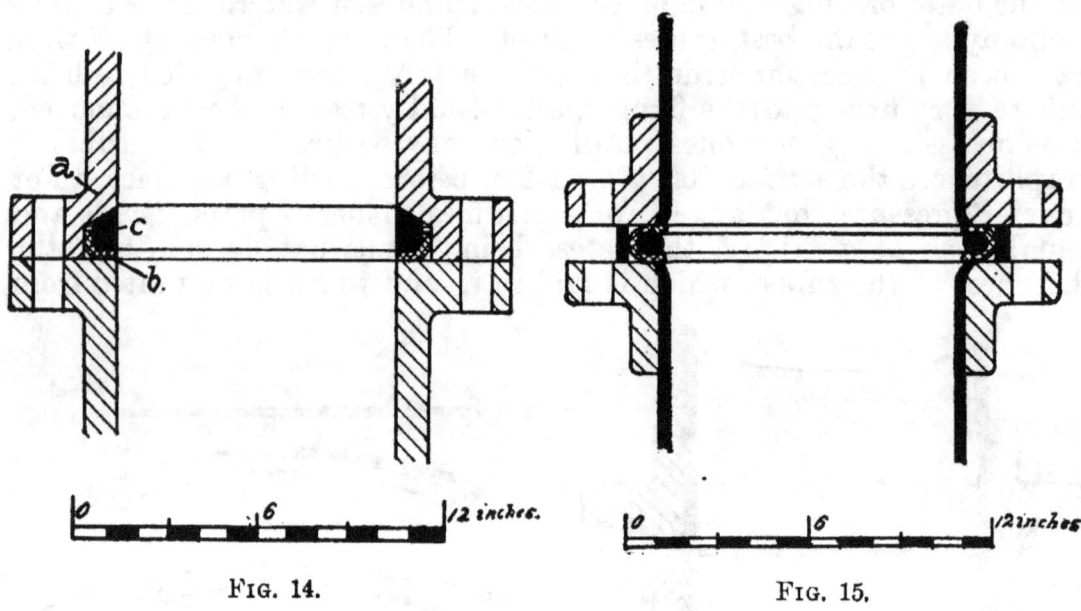

FIG. 14.   FIG. 15.

will readily suggest themselves; for example, that in Fig. 15, which could be used with ordinary flanges by inserting a forged distance-ring between them so as to form the space for the packing.)

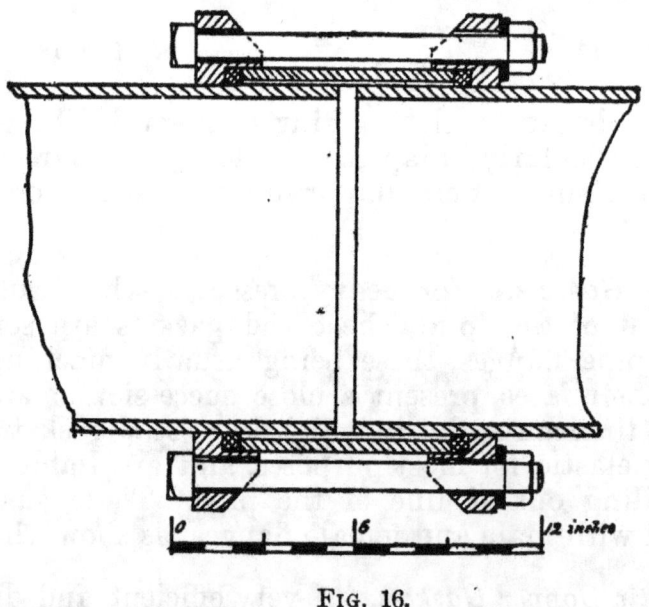

FIG. 16.

1.2.27. In the Paris compressed-air power transmission system, plain cast-iron pipes without flanges or spigot ends are used for the mains. The sections are connected by the Normandy joint, which consists of a sort of double stuffing-box, and is shown in Fig. 16. It is very flexible, and almost absolutely tight under the 80 lbs. pressure used. The pipes are not turned at the joint, but are put in as they come from the foundry. With some modifications this joint is also suitable for higher pressures in column-pipes.

1.2.28. Expansion Joints.

For long pipes, particularly in shafts, inclines, and levels, and for pipes rigidly fixed at the extremities, expansion joints must be used. The most common form of expansion joint consists merely of a stuffing-box, as shown in Fig. 17, or of a recessed spigot end containing hydraulic packing, as in Fig. 18. The end of the pipe entering the stuffing-box must be smooth, and is best made of brass. For steam-pipes in shafts, expansion joints are particularly necessary, and for these the one shown in Fig. 17 is the proper form.

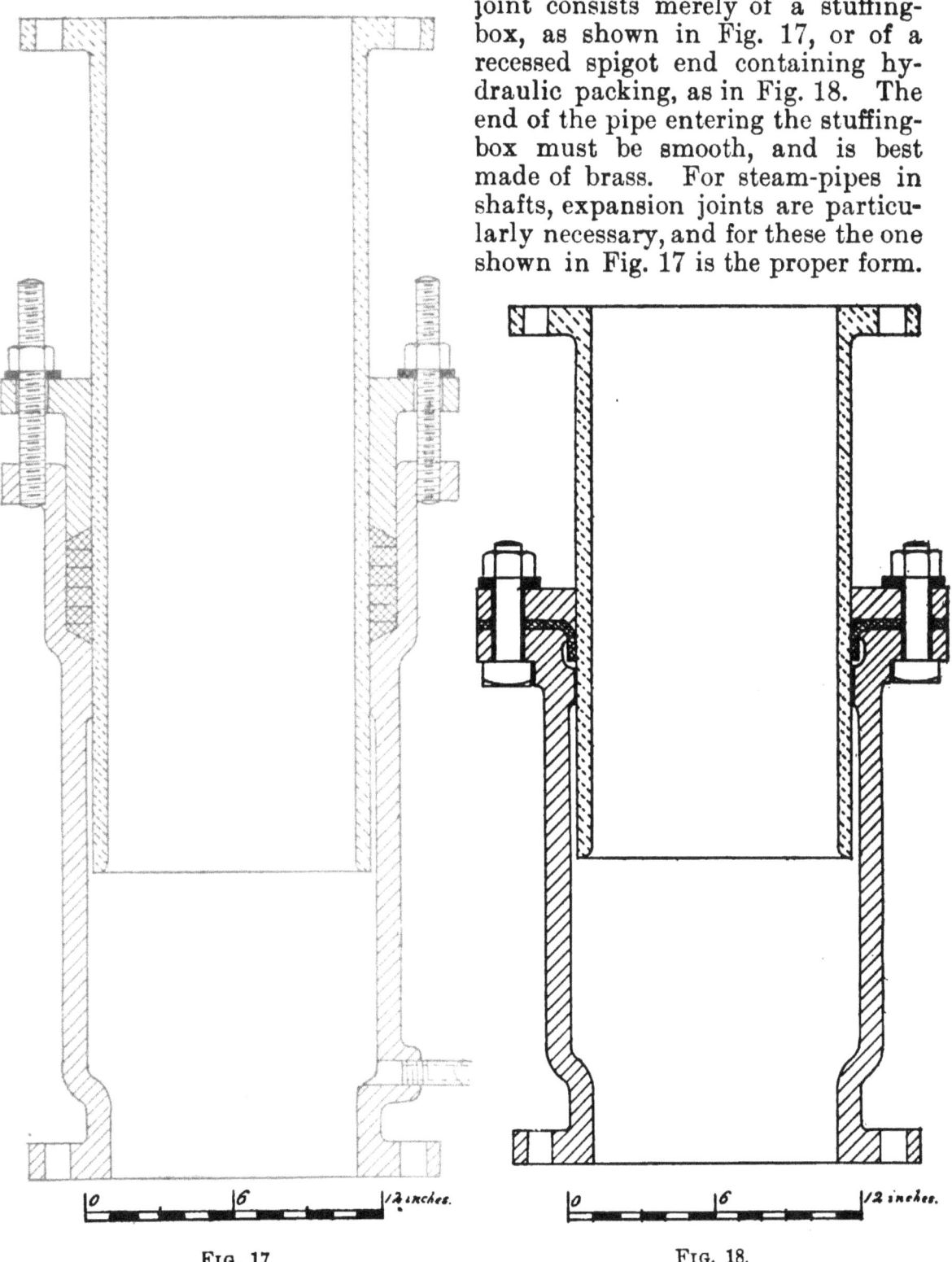

Fig. 17. Fig. 18.

The form Fig. 18 is only adapted for water-pipes. Stuffing-boxes in shafts and inclines should always be placed so that the gland is on top, since, if placed otherwise, they are almost sure to leak. Expansion joints are usually troublesome, and should be carefully looked after.

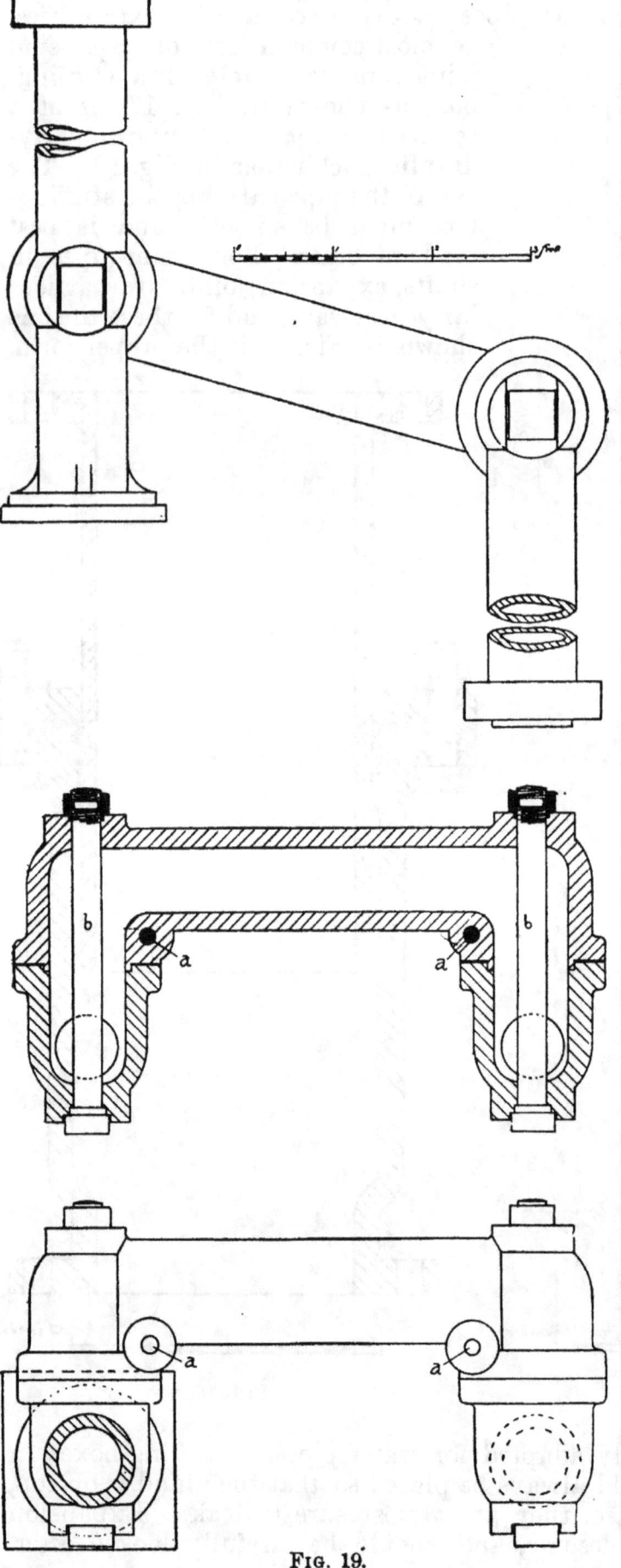

Fig. 19.

For steam-pipes, or where the water is hot, the expansion will necessarily be greater than for the ordinary variations due to climatic temperature. Both these variations can always be calculated; those due to settling of ground or timbers cannot. Ample range should therefore always be provided, so that the expansion joints will not pull out of the stuffing-box, which would be a serious matter with a steam-pipe under ground.

1.2.29. An expansion joint composed of a double swiveling pipe-section, shown in Fig. 19, was used at the "Combination Shaft," Virginia City, Nevada, on a cast-iron pipe under very heavy water-pressure, and gave good satisfaction. One of the pipes rests with a pedestal on a support in the shaft, the other being free to move. The bolts at a are inserted to reinforce the casting at the dangerous section. No packing was used between the faces of the casting. The threads of the swivel bolts b were packed by winding wicking around a groove, cutting part of the thread away. Where the range of expansion is not great, U-shaped pipes are sometimes used in steam-pipes to give them a certain amount of elasticity, or a corrugated section of pipe made of copper, brass, or wrought-iron is used. Such joints are, however, not suitable for long pipes, on account of the large number required to allow for the variation in length. Water-pipes having slip joints usually do not require any expansion joints.

1.2.30. Water-pipes laid in trenches at the surface do not require expansion joints. These are needed where pipes are laid over long bridges or trestle-work, as they are there exposed to changes of temperature. Large pipes should not be carried by wire cables or suspension bridges, as both of these sway the pipe and cause strains and leakage.

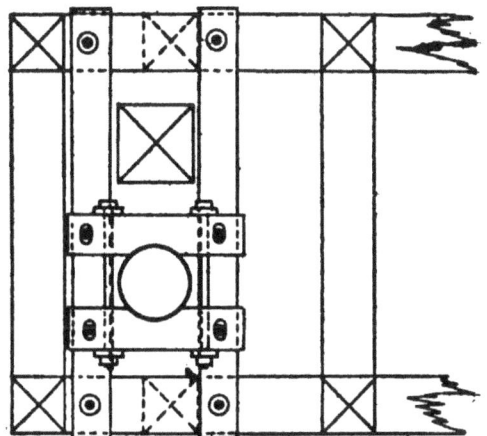

1.2.31. *Pipe Supports.* Pipes in a vertical shaft should have their weight well supported, and they must also be stayed laterally to be kept in line. In the Cornish system, with pumps not over 250' apart, the columns are usually stayed at intervals of about 50' by clamps of wood or iron. Generally, these rest on beams laid across the wall-plates of the shaft timbering, the beams often serving at the same time as supports for pumprod guides. Such a stay is shown in Fig. 20.

1.2.32. Posts are frequently inserted between several sets of shaft timbers below the pipe supports, so as to distribute the weight of the pipe on a number of wall-plates. Sometimes the pipes are clamped directly to the wall-plates with an intervening saddle-piece, as shown at *a*, Fig. 21, which represents a heavy form of such a fixture used where a goose-neck or offset-pipe on the top of the pump clack-chamber connects to the column-pipe. The weight of the column-pipe is sometimes also carried rigidly by an adjustable bolt support (Fig. 22)

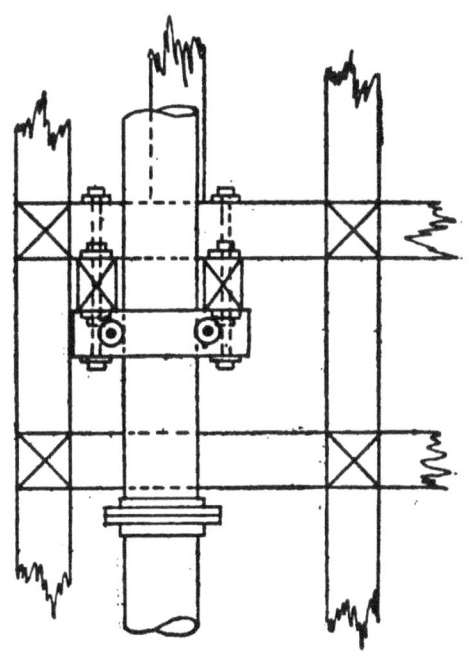

Fig. 20.

clamped to the pipe below a flange above the offset-pipe over the clack-chamber. There should be only one such rigid connection on the pipe, so that the latter can expand and contract. All supports and stays should be frequently looked after, particularly where the shaft is in bad ground and liable to be crowded out of line.

1.2.33. Water-pipes in inclines are usually laid along the lower side, resting simply in wooden saddle-pieces, which serve both as weight support and lateral stays. Steam-pipes are usually hung from the roof of inclines.

1.2.34. *Bends and Elbows.* Pipes should be well supported at bends and elbows, because, in addition to the effect of the weight, the unbalanced pressure of the water tends to crowd the pipe toward the convex side. Short bends in riveted pipe are often made up of sheets riveted up like the pipe. In flanged welded pipe, short bends are made of castings. Where the velocity is great, the bends should have as large a radius as possible, especially if the bend be through a considerable arc. Slight bends in flanged pipe are often made by inserting between the flanges of two sections of pipe a ring with inclined faces, on each of which packing is placed, as in Fig. 23.

1.2.35. Elbows used with the ordinary screwed pipe have too short a bend and offer too much resistance for high velocity of flow. In case of high velocity, it is advisable, therefore, to use special fittings. The ordinary malleable iron pipe-fittings are

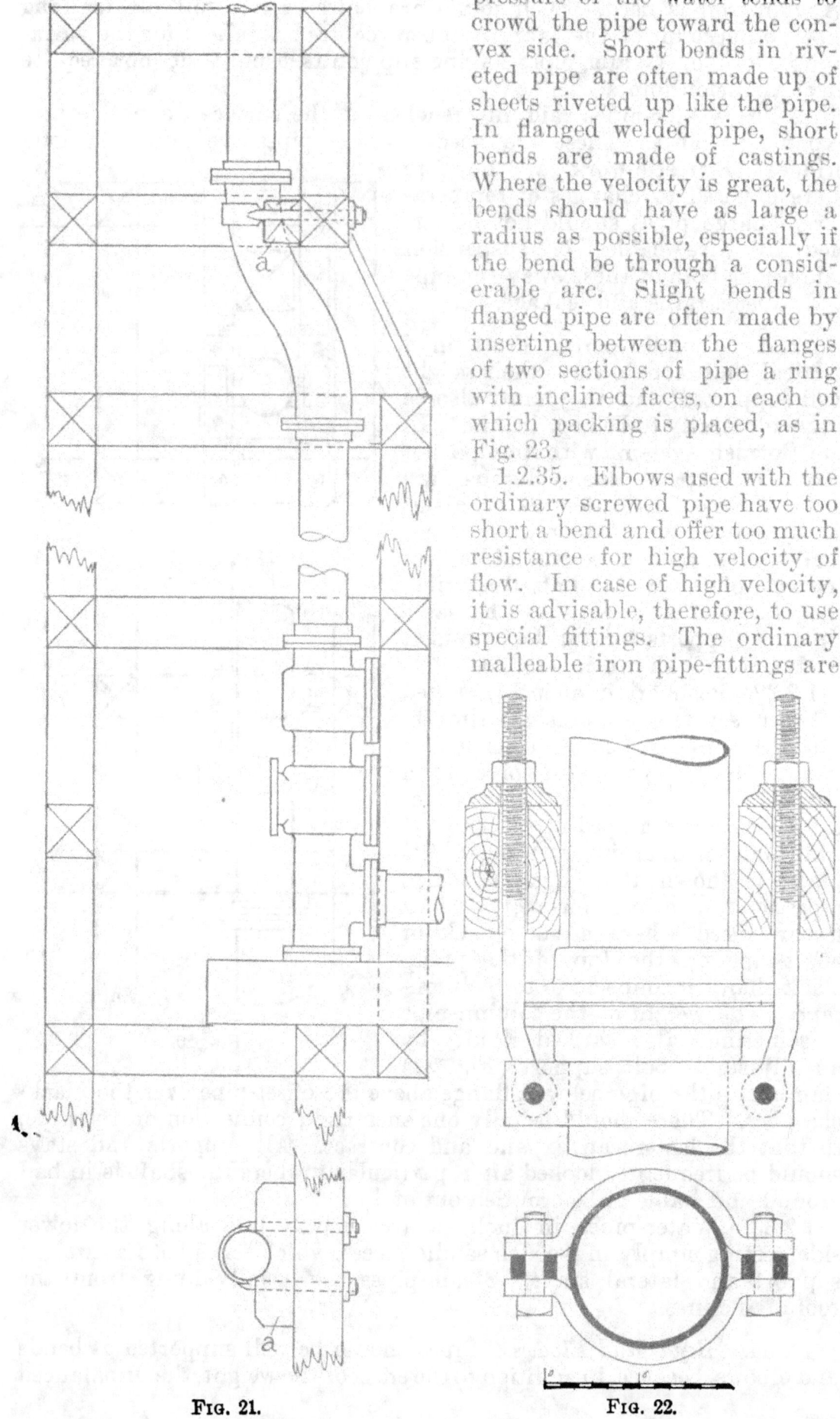

FIG. 21.   FIG. 22.

also unsuitable for many cases, and special cast-iron ones, which are less liable to split, are used for work requiring special care. Some machine shops that make a specialty of screwed-pipe work manufacture fittings of this kind, particularly elbows of larger radius than the ordinary trade fittings.

1.2.36. *Diameter of Water-Pipes, and its Relation to Velocity of Flow.* The diameter of the suction-pipe of a pump should always be such that the velocity of flow required by the speed of the pump can be maintained by the excess of atmospheric pressure plus any available head on the suction-pipe over and above the resistance due to valves and pipes. The suction-pipe, for single pumps particularly, should be as short as possible, making the mass of water which must be put in motion from rest at each stroke a minimum, so that its motion will be accelerated in the shortest possible time. Where a number of pumps operate through the same pipes in rotation or regular succession, so that the water in the suction- and discharge-pipes is always in motion, the size of the pipes may be reduced. Where the height from the suction level to the highest part of the space, the volume of which is affected by the pump-displacement, is great, the suction-pipe must be larger than where this height is small, because the available acceleration due to excess of atmospheric pressure is less. Since the mass of water to be accelerated is greater in the former case, the admissible pump speed will in general also be reduced. It is evidently necessary that all pipes be tight against leakage, but with suction-pipes this is particularly so, in order to prevent air from being drawn in, which would reduce the efficiency of the pump.

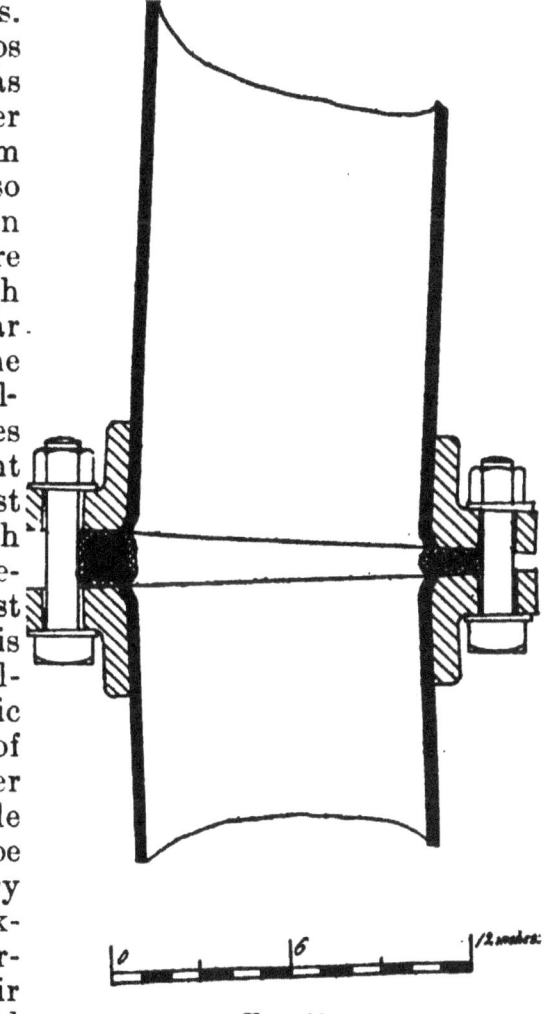

FIG. 23.

Where water is forced through a line or column of pipe by a reciprocating pump, and where, therefore, the water in the pipe is alternately started and permitted to come to rest, the velocity of flow cannot be allowed to be great; otherwise, the column of water will continue its motion for a short interval after the pumps have reached the end of their stroke, and will then fall back when the pump-piston is already on its return-stroke; the effect being to close the discharge-valve with a blow, whereby the entire column of water is arrested more or less suddenly. This is very liable to occur in the Cornish system, where air-chambers are rarely used, on account of the difficulty of applying them of proper size. In direct-acting pumps, which make a greater number of strokes per minute, air-chambers correct this evil to a great extent by equalizing the flow of water and making it continuous.

1.2.37. The least size of pipes is sometimes determined by other conditions; as, for instance, in Cornish sinking-pumps, where it is desired to remove the bucket through the column-pipe.

1.2.38. In general, the discharge-pipe need not be larger for double-acting pumps than for single-acting ones of half the capacity, because the velocity of flow is the same, the water being, in the latter case, at rest half of the time. Greater velocity may, with the same freedom from water-ram, be given to a short column of water than to a long one. For example, where a pipe is longer than the height vertically pumped, as in inclines, or where the pipe is partly horizontal in its course, the velocity of flow should be less than for an entirely vertical pipe, and the diameter therefore greater for the same capacity, because in that

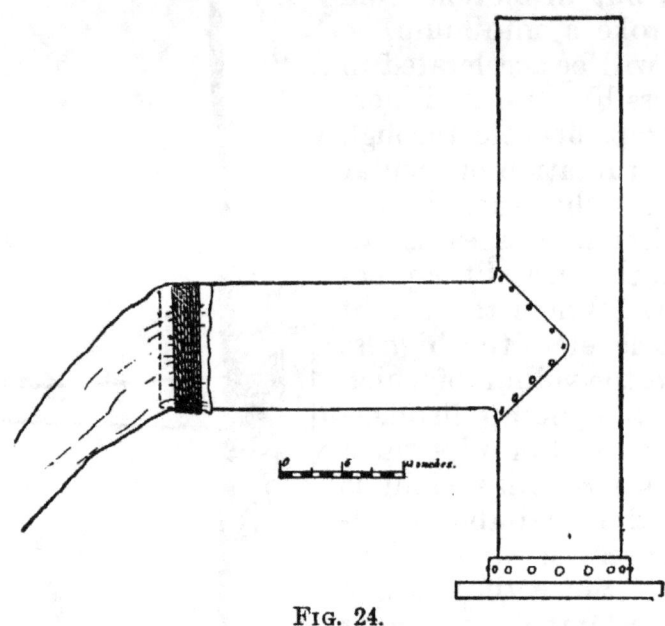

Fig. 24.

case the energy of the greater moving mass has less proportional retarding force due to gravity, and shocks are more liable to occur. Velocities over 5' per second should not be allowed in discharge-pipes, unless a number of pumps are arranged to come to the end of their respective strokes in rotation, so that the water in the pipe will be continuously advanced. In the Cornish system the diameter of the column-pipe is frequently the same or nearly the same as that of the plungers, and, for a double line of pumps, a separate column is used for each plunger, except where the pumps act alternately on independent rods, in which case only one column need be used.

1.2.39. *Discharges and Inlets of Water-Pipes.* With Cornish pumps, particularly, the discharge from vertical or column-pipes into the station-tanks should not be by means of ordinary elbows or short bends, because the intermittent flow of the water will cause a jar by striking against the side of the elbow. It is best to carry the pipe up vertically for a few feet above the outlet-pipe, because then the water can rise freely without shock, and flow gradually from the outlet. Fig. 24 shows the usual discharge-top for column-pipes of Cornish pumps.

It is generally made of galvanized iron, for the sake of lightness in handling, and has a short piece of canvas hose attached to the outlet to prevent splashing.

1.2.40. In order to reduce losses due to resistance, inlets to pipes should be flaring (or bell-mouthed), if the velocity be great. It is also economy to gradually enlarge the outlet, and submerge the end in the discharge-reservoir, in case of high velocity, because thereby the energy of motion is changed into pressure or lift, and, in case of pumping, less of the pump work is lost. These remarks apply particularly to low lifts and considerable velocities, and where the additional lift gained is an object, on account of its considerable proportion of total lift.

1.2.41. *Thickness of Water-Pipes.* Pipes subject to uniform, constant water-pressure can be made much lighter than those subject to water-hammer, and to varying pressures due to starting and arresting the column of water, as in the discharge-pipe of a single reciprocating pump. Again, pipes which lie on the ground, and which are not liable to be disturbed, can be made lighter than those which, like the column-pipes in vertical shafts, are subject to strains from being forced out of line by moving ground. Corrosive action of the mine-water may also require extra thickness. All strains and destructive influences must be taken into consideration, in designing a line of pipe, especially in mines where delays are nearly always expensive. The column-pipes for underground pumps are therefore usually made several times the strength that would be required for a pipe-line operating under constant pressure. What applies to strains in discharge-pipes of pumps, applies, however, with greater force to such power-pipes as are used for operating reciprocating hydraulic engines, because here the shocks are liable to be even more severe than in the case of pumps. The discharge-pipes of centrifugal pumps are not so liable to water-hammer, and can therefore be considered in the same category as pipes subject to uniform pressure.

1.2.42. *Air-Chambers on Water-Pipes.* Air-chambers are frequently used along a line of pipe and at sharp bends to reduce shocks, such as occur when valves are suddenly closed or when the flow in pipes supplying water to power-wheels is suddenly arrested by obstructions finding their way into and closing the nozzle.

1.2.43. Air-chambers under pressure usually require some charging device, as the air is absorbed by the water. This device may be a small air-compressor operated by hand at long intervals, or whenever a try-cock or gauge-glass on the air-chamber shows that the air-space has become too small. Air-chambers should be so tight that no air can escape. It is well to coat them inside with paint or asphalt, for heavy pressures, as the air is liable to leak through the pores of the metal.

1.2.44. Air-chambers on pumps perform the functions of equalizing the flow in the discharge- or in the suction-pipe, and of reducing shocks on the valves. They will be considered more in detail in connection with direct-driven pumps. Spring-loaded pistons or plungers are sometimes used in place of air-chambers of small capacity.

1.2.45. *Relief-Valves.* Spring-loaded or weighted relief-valves or pistons are also used on pipes liable to sudden stoppage of the water column, so as to afford an escape for the water under excessive pressure.

Weighted valves are not so good as those loaded by springs, because they are slower to act, on account of the greater mass to be moved.

1.2.46. *Protection of Water-Pipes against Corrosion.* Pipes conveying water, and particularly those used in mines where the water is acid, are either made of material to resist corrosion, or, if the corrosion be slow, as is usual, of greater thickness, so that they will stand a reasonable time with such protection as is afforded by a coating applied to their surface. The use of copper pipes in exceptional cases has been

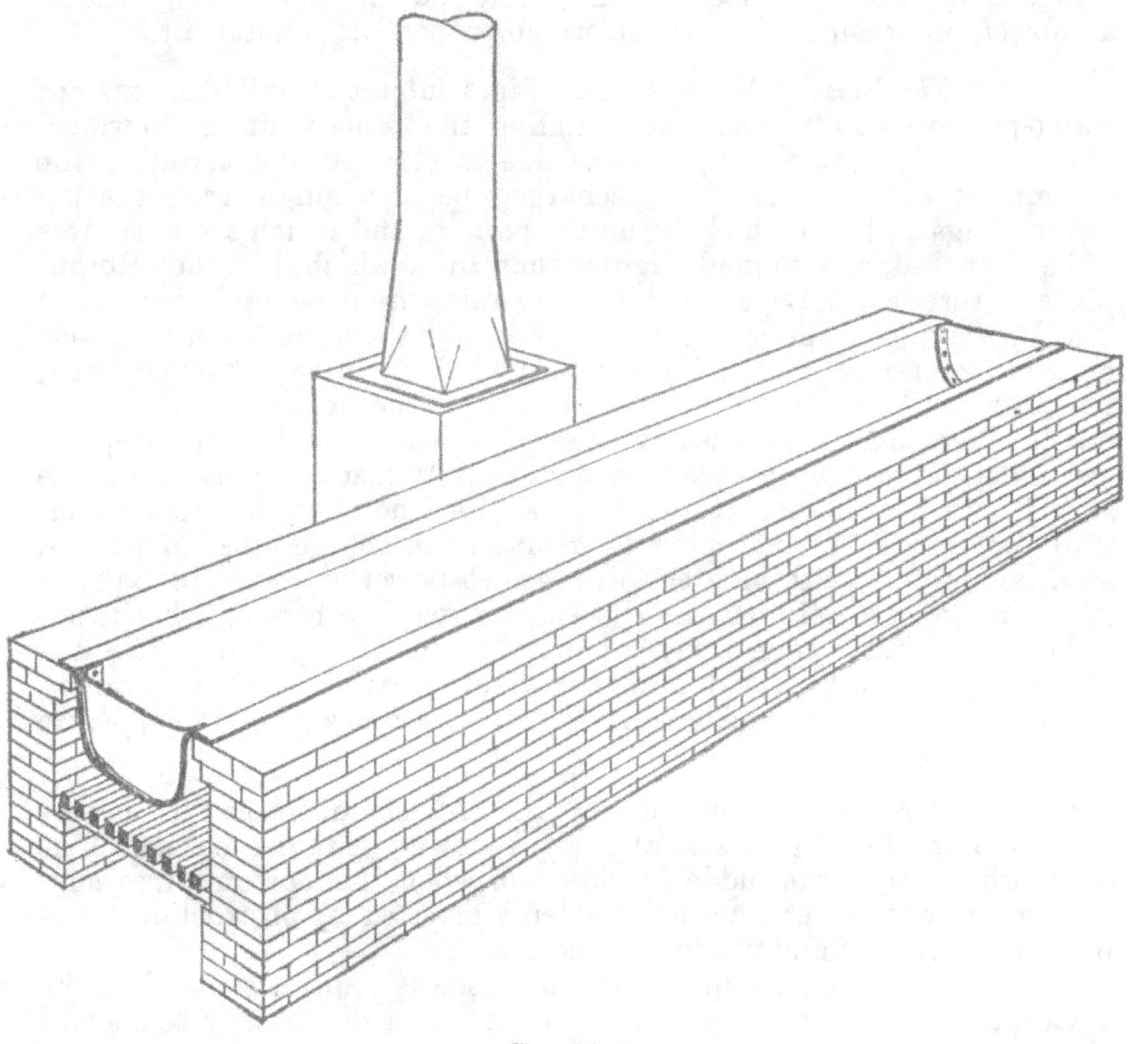

Fig. 25.

previously mentioned. They are rarely used, on account of their cost. Coatings of asphaltum, or paints prepared from the resinous part of oils, constitute the usual method of protection. The asphaltum coatings are applied by dipping the pipes into a melted bath of it. The pipes should be thoroughly heated to the temperature of the bath, and the latter must be maintained at uniform temperature. Where pipes have to be transported great distances over rough roads, the asphalt coatings are liable to be injured, and it is therefore sometimes better, if the appliances be available, to dip the pipes at the mine. Fig. 25 is a form of asphalt bath for dipping pipes. The illustration shows the apparatus arranged with a double-end fireplace under the pan. By this construction, with the chimney at the middle, more uniform heating is secured. In out-of-the-way places a pan is generally made from a spare length of pipe by cutting it open lengthwise and riveting pieces to the ends.

1.2.47. It requires some experience and attention to maintain a uniform temperature throughout the bath with this arrangement. Where steam is available the bath can be heated very uniformly by placing steam-pipes in the bottom, in which case a wooden trough will answer as a makeshift.

1.2.48. To avoid the difficulties attending the hot coating of pipes in out-of-the-way places, the pipes are often painted or dipped cold with some of the so-called paraffine paints. The dipping should in this case be done vertically, the coating fluid being contained in a vertical pipe sunk into the ground, and only slightly larger than the pipe to be dipped, so that a minimum of surface is exposed to the atmosphere and for evaporation of the very volatile solvent. In applying any coating to pipes they must first be thoroughly cleaned, and every particle of rust scraped off, as otherwise the coating will not adhere well at such places. The asphalt coating costs generally about half a cent per foot per inch diameter of pipe, so that a 3" pipe would cost $1\frac{1}{2}$ cents per foot to dip.

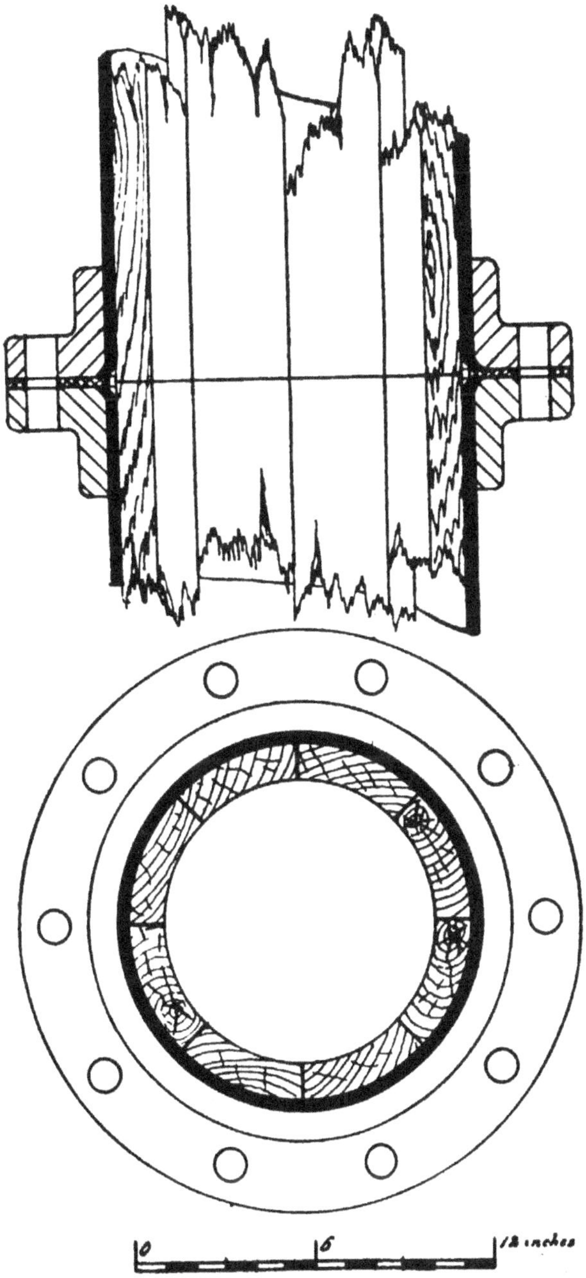

1.2.49. Where the water in the mine has a high temperature, as in the Comstock mines, coatings of the kind described are of no value in protecting the pipe. Galvanizing the pipes will protect against some waters. Some pipe manufacturers use an alloy consisting of lead, tin, and nickel, lead being the chief constitutent. This is a better coating than the zinc of the galvanized pipes, and also has the advantage that the pipes can be bent cold without cracking the coating. To bend galvanized pipes and not injure the coating, they should first be carefully heated to a moderate temperature.

FIG. 26.

1.2.50. Iron pipes have also been protected inside by wooden linings. At the New Guston Mine, Montrose, Colorado, the lining shown in Fig. 26 was used. The pipe in this case should be asphalted or painted with a protective coating before introducing the lining. Redwood is the best material for the latter. The staves should be cut off slightly longer than the lengths of the pipe, so as to secure contact of the ends of the staves and also allow for the packing between the flanges. The pipes are necessarily larger for wooden linings, and this is perhaps the main objection to their more common use. A thin coating of cement has in some cases been a good protection.

1.2.51. In the greatest number of cases the best plan will be to use heavier pipe and protect it as well as possible by coatings.

1.2.52. *Air in Water-Pipes.* Frequently pumps take in a small amount of air on the suction-stroke, either by leakage or intentionally, in order to keep the air-chambers filled; and this air will accumulate not only in air-chambers, but also, when these are filled, at any high places along the discharge-pipe. Besides contracting the free passage of the water, such air is liable to be carried along in a body when the overpressure necessary to force the water through the contracted space has become sufficiently great, and then to cause water-hammer by rising

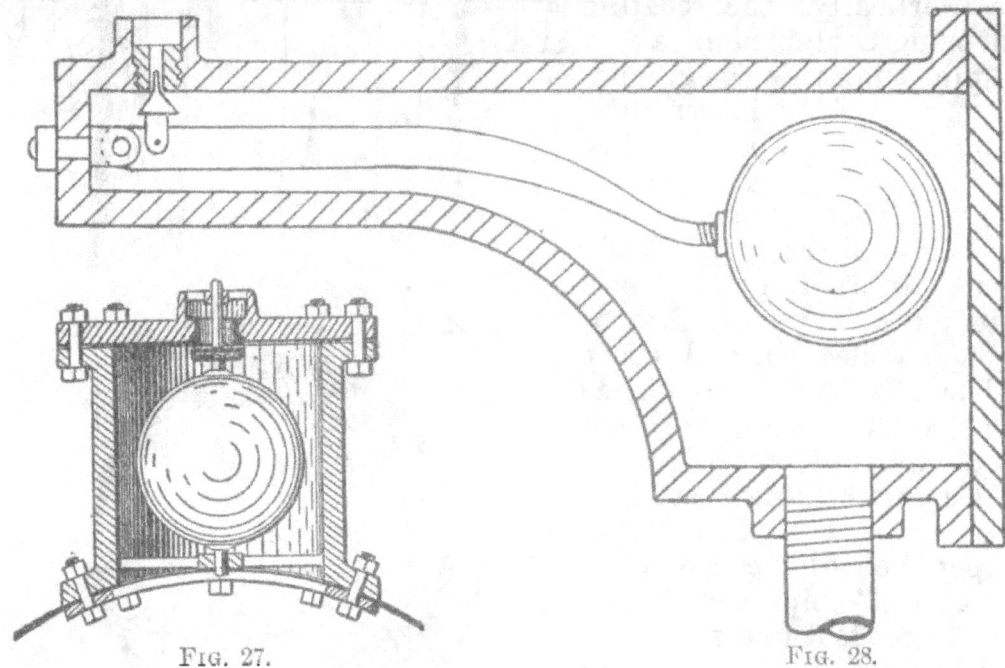

Fig. 27. Fig. 28.

back through the descending pipe; or, if carried far enough, by entering the next rising part of the pipe, where it is in a position still more dangerous to the pipe. Therefore, wherever possible, discharge-pipes of pumps should rise all the way toward the discharge end, so that the air may be continuously expelled. Where this is not possible, it is necessary to use either some form of automatic air-valve, or a vertical pipe connected to the high part of the pipe-line (the vertical pipe rising to an elevation equal to the pressure-head at that point). A small adjustable opening or a cock, placed at the highest point to permit the air to escape with a small waste of water, would in some cases serve the same purpose. For all air-escapes it is necessary to have a pocket or chamber at the highest part of the pipe-line, to permit the air to accumulate, as it would, for the greater part, run past any small opening without being diverted into it. Automatic air-valves for letting accumulated air out of pipes must have sufficient weight in air to open the valve against the overpressure in the pipe. They must also be so constructed that they will close by the combined effect of buoyancy and the pressure due to the rush of water. Figs. 27 and 28 show air-valves of this type.

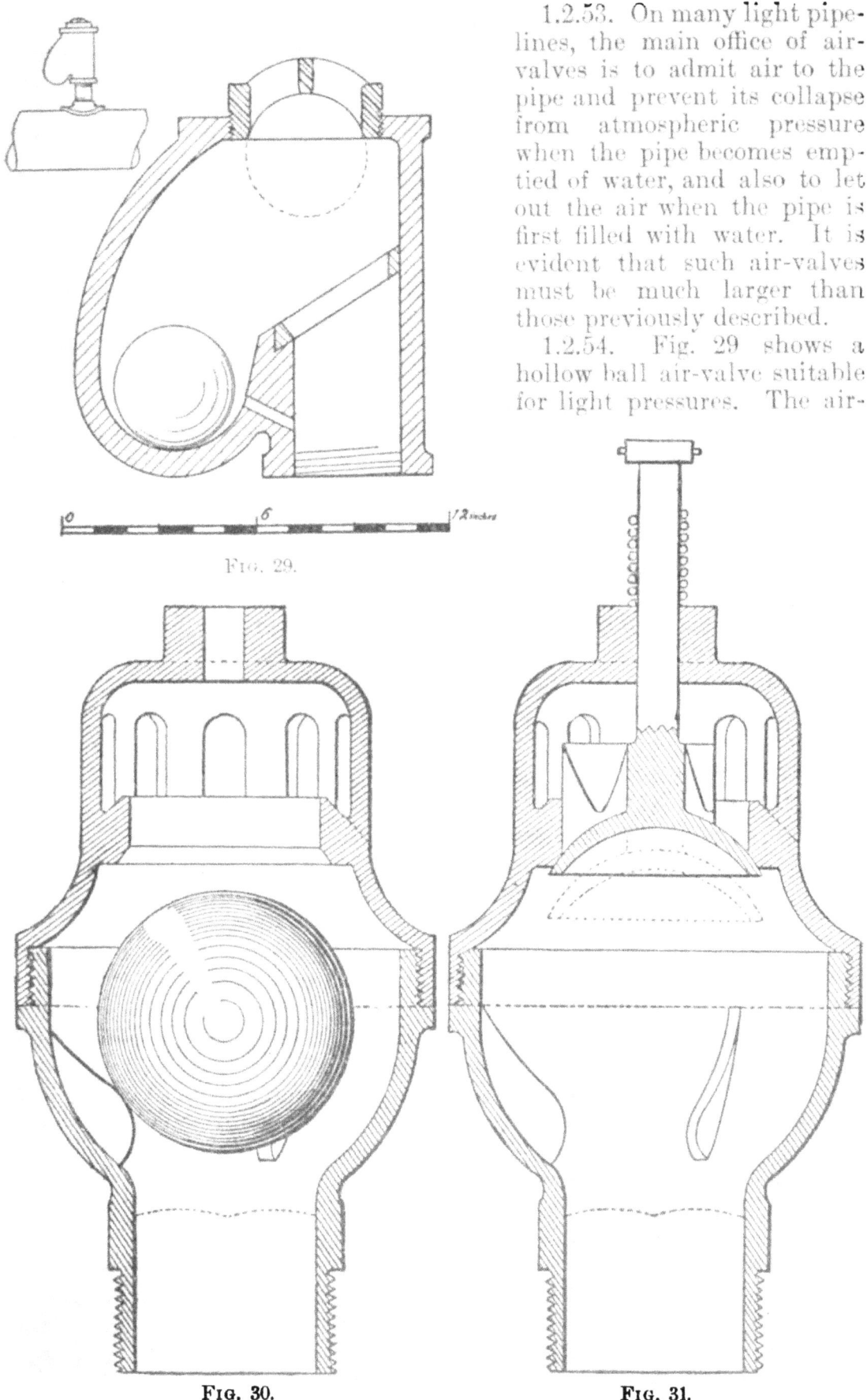

FIG. 29.

FIG. 30. FIG. 31.

1.2.53. On many light pipe-lines, the main office of air-valves is to admit air to the pipe and prevent its collapse from atmospheric pressure when the pipe becomes emptied of water, and also to let out the air when the pipe is first filled with water. It is evident that such air-valves must be much larger than those previously described.

1.2.54. Fig. 29 shows a hollow ball air-valve suitable for light pressures. The air-valve in Fig. 30 has a wooden ball covered with rubber, and is, therefore, more rigid and not liable to be pressed out of shape and remain stuck in its seat. For high pressures, the same make of valve is constructed with a bell-shaped metal valve, as in Fig. 31. The bell-shaped valve (Fig. 31) is closed by the rush of escaping water. In all the forms of ball valves, the ball is the valve and float in one. They do not operate to let air out of the pipe, unless the pressure falls very low, as in case of a break in the pipe or its emptying.

1.2.55. Besides the air taken in by pumps, there is always air contained in the water. It is not generally possible to predict under what pressure such air will be liberated from the water. It is, however, almost certain to be liberated if the pressure falls below that at which it has entered from the outside, where it was under atmospheric pressure.

1.2.56. Air is generally absorbed under pressure in an air-chamber, and such air will be released when the water which contains it reaches a high point at a lower pressure. Air will also be more readily released when the temperature increases, so that air may be looked for in the elevated parts of long pipe-lines which are exposed to the heat of the sun.

1.2.57. *Notes on Steam- and Air-Pipes.* In this class of pipes the first care next to safety and preventing leaks should be to keep as much of the heat in the steam or air as possible. It is advantageous, therefore, to locate such pipes in upcast shaft compartments. In the case of steam- and reheated-air-pipes further protection against radiation must be afforded by non-conducting coverings. The latter should in turn be protected from moisture in order to be efficient. This can often be done by wrapping the non-conducting material with tarred canvas. The pipe connections should not be covered, as leakage from them might enter the non-conductor, and they should also be accessible for repacking. It is a good plan to provide small conical rings at intervals, to act as "umbrellas" for shedding off the drip. These are best placed just below pipe connections, so as to carry off any leakage drip and prevent its soaking into the non-conductor.

1.2.58. Steam-pipes, and generally air-pipes, should be provided with traps at low points, for the purpose of draining off the condensed or entrained water, which must be prevented from getting into the motor cylinder of the pump engine, and which, besides contracting the passage at points where it accumulates, and thereby causing resistance to flow, is also liable to produce water-hammer and endanger the pipe. For this reason, as soon as a steam-pipe is shut off for a time, the drains should be opened to let out all the condensed water.

1.2.59. A break in a large steam-pipe underground is a serious matter. Where such an accident is liable to occur, as in some shafts in moving ground, provision should be made either to have the increased rush of steam automatically operate a self-closing device, or to connect a throttle at the surface, or valves at intervals, with a handrope passing down the shaft or other parts of the works containing the pipe.

1.2.60. Where steam or air is conducted a long distance to drive a reciprocating pumping-engine underground, it is best to connect the pipe to a receiver from which the engine takes its air or steam. The receiver, from which the engine draws intermittently, acts as an equalizer of pressure and flow in the pipe, so that a somewhat smaller pipe can be used with the receiver than without it, because the flow in the pipe is practically uniform.

1.2.61. It is better to use first-class gate-valves on steam- and air-pipes as well as on water-pipes, as they cause less obstruction to the flow than globe valves, which, if used, should be so placed that water cannot accumulate in the globe. Tightness against leakage is important in steam- and air-pipes, for economical reasons. In long pipes the loss from leakage is often enormous. These should, therefore, be carefully designed and erected.

1.2.62. Steam- and air-pipes should have stop-valves, not only at the pump engine, but also at the boiler or air-receiver, so that the pipe can be repaired without shutting down the boiler or exhausting the receiver. Before connecting steam- or air-pipes to the engines to be operated through them, they should be thoroughly blown out to remove any loose scale or dirt which might afterwards get into the engine.

1.2.63. The heat generated by steam-pipes has a tendency to cause vapor to form, which rots the timbering of the mine.

1.2.64. *General Remarks.* All pipes (water, steam, or air) should be larger when their length is great, to compensate for the additional resistance to flow.

1.2.65. Elbows and bends for the same reason should be formed to a large radius, where economy is desired and where space permits.

1.2.66. All shut-off valves and gates on water-pipes should be so arranged that they can only be closed slowly; then the water flowing in the pipe will be brought to rest gradually and without shock. The longer the pipe and the swifter the flow the more slowly should the gate or valve be closed.

1.2.67. Joints in pipes should be accessible. In underground workings they should stand some crowding out of line without leaking, and should remain in good condition for a long time.

1.2.68. It is of the greatest advantage to have as much as possible of the supporting arrangement for pipes, pumps, and rods in a shaft designed to be made of wrought-iron and timber, and the iron work of simple form, so that breaks can be quickly repaired by the mine blacksmith and carpenter. For large pumping-plants, a small machine shop is almost a necessity. Extra flanges for pipes, elbows, and other parts should be kept on hand.

1.2.69. If a line of pipe be properly designed and carefully put up at the start, much annoyance, repair work, and stoppage of machinery will be avoided, and the expenses of these in a year's run will almost more than equal the increased first cost.

CHAPTER III.

Pump-Valves.

1.3.01. *General Types.* Valves for pumps used in mines are of various types, their design and construction depending upon the conditions under which they are intended to operate. They may be divided roughly into hinged valves, commonly called clacks, which open by swinging about an axis parallel to the face of their seat; straight-lift valves, which rise evenly, and generally vertically, off their seats; and flexible valves, which alter their form on opening.

1.3.02. The pumps of the so-called Cornish system have usually hinged or clack valves, although single- and double-seated straight-lift valves are also often used, particularly in Europe.

1.3.03. In direct-driven pumps, straight-lift valves are almost entirely used. These are usually simple, often practically rigid, rubber disks, the seat being in the form of a grating. Flexible valves of rubber or leather are suitable only for very low lifts.

1.3.04. *Requirements.* The points to be aimed at in the design and construction of a pump-valve are:

First—It should close tightly against its seat, which latter is usually made so that it can be readily removed for the purpose of truing up and repairing. Tightness of valves and plungers or buckets is particularly required in a pump which has to raise water partly by suction, and where reduced inflow of water necessitates slow running of the pumps.

Second—It should open easily, and remain open with a minimum of overpressure on its lower side.

Third—It should, when open, present very little resistance to the flow, and divert the current as little as possible from a straight course.

Fourth—It should close as promptly as possible, immediately on the completion of the stroke, or when the forward motion of the water ceases, because, if the valve is still open during the commencement of the return-stroke, the water flows back and acquires a velocity which is suddenly checked with a blow by the closing of the valve. The blow is the more severe the more tardy the valve is in starting to close and the longer the column of water above it.

Fifth—It should be simple in construction and not liable to get out of order easily.

Sixth—It should be readily accessible for purposes of repair and interchange.

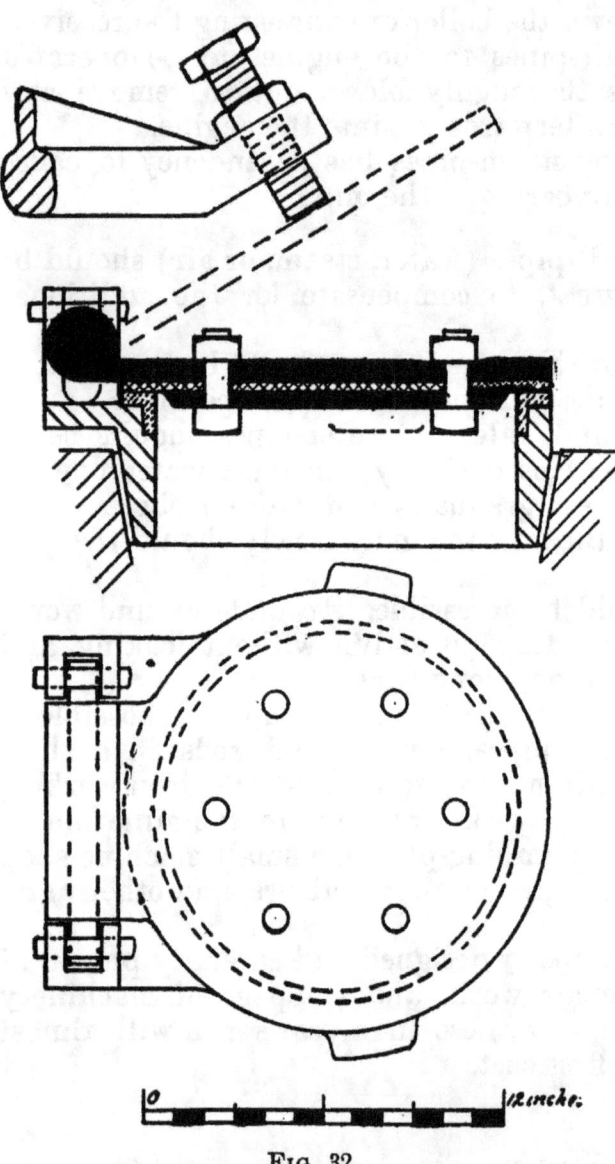

Fig. 32.

1.3.05. *Valves and Valve-Seats.* In mining pumps, which have nearly always to deal with water carrying sand in suspension, the tight closing of the valve is, by this cause, often prevented. The valve-faces, or the whole valve, are usually made of some elastic material, so that any particles lodging on the seat will be pressed into the valve-face and not prevent its coming in contact with the metal seat, as would be the case if the valve-face were also of metal. When the water permits it, leather is much used for facing the valves. Where the water is very acid, rubber must be used. Hot water requires rubber-composition. This material has long been used for the disk-valves of direct-acting steam-pumps. It is said to have been first used for the faces of clack valves of Cornish pumps by Mr. Deidesheimer when Superintendent of the Hale and Norcross Mine, in Virginia City, Nevada. The composition disks are usually $\frac{5}{8}''$ thick for clack valves. Fig. 32 shows a hinged clack of common form, with composition-rubber facing. When the valves are large and the water very hot, it is better to bore out the central portion of the disk in order to reduce liability of cracking from unequal expansion. Hinged valves are more liable to leak than straight-lift valves, as they generally wear unequally by striking first either at the edge nearest to or at the edge farthest from the hinge. When the hinges are made of metal the pins should be very loose, so that they will not become clogged and by their friction retard the valve. The leather faces of clack valves are often extended to serve as hinges for the valves, as in Fig. 33, which shows a double valve of this kind.

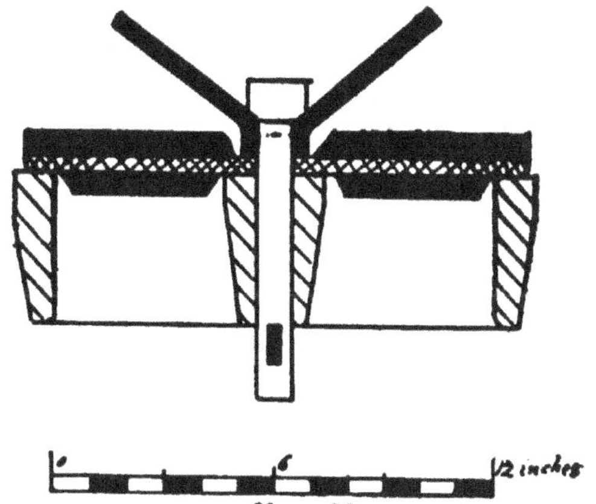

FIG. 33.

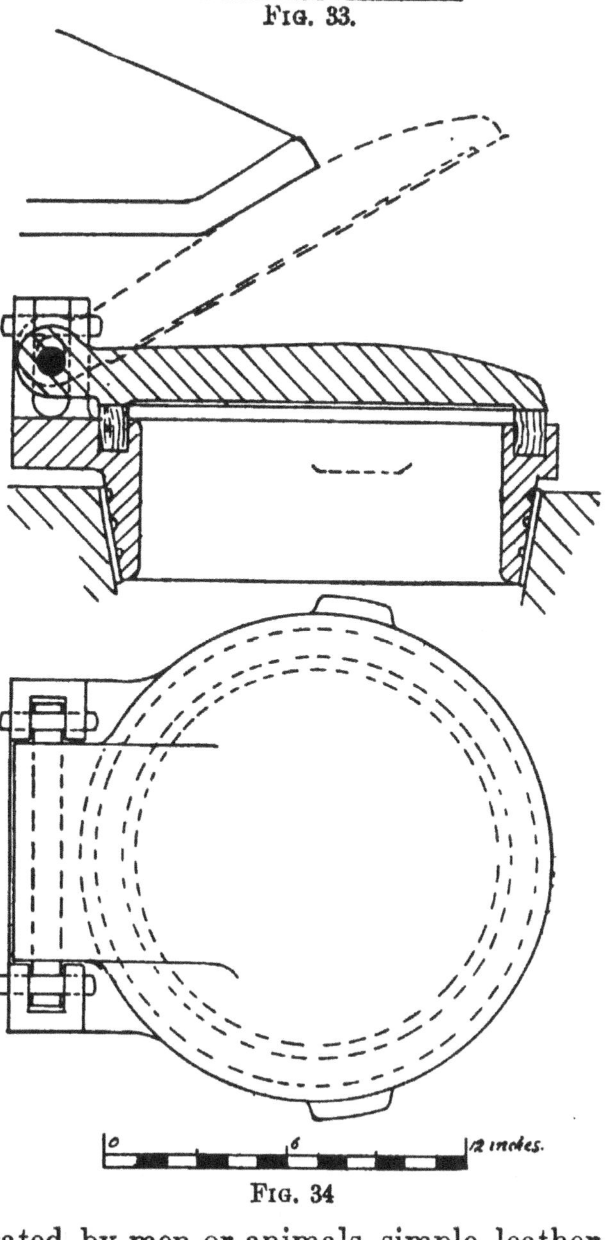

FIG. 34.

1.3.06. For low heads and small pumps, such as are operated by men or animals, simple leather flaps, reinforced by a couple of washers held together by a bolt, are often used. Sometimes they are nailed to one side of a bored wooden block, which serves as a seat.

1.3.07. Boxwood, maple, beech, and even pine, have been used for valve-seats of metal-faced valves, and they are very durable, but always leak, as the grit in the water cuts out the soft part between the fibers of the wood, and this also retains particles of sand, which cut out the valve-face. The small blocks of wood are pressed into a groove in the valve-seat, the end of the grain being presented to the valve-face. Fig. 34 shows a hinged valve with its seat constructed in this manner.

1.3.08. For valves with elastic faces, brass seats, or seats faced with brass, are advisable with acid water. The last Cornish pumps operated on the Comstock had brass-faced valve-seats, constructed as shown in Fig. 32.

1.3.09. Fig. 35 is a straight-lift valve like those used in direct-acting pumps. It is simply a thick rubber disk, supported, when closed, on a metal grating which forms the seat. For higher pressures the openings in the grating must be made very small, and rubber-composition used for the valve. Such valves have been used for pressures of 500 lbs. to the square inch, but for such pressures the valves are usually held in brass cages, as in Fig. 36. Straight-lift valves are also made of metal, with leather or rubber facing, as in Fig. 37.

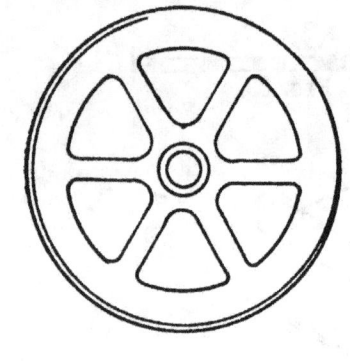

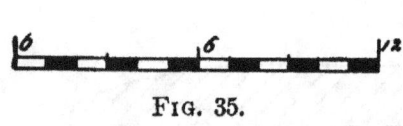

Fig. 35.

Fig. 36.

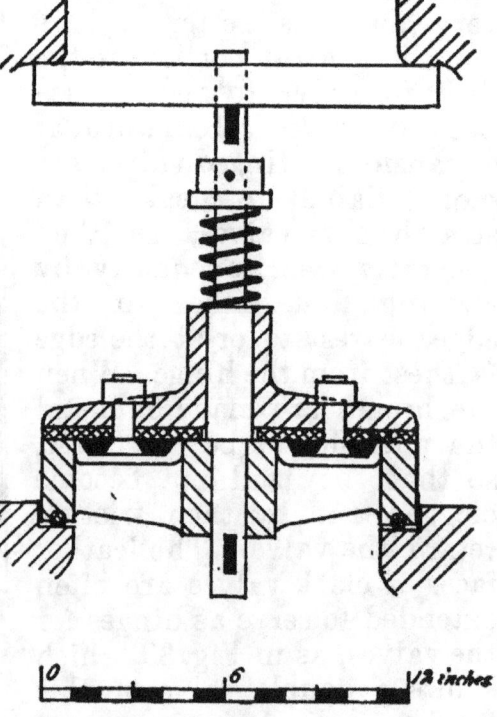

Fig. 37.

1.3.10. The guides of straight-lift valves must be arranged so that they will not cause friction or binding, and thereby retard the action of the valve. The width of the bearing of the valve on its seat must be such that the material will not be destroyed too rapidly by the repeated and more or less heavy blows on the closing of the valve. On the other hand, the bearings should not be too wide, otherwise greater overpressure will be required below the valve to open it. This overpressure is, however, not greater in the ratio of the areas exposed to pressures above and below the valve, because there is always a film of water between the valve-face and its seat, through which the pressure is transmitted, and to a considerable extent balanced.

1.3.11. Often the water of a mine is corrosive in its action, or contains much gritty material which cuts the valve-seats and valve-faces, so that great difficulty is experienced in finding a proper material or construction by which the valves can be kept tight.

1.3.12. Flexible valves are generally made of rubber. They are suitable only for moderate lifts. Round and rectangular forms exist. Their seats form a grating, which supports the rubber at many points. Fig. 38 illustrates a type of round flexible valve, such as is used for air-pumps of steam engines. The seats of all valves having flexible faces must have all the sharp corners of the edges of the seat rounded off, so

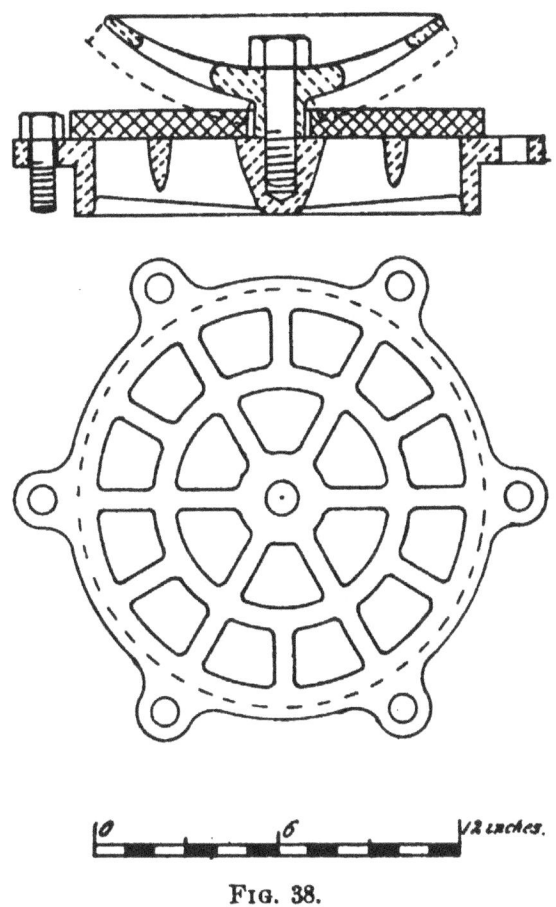

Fig. 38.

as not to cut the flexible material. Flexible valves open with very little overpressure beneath them, because the least excess of pressure bulges up the exposed part of the valve and lifts it a very little at the inner edge of the seat, where the water enters, and thus communicates the pressure to a greater area, which is again increased, and the valve thereby rapidly peeled off its seat.

1.3.13. *Area and Lift of Valves.* Quick, easy opening and closing, with a minimum of obstruction to the current passing the valve, were mentioned before as requisites for all pump-valves. As it is generally desirable (for reasons stated farther on) to keep the lift of valves as small as possible, it is necessary to make them of a correspondingly larger area, so as to keep down the velocity and consequent resistance to the flow past the valve. Such enlargement of area must, however, be kept within limits, as the leakage is liable to be greater with larger valves when closed, and also because the valve, during its closing stroke, permits some water to flow back, so that the decrease of this back-flow due to the lower lift and shorter time of closing is counteracted more or less by an increase due to the greater circumference exposed to back-flow. The higher the piston-speed of a pump, the greater should be the area of the valves, in order to insure small resistance as well as quick closing. Suction-valves should be of ample area in order to reduce the resistance, particularly where the suction-lift is considerable; also in high altitudes and where the water is warm.

1.3.14. In order to keep the diameter and also the lift of valves within bounds, straight-lift valves are quite often constructed with double or multiple seats, as in Figs. 39 and 40, the valves and seats being annular with inner and outer discharge-edges.

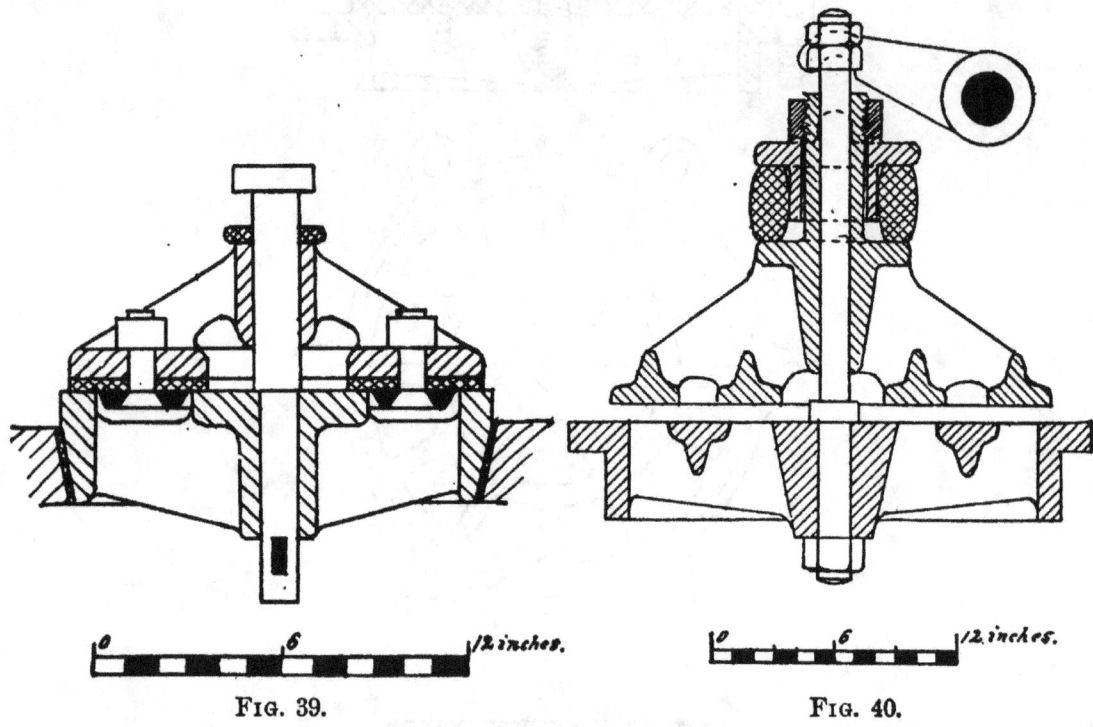

FIG. 39. FIG. 40.

1.3.15. Where valves are placed in buckets, as in the ordinary lift and jackhead pumps, the valves can naturally not be made of the requisite area, and the resistance introduced by the contraction acts to reduce the speed at which such pumps may be operated. This defect is, however, to a great extent counteracted by the uniform direction in which the water moves, as it is not reversed in its course in this class of pumps. Such valved buckets will be described in the chapter on Cornish sinking-pumps. (See 2.3.27.)

1.3.16. *Action of Valves.* Both suction- and discharge-valves should open and close as nearly as possible coincident with the ends of the pump-stroke. Tardy closing produces back-flow and increased intensity of shock. Tardy opening of the suction-valves is due to their excessive resistance, and indicates that there is liability of a reduced fill of the pump for the suction-stroke, and a shock when the plunger or piston strikes the water on the return-stroke. Promptness of closing is particularly desirable for the discharge-valves where the head is great. A slightly reduced lift and increased resistance due to it in the discharge-valve is not so great a detriment. Promptness of closing can be secured by making the valves heavy, or by using the pressure of springs. Stops must be used in all cases to keep the valve-lift between limits, and it is well to make these so that the extreme lift can be adjusted to suit the best working of the pump. Clack valves for Cornish plunger pumps, Fig. 34, are usually made heavy, of cast-iron, and the stops are cast on the clack-chamber doors. Sometimes spring-stops, which are compressed by the valve when the overpressure beneath holds it open, are used. Such springs also serve to accelerate the closing of the valve at that point where its weight is least effective. Fig. 41 shows such an arrangement, which was designed by Mr. S. N. Knight, of Sutter Creek, Cal. The closing of the rubber disk-valves commonly used in direct-acting steam-pumps is accelerated by springs. (Fig. 35.)

1.3.17. *Number of Beats of Valves.* The admissible number of beats per minute of a valve, and therefore the number of strokes of the pump,

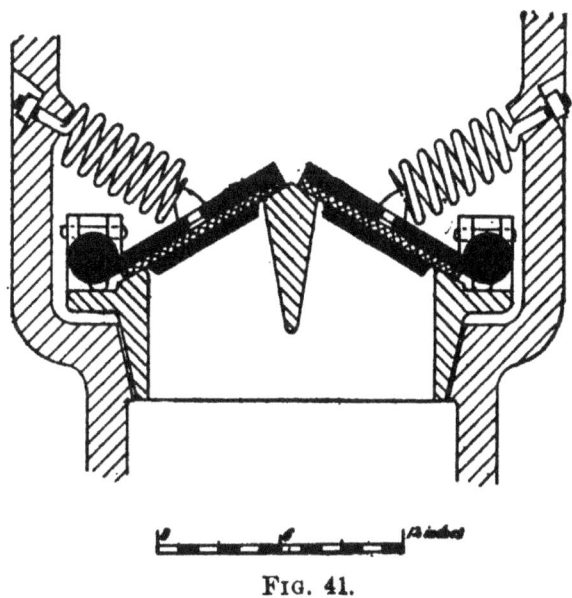

FIG. 41.

depends upon many conditions. Among these are: design, size, weight, and lift of the valve; length of the pump-stroke; velocity of flow at each part of the stroke; the head pumped against; length of the discharge-pipe; the height of suction-lift; and also whether a single pump does the work, or whether two or more, operating in rotation, force the water into the same pipe. All these conditions influence the motion of the valves to such an extent that they must all be considered and calculated or otherwise determined as far as possible, in order to decide at what rate a pump can be allowed to run under different conditions.

1.3.18. *Mechanically Actuated Valves.* A modern method of securing perfect action of pump-valves is to aid their movements by mechanical means, as in the pump valve-gear of Prof. Riedler, a form of which is shown in Fig. 42. The valve is here constructed so as to open as freely as possible without the assistance of mechanism. A little before the time when the valve should close entirely, and when the velocity of flow is already considerably reduced, so that a partial closing will offer no appreciable obstruction, a lever or rod operated by valve-gear from the

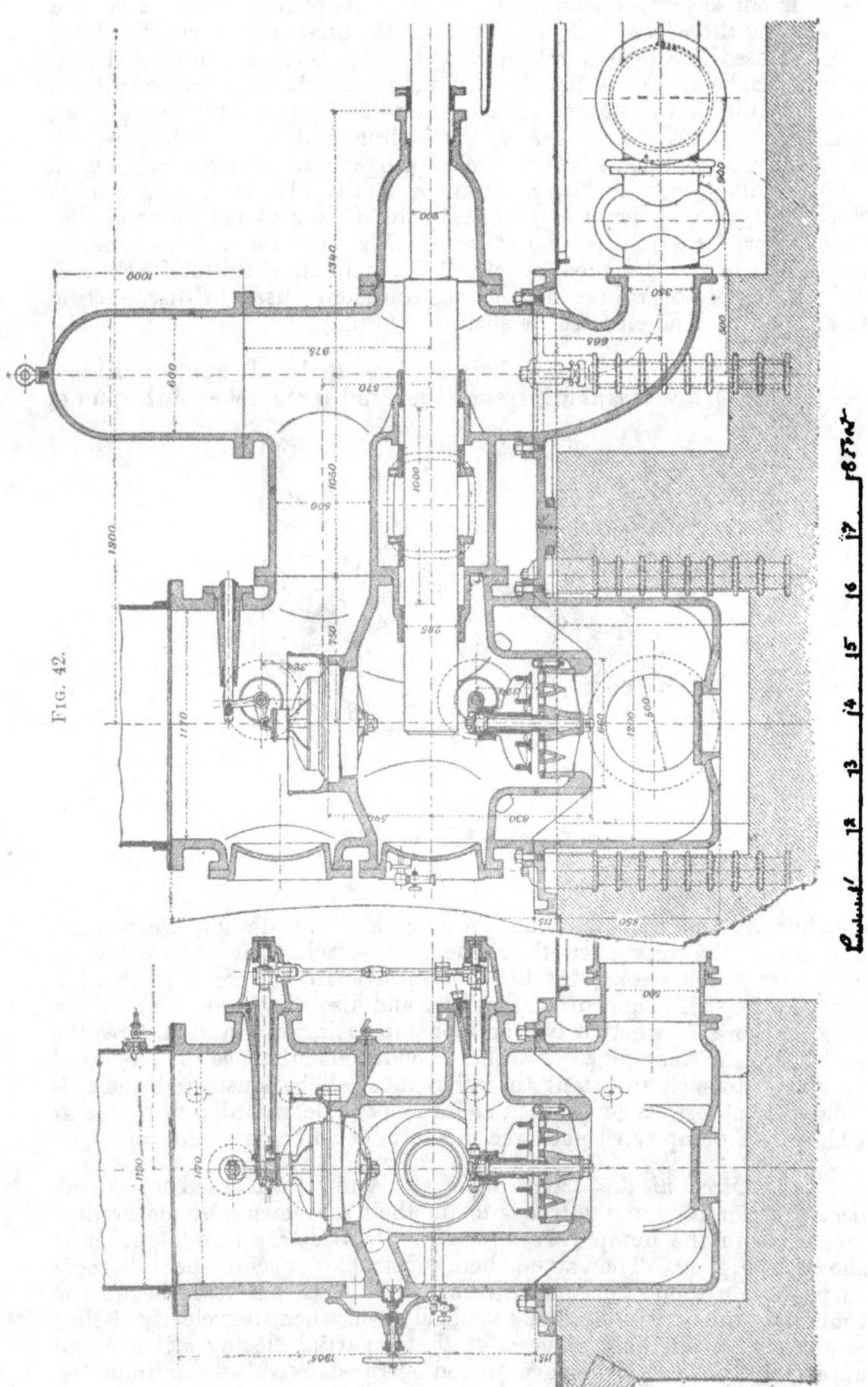

Fig. 42.

crank-shaft moves toward and closes the valve; the arm then recedes, and removes all pressure from the valve before the time for its opening arrives. With non-rotative pump-engines this arrangement is not applicable, but it is used successfully for steam-pumps driven by rotative engines. Riedler has constructed his pump valve-gear in various ways; some are operated by cams, others by eccentrics; in some the closing levers are used to remove the pressure of a spring from the back of the valve before its time of opening; in others the lever is armed with a spring; and still in other constructions a small hydraulic plunger is used instead of a spring.

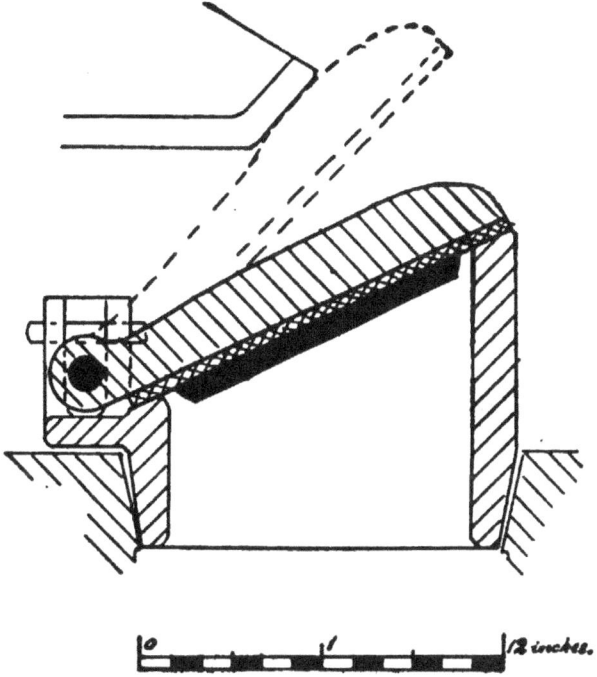

Fig. 43.

1.3.19. *Inclined Valves.* Clack valves with inclined seats, as shown in Figs. 41 and 43, permit a more direct path for the water than the type shown in Figs. 32, 33, and 34; but in vertical pumps the angle which the valve makes with a vertical line is less for the wide-open position than for the valves with horizontal seats; there is therefore less acceleration, tending to close the valve by its own weight, and the use of a spring at the back of the valve is indicated. When used in inclined or horizontal pumps, having the clack-chambers placed parallel to the pump, a single valve, inclined to the axis of the chamber so as to be more nearly horizontal, works very well, even without a spring, because the weight-acceleration tending to close the valve is greatest at its wide-open position. (Fig. 44.) Particles are not so liable to lodge on inclined valve-seats.

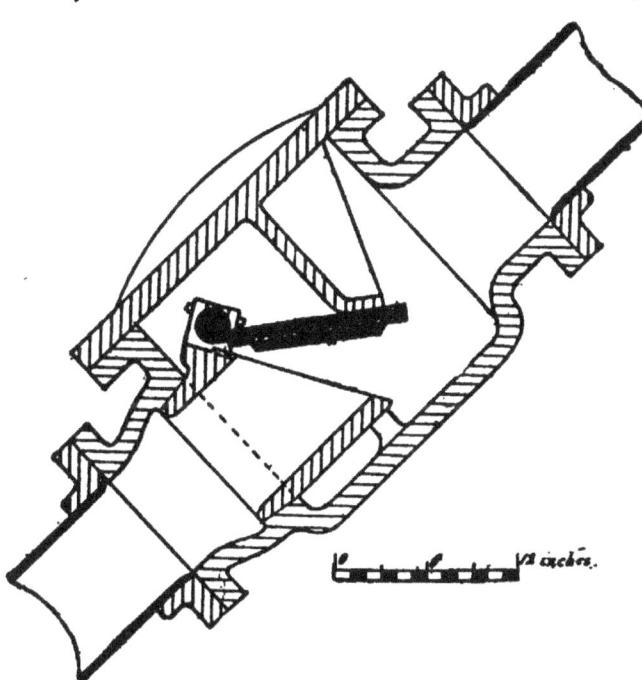
Fig. 44.

1.3.20. *Multiple Valves.* Several valves in a set are frequently used. This is the usual method in large direct-acting steam-pumps. In the Cornish system double valves, as in Figs. 33, 41, and 45, are often employed. The use of a number of smaller valves, instead of one large one, is generally necessary for high pressures. Multiple valves also present the opportunity of making the weight, lift, or spring-tension of the different valves unequal, so that they will seat successively and not all together, thereby causing a more gradual arrest of the water-column as it falls back, and thus more efficiently reducing the chance for blows than could be done by a single valve, unless the single valve is operated by mechanism, as in Riedler's construction. Fig. 46 shows a multiple valve of a type much used for waterworks pumps. Fig. 47

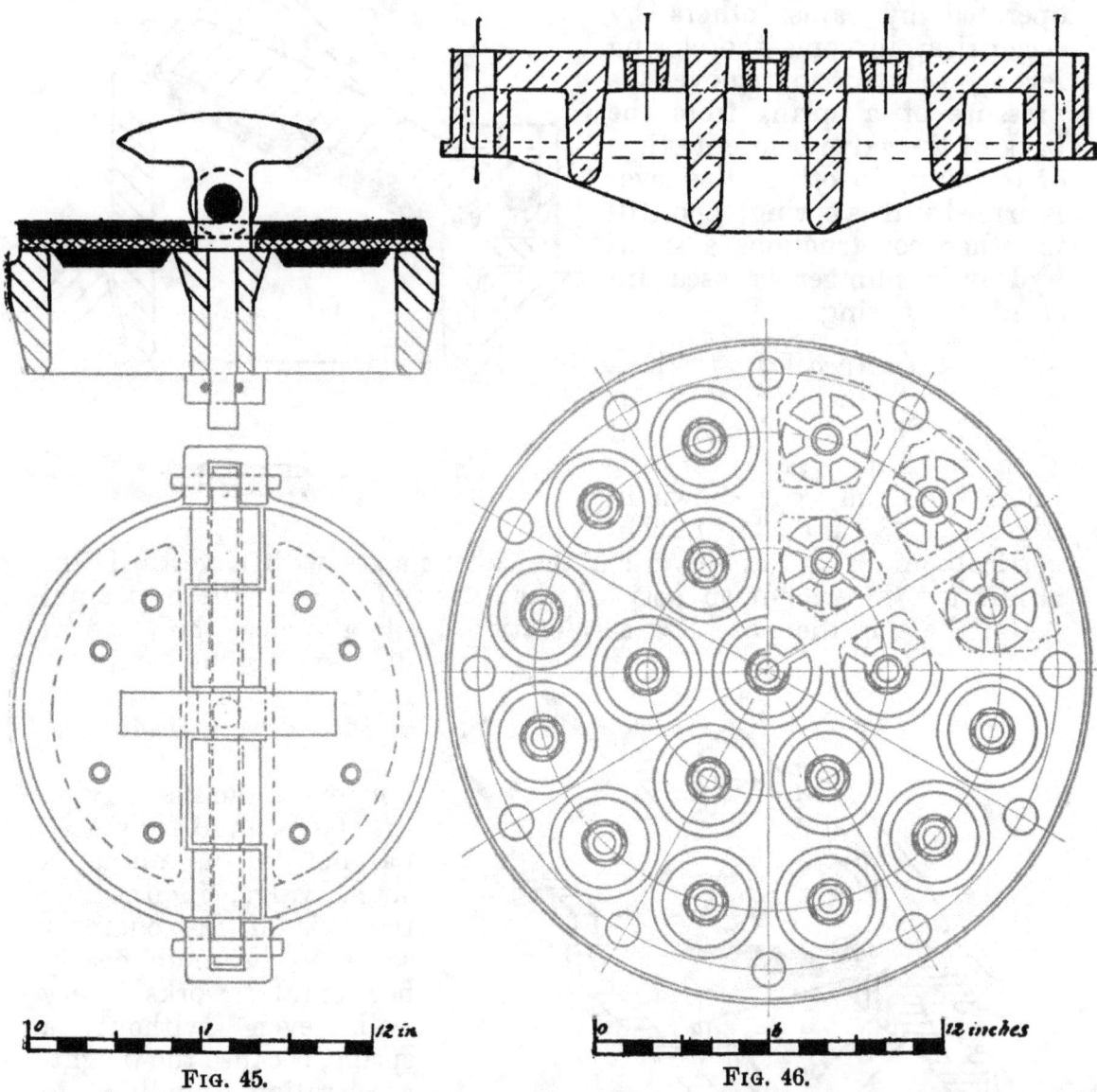

FIG. 45. FIG. 46.

is a form of valve-support which permits of getting a large number of valves into a comparatively small valve-chamber.

1.3.21. *Spring-Loaded Valves.* If a valve be made heavy in order to assist in its rapid closing, its resistance to opening is increased, and such increase is twofold: first, the heavier valve must be balanced by a greater force beneath it; and, second, there must be an additional increase of force in order to move the greater mass of the valve into its full-open position in the same time that a lighter one would be moved. If, on the other hand, a spring, the mean tension of which is equal to the increased weight for which it is substituted, is used, there will be less resistance to moving the valve to its full-open position, and the valve

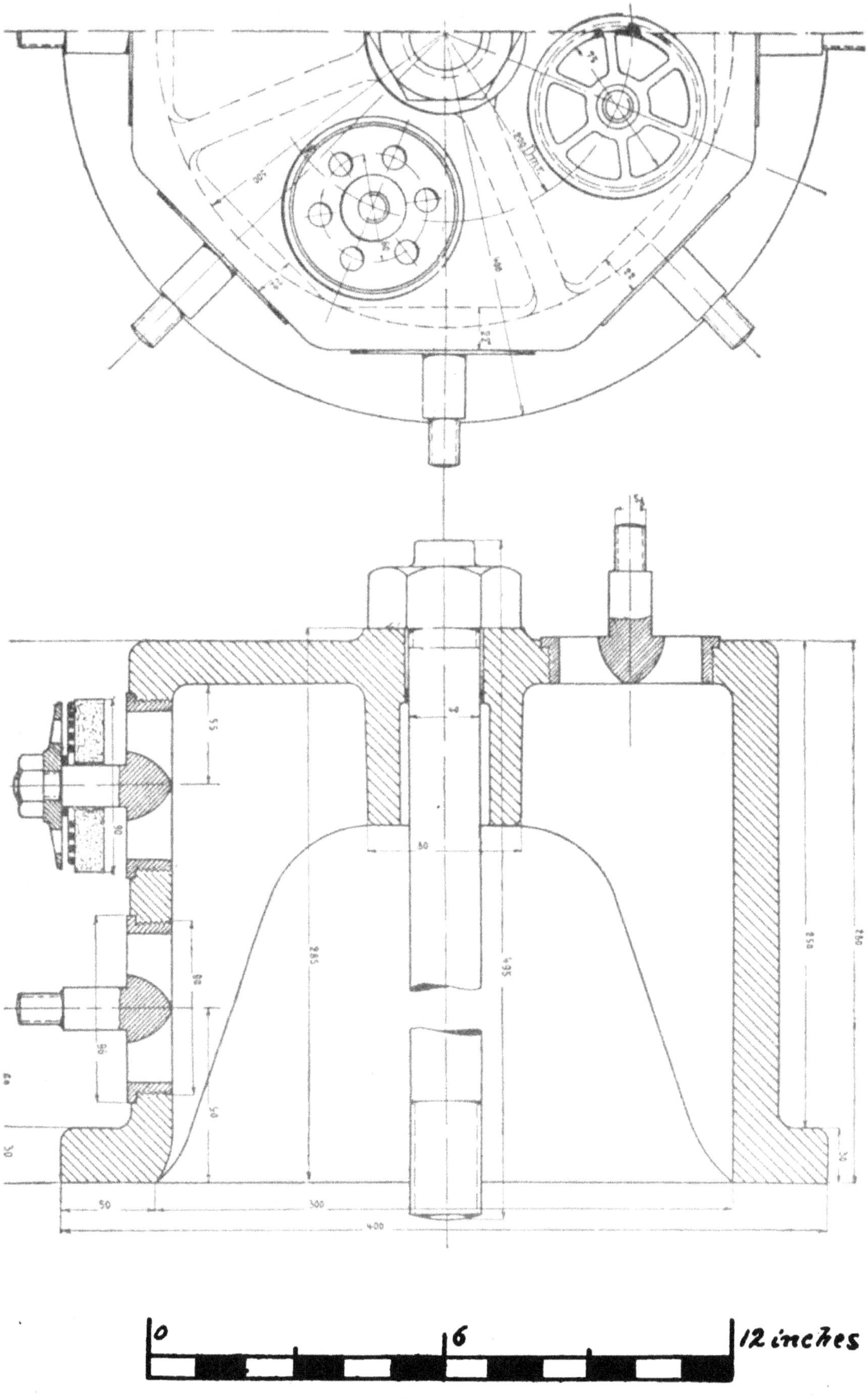

MINE DRAINAGE, PUMPS, ETC.

Fig. 47.

will also close more rapidly than by means of an equivalent weight, because, in the latter case, the increased weight or force has also to move an increased mass, while the same force exerted by a spring has less mass to move, and will, therefore, move it the same distance in less time. This argument shows that it is better to make the valve as light as is compatible with strength, and to accelerate closing by means of proper springs. The springs should be made adjustable in tension, and, to secure easy opening of the valve, they should not bear appreciably on the latter when closed. By using a number of smaller valves equivalent to one larger one, their aggregate mass can be less than that of the single one, because their thickness can be reduced with their area. A multiplicity of valves favors the application of springs, because these can be made light for small valves. Spring brass is the proper material for valve-springs, as steel would soon rust away and is also more liable to break from shocks. Springs are very extensively used for the rubber disk-valves of direct-acting pumps (Fig. 35). For pumps of the Cornish system, their application has been limited, but there seems to be no reason why, if properly constructed, their use should not be advantageous.

1.3.22. *Valve-Chambers and Valve-Seat Fastenings.* The seats of valves are usually separate from the valve-housings, or clack-chambers, and are held in place either by bolts, or simply by their own weight aided by the friction of a conical recess into which they are forced by the pressure of the water upon the valve when closed. For small valves, the seats are frequently secured by screwing them into the body of the chamber. Fig. 37 shows a single disk-valve with its seat secured by a central bolt. The suction-valves of Cornish lift pumps are now, less often than formerly, arranged for drawing up through the column-pipe. This arrangement is advantageous only where a shallow mine, involving the use of only a single lift, is drowned out, or where a deep mine is liable to be flooded up to, but not above, the level of the next higher set of pumps.

1.3.23. The clack or valve-seats of Cornish pumps are usually formed with a tapering ring or spigot, which fits into a boring of the clack-chamber. The spigot is wrapped with canvas, or similar material, before putting in place. If this is not done, the seat may jam so tight, or rust fast in the conical bore, that it cannot be got out without risk of breaking. The projecting part of the seat should have lugs cast on at opposite sides so that a bar can be inserted under them, and the seat pried up. Fig. 48 shows a common form of clack-seat in its chamber. For inclined pumps having also inclined clack-chambers, it is well to have some additional bolt-fastening for the valve-seat, as the latter has little tendency to fall back into its place, if by accident forced therefrom.

1.3.24. In order to gain access to the valves of pumps, the chambers are provided with doors or covers held in place by bolts. Fig. 48 is the clack-chamber of a Cornish plunger pump; the bolts for securing the door are hinged to the chamber-casting by an eye on one end, and fit into slits extending from the edge of the flanges. This arrangement has the advantage that the cover can be removed very rapidly by simply slacking the nuts sufficiently to permit the swinging aside of the bolts, and also that the nuts cannot be lost easily, or fall down the shaft, to the peril of men working below. In a shaft there should be the fewest possible number of loose pieces or tools placed where there may be danger of their falling.

1.3.25. The covers of large valve-chambers are heavy, and the doors must be cast with lugs, and have a ring by which to lift them, and they should have starting-bolts to break the joint when it is necessary to take them off.

1.3.26. In order to compensate, in part, for the weakness of the chamber, due to having an opening in one side, it is well to have the

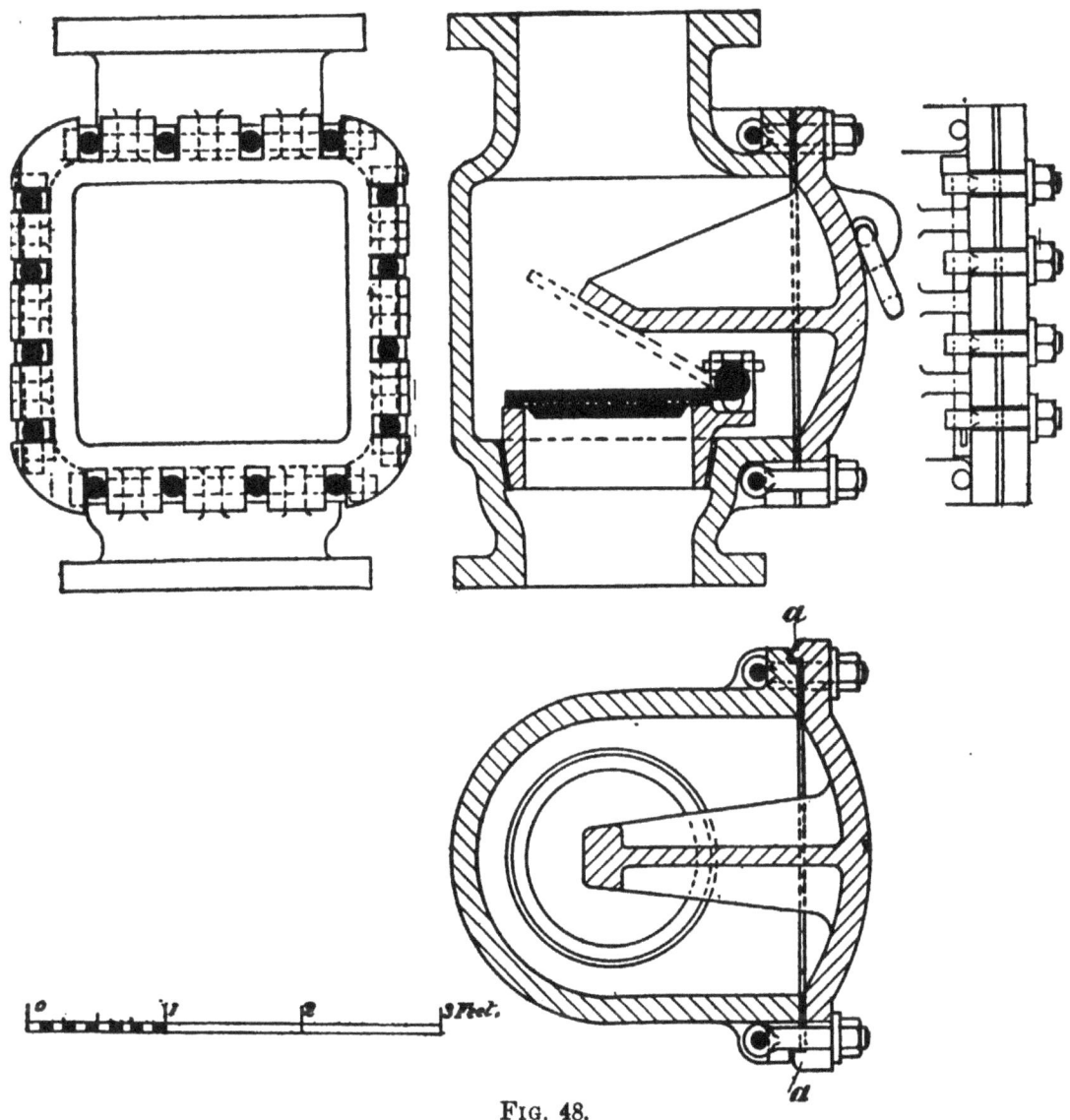

Fig. 48.

covers formed with a projecting ledge at the two vertical edges, which are fitted over the outer edges of the flange on the chamber, and serve to bind them together, as at a, Fig. 48. The doors of valve-chambers should be placed so that they are readily accessible, as the valves usually require frequent changing or repairs. Besides the large, heavy door for the removal of the valves, it is sometimes an advantage to have another small door on the side or in the main door, which can be quickly removed, and which is just sufficiently large to admit of inspection when anything is wrong with the valve, such as chips or gravel on the seat, which can be removed by the hand.

1.3.27. In horizontal pumps the valve-chamber covers are often on top, and admit of easy access to the valves.

1.3.28. *Stop-Valves.* In order to get at the discharge-valves of pumps, without draining the entire column-pipe above, a gate-valve is sometimes placed above the valve-chamber, which can be closed when it is necessary to get at the discharge-valve. The ordinary gate-valves in the market are generally of too light construction to bear the weight of the column-pipe, and also the lateral strains that are liable to be thrown on their casings. Special, heavy valves should be used for this purpose. A very good plan with Cornish pumps is to put above the discharge-clack an additional clack, which remains open and inoperative during the working of the pump, and swings entirely out of the way so as not to obstruct the flow. The valve must be arranged so that it can be closed by a handle from the outside, which is done without a shock when the pump is stopped. If the water be let out of the column-pipe, the pump-work will usually be out of balance on starting up again until the column-pipe is filled.

1.3.29. *Spare Gear.* In order to avoid delays, there should always be a number of valves and seats on hand, and ready to immediately replace others taken out. The number of parts necessary to be kept on hand can be reduced by having the valves, or at least such parts that cannot be made or repaired at the mine, of the same pattern so that they will be interchangeable.

SECTION II.

PUMP SYSTEMS OPERATED BY RODS.

CHAPTER I.

General Description of System.

2.1.01. Notwithstanding the fact that other more recently developed methods of transmitting power to operate pumps underground, such as direct steam, compressed air, water pressure, and, to some extent, electricity, are in most instances, particularly for great depths, more suitable and economical, the method of operating pumps in shafts by means of rods has still a considerable range of application for moderate depths.

2.1.02. The name "Cornish System" applies to an arrangement whereby a rod simultaneously operates a series of pumps, all of which are plungers, except the lowest, which is a lift pump. Each pump delivers the water into a tank, from which the next higher pump draws its supply. This system is said to have been first applied in 1801 by Captain Lean at one of the mines of Cornwall. The reason for using plungers is, that these, where they are not required to operate under water, can run uninterruptedly for a much longer time than lift pumps. Where submersion is liable to occur it requires a pump which can be operated and repaired under such conditions, and for that reason the older lift pump was retained as the lowest of the series. The kind of power used to operate the rods may be either steam or water. Originally the only method of working the rod was by means of a single-cylinder, single-acting engine, which lifted the rod and the water in the lift pump, and then allowed the rod to sink back, its weight driving down the plungers. The single cylinder of this Cornish engine did not admit of an economical degree of expansion of the steam, because the excess of pressure at the beginning of the up-stroke produced excessive strains in the pumprod and effected too great an acceleration of it and its attachments, causing shocks and frequent breakdowns. The introduction of compound, or Wolf, engines secured a higher and more economical rate of expansion with less variation in the extremes of pressure. These engines were, however, still single-acting, and therefore of large size in proportion to the work. Double-acting, non-rotative engines were introduced about the latter part of the "sixties." Later, double-acting, rotative engines, with crank and flywheel, came into use. A defect of this kind of direct-coupled, rotative engine is, that it cannot be operated at very slow speed, as it may then stop on the center, and it is therefore not suitable for the same variability of pump-work as the non-rotative engine, which operates for any length of pause between the strokes. Kley, of Bonn, remedies this defect of ordinary rotative engines by arranging the valve-gear so that the engines can rotate in either direction, and therefore they can be reversed automatically before the end of stroke for slow speed, and in that case be operated similarly to non-rotative engines, while at a greater speed they turn the centers and rotate the crank in one direction. A recent arrangement of rotative pumping-engine is that of Regnier, in which the dead points are overcome by a smaller engine coupled to a crank at right angles to the main crank. These engines require only a comparatively light flywheel. They are, at present, the most perfect rotative engines for working pumps through rods, and a number of them are operating at mines in Germany.

2.1.03. Ordinary steam engines, geared to a crank operating the pumprod through a bob or beam, form one of the oldest applications of the rotative principle, and are much used on this coast. Probably the largest examples of this type are found on the Comstock. The geared arrangement is also the one suited to driving Cornish pumps by waterwheels. Reciprocating hydraulic-pressure engines began to be used for operating pumps about the middle of last century. Many examples of this class of engines exist in Germany, France, and England. On this coast, Mr. S. N. Knight, of Sutter Creek, Amador County, has been prominent in introducing a type of his own design.

2.1.04. In all double-acting arrangements for operating Cornish pumps by engines or other motors, part of the work is done on the up- and part on the down-stroke. For rotative engines and motors the work on the up- and down-stroke should be approximately equal. On the down-stroke the main work of pumping is accomplished by the plungers, while on the up-stroke the weight of the pumprod is lifted with the water in the lift-pump column. It is evident, since the weight of the rod aids the plungers in lifting their water, that, if the weight of the pumprods, plus half the total pressure on the lift-pump-bucket, equals half the total pressure on all the plungers, the work on both strokes will be equal. Unless balance-bobs are used, this leads to a very light pumprod, which, in order to be sufficiently strong to resist compression, must be made of iron girders, channels, or tubes. Such pumprods are expensive, and the connection of the sections presents difficulties and requires first-class workmanship. Pumprods are usually made of wood in this country, where there is an abundance of excellent timber. In order to secure proper strength, wooden rods require to be heavier than iron ones, for which reason they have to be equipped with balance-bobs, so as to equalize the resistance on the up- and down-strokes. Wooden rods, with balance-bobs, are almost invariably used where the Cornish system is applied on this coast. The maximum stroke used with Cornish pumps is about 10'.

2.1.05. A system allied to the pumprod system (inasmuch as the same kind of engines are used as motors, and the water raised by successive lifts) is that in which the column-pipe is made to serve as the pumprod, thereby saving some room in the shaft.

2.1.06. Owing to the weight of the pumprod, which has to be balanced, there is little virtue in double-acting Cornish pumps. A double-acting pump might be warranted only where a large quantity of water is to be raised from a depth of a few hundred feet.

2.1.07. Most of the proposed double-acting constructions have opposite plungers, one stuffing-box being a hanging one, and therefore exposed to all the sand and mud contained in the water. The only double-acting pumprod system which, in the opinion of the writer, has any merit, and which also has found some application in Europe, is the Rittinger telescope-pump system, in which the column-pipe, like that mentioned in 2.1.05, serves the purpose of the pumprod. Such systems are, however, too complicated, and the pumps too inaccessible for our purposes here, and the writer knows of no case of their application.

2.1.08. The pumprod system has been used for depths of over 3,000', but it is unquestionably unsuited for economical work under such extreme conditions. On account of the elasticity of the pumprod, the lower pumps do not get their full stroke, and their action at the end of the stroke becomes uncertain.

CHAPTER II.

Pumprods.

2.2.01. *General Arrangement.* The pumprods which serve to transmit motion from the engine or other motor to pumps of the Cornish system, were formerly sometimes several thousand feet long. Owing to the elasticity of the rods, referred to at the conclusion of the preceding chapter, and their great mass, both of which affect the working of the pumps and produce severe strains and frequent breakages at speeds that would be admissible with shorter rods, the working-speed of such long rods must be kept very low, and the pumps and entire working-plant must be larger for the same capacity.* The Cornish system should properly, therefore, not be applied for such depths, particularly as other methods of transmission are to-day available to give equal commercial efficiency.

2.2.02. Pumprods are composed of pieces or sections joined at their ends by very strong connections, which must be capable of bearing the continual reversal of heavy strains to which they are subject. The rods must be securely guided in the direction of their motion in the shaft or incline. As constructed in this country, they generally require to be arranged with counterweights to balance the excess of weight over that required to equalize the work on the up- and down-strokes. The plungers and sinking-pump rods are usually attached to the side of the main rod. In Europe, the main rod is frequently forked or made double to enable a single line of plungers to be placed in the axis of the rod.

2.2.03. Even where the sinking of a shaft has been completed, the sinking-pump often remains in place, no pump-station being put in at the bottom, where it might be flooded.

2.2.04. In order to enable sinking to be carried on easily, the sinking-pump rods must be capable of being readily disconnected and hauled up with the bucket of the pump. For this reason the sinking-rod is usually not in line with the main rod, but offset to one side and clamped to it, or to an intermediate distance-piece, in such a manner that it can be let down to suit the increasing depth.†

---

*At the Combination Shaft, Virginia City, Nevada, a vertical pumprod, 15" square and over 3,000' feet long, operated a double line of 15" plunger pumps, at a maximum of 6½ strokes per minute, the stroke being about 7' 6" at the surface.

†When Cornish pumps were still in operation on the Comstock, the sinking lift-pump was discarded in several mines and a direct-acting steam sinking-pump of the Blake, Knowles, or Dow type was employed.

2.2.05. *Material of Rods.* Pumprods are made either of iron or wood. The sections of the latter are usually connected by iron strapping-plates, though wooden plates are also used. Owing to the excellent quality of the timber, and the long pieces of it which can be obtained free from blemishes, wooden main pumprods are almost exclusively used on this coast. Iron pumprods of hollow rectangular cross-section, as in Fig. 49, are lighter than the wooden ones, and if properly designed and constructed, no balance-bobs or other counterbalancing devices will be required with them, so that the extra cost of such a rod might be compensated for by the saving in cost of balance-bobs and their stations. The moving mass being much reduced, not only by the lesser weight of rod, but also by the absence of counterweights, higher speeds and greater depths are admissible by the use of iron rods. Tubular iron rods can be made still lighter than those composed of I-beams.

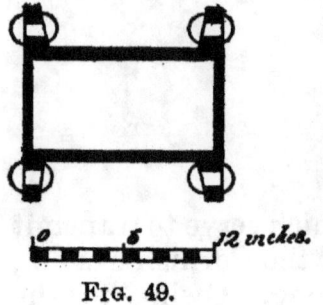

FIG. 49.

2.2.06. Iron rods require very careful workmanship, especially in the connections, which have to be very strong and rigid. Moreover, the exact length of pumprod sections cannot always be determined long in advance, and while wooden rod sections have the advantage that they can be cut to length conveniently, iron rods require careful and costly machinist work to effect such changes.

2.2.07. The only instance of a hollow iron main pumprod known to the writer on this coast is one at the Grand Central Mine (Arizona). This rod gave out in a very short time by becoming loose in the joints. Sinking-rods of iron, of solid section, are, however, extensively used, because, being only subjected to tension, they are better adapted to lift-pump work than wooden rods. They are also cheaper, consisting simply of solid rods with ends suitable for connecting by keys or other fastenings.

2.2.08. Wooden pumprods are usually of square section, except where two connected rods work a single line of plungers between them, in which case they are oblong in section. The so-called Oregon pine (Douglas fir, or *Pseudotsuga Douglasii*) is the best material for wooden pumprods.

2.2.09. Data relating to this wood were published in the 10th Census of the United States Government, of 1880, at Washington, containing the Report on the Forest Trees of North America, by Charles A. Sargent, Professor of Agriculture at Harvard College; pages 255, 259, 264, 410, 412, 476, 478.* Experiments for determining the tensile, compressive, and shearing strength of Oregon pine were also made by Arthur Brown for the Southern Pacific Co., by W. A. Grondahl for the Oregon & California Railroad Co., and by Prof. F. Soulé, of the State University, Berkeley, Cal.

2.2.10. All the experiments confirm the excellent qualities of the timber, but they also show that its resistance to longitudinal shear, or sliding of the fibers upon each other, is very slight. This fact must be taken account of in constructions, and such connections as hook-splices, and similar fastenings depending upon the resistance to longitudinal shear, should be avoided with this material. The following table, taken

---

*A very excellent pamphlet, giving the important results of some of the experiments mentioned, has been published by the Pacific Pine Lumber Co. of San Francisco.

from a paper read by Prof. Soulé before the Technical Society of the Pacific Coast, shows the results of different experimenters:

TABLE II.

Tests of Oregon Pine (Douglas Fir—Pseudotsuga Douglasii).

	University of California.	United States Government, Watertown.	Arthur Brown, for S. P. Co.	W. A. Grondahl, for Or. & Cal. R. R.
Ultimate crushing strength	5,055	8,496	6,000	------
Parallel to grain	------	10,685	------	------
	------	5,772	------	------
Ultimate shearing strength, parallel to grain	635	442	600	689
	------	356	------	------
Ultimate tensile strength	------	13,810	15,900	16,600

2.2.11. The pieces selected for rods must be straight-grained and as free from knots as possible. Pieces 16" square and 70' long were obtained of the requisite quality for the pumping-plant of the Ontario Mine, Park City, Utah. Sections of such length are transported by rail on two flat cars, being supported on swivel frames placed on each car.

2.2.12. *Lengths of Pumprod Sections.* The sections of pumprods should be chosen as long as they can be conveniently handled, because the number of connections is then reduced and more free rod-length available for attachment of pumps or balance-bobs. The long strapping-plates often required for wooden rods generally necessitate considerable length of the sections.

2.2.13. The admissible length of pumprod sections for vertical shafts is sometimes, however, limited by the height of gallows-frames or buildings, which do not permit raising the rods vertically prior to lowering them down the shaft.

2.2.14. *Connections of Wooden Main Pumprod Sections.* The iron strapping-plates generally used for connecting the sections of wooden main pumprods are usually four in number, and are frequently over 30' in length. The 16" pumprod at the Ontario Mine, previously referred to, had strapping-plates 33' long of 1"x 10" iron. Two opposite strapping-plates are secured to the sections by bolts passing through the wood. The bolts are usually square in section where they pass through the wood and through the plate under the bolt-head. The bolts should be sufficiently numerous so that the plates will hold to the rods by friction, independent of the shearing resistance of the bolts. It is, however, generally customary to utilize also the strength of the bolts of one pair of plates by driving hard-wood keys between the square ends of the rod sections after two plates have been bolted in place, because the plates are liable to become loose through shrinkage of the wood. The keys are then sawed off and the other pair of plates put on. The bolt-holes for these in one of the sections have then to be bored in the shaft. Where an entire new line of rod is put in, it is well to permit it to hang, if the time can be spared, after one pair of plates have been put on each joint. This permits the rod to straighten by its own weight, and stretches the

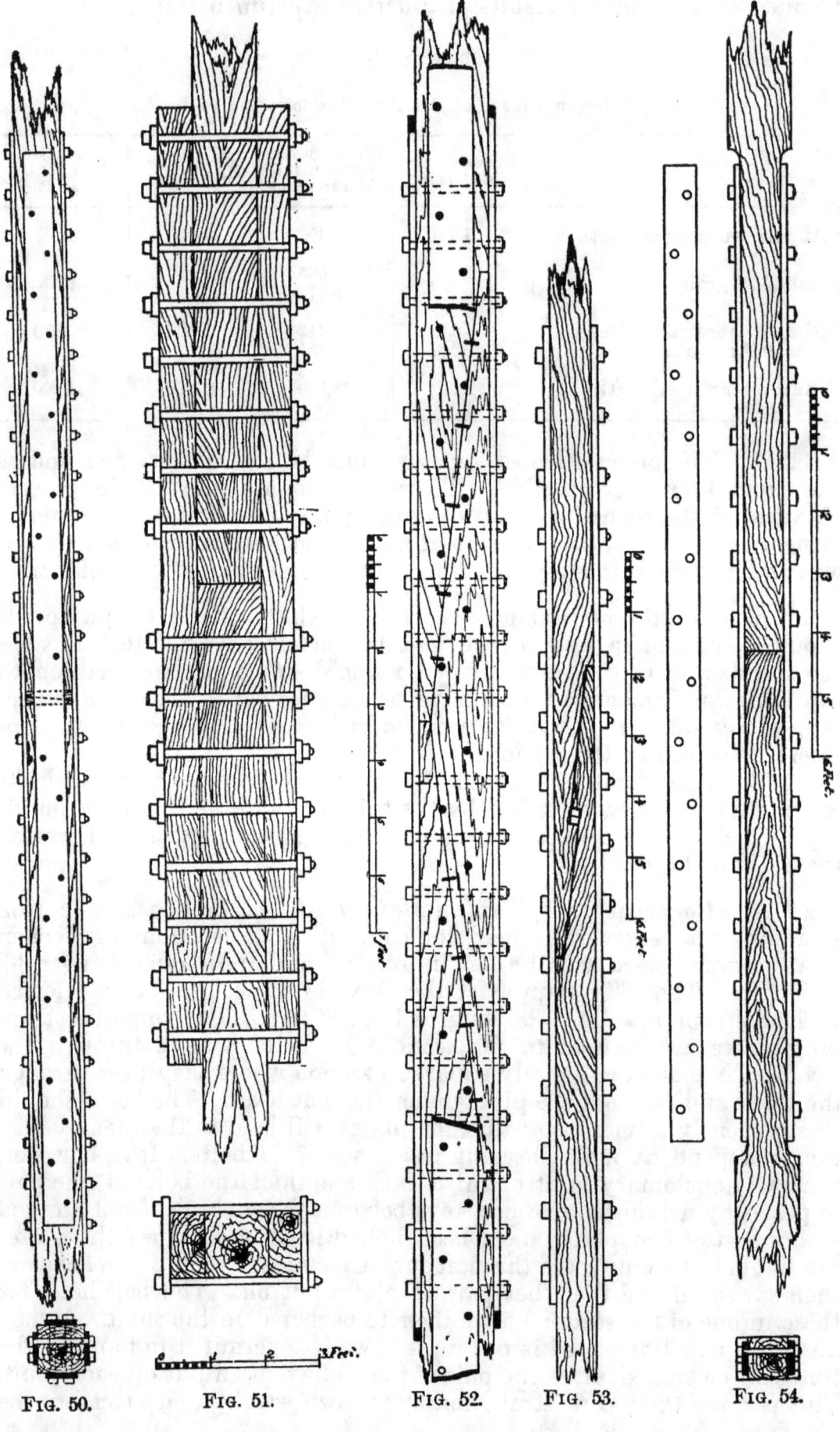

Fig. 50.   Fig. 51.   Fig. 52.   Fig. 53.   Fig. 54.

joints so that the keys can be driven home with better effect. The keys between the ends of rod sections are sometimes omitted. Fig. 50 shows a usual form of connection. The hook-splices sometimes used to additionally connect the ends of the rods, cannot add much to the strength of the joints. Where only one pair of strapping-plates is used on a rod, they are more liable to split through the line of bolt-holes. For this reason, also, the bolt-holes in the plates are not all in one line, but are placed in zigzag order.

2.2.15. Wooden strapping-plates, as in Fig. 51, are also occasionally used. As the plates hold the rod by friction, it is simpler to clamp the plates to the rod, and then also the rod and plates will not split through bolt-holes. Wooden strapping-plates have a better hold on the rod, because the coefficient of friction between wood and wood is greater than between iron and wood. The only trouble with wooden plates is the shrinkage of the wood, which loosens the clamp-bolts, and therefore these require frequent screwing up. Wooden strapping-plates, being of the same material as the rod, contract or shrink and expand equally with the latter, while iron plates are liable to severe strains, their expansion being different from that of the wooden rod. An objection to wooden strapping-plates is the space they occupy in the shaft, as they are naturally much thicker than iron plates.

2.2.16. Additional strength of joints is sometimes aimed at by overlapping the ends of the rods under the strapping-plates by a separate piece holding the rod ends by means of steel keys, as in Fig. 52. Such a connection was used at the Ontario Mine, previously referred to.

2.2.17. Main pumprods of iron or steel are generally so constructed that the joints in the different lines of channels, I-beams, or plates composing the rod shall alternate or break joints, so that no two joints fall at the same cross-section. The joints are secured by short strapping-plates held to the sections by tapered steel bolts well fitted into the straps and rod irons. The ends of the beams and plates composing the rod should be planed true. Keys of steel running through the rod and strapping-plates serve to bring the sections hard together, so that the tapered bolts can be inserted and screwed up. It is important that the workmanship of the joints of iron and steel rods be perfect, otherwise they will get loose.

2.2.18. *Connections of Sinking-Pump Rods.* Where wooden sinking-pump rods are used, they are often connected by hook-splices and one pair of iron strapping-plates, as in Fig. 53. The connection shown is objectionable, not only on account of the hook-splice, but also for operating inside of the sinking-column, because the projecting nuts and boltheads wear against the inside of the pipe.

2.2.19. By making the rods of oblong section greater than needed, the strapping-plates can be let into the wood deep enough to keep the nuts and bolt ends below the surface of the wood, and thereby prevent their wearing on the pipe. (Fig. 54.)

2.2.20. Iron pumprods of round or square sections are much used for lift pumps. Being always in tension, the joints are not so liable to get loose as with main pumprods. Fig. 55 shows a form of connection, the end of one section being fitted and keyed into a socket formed on the other section. A split pin through the projecting smaller end of the

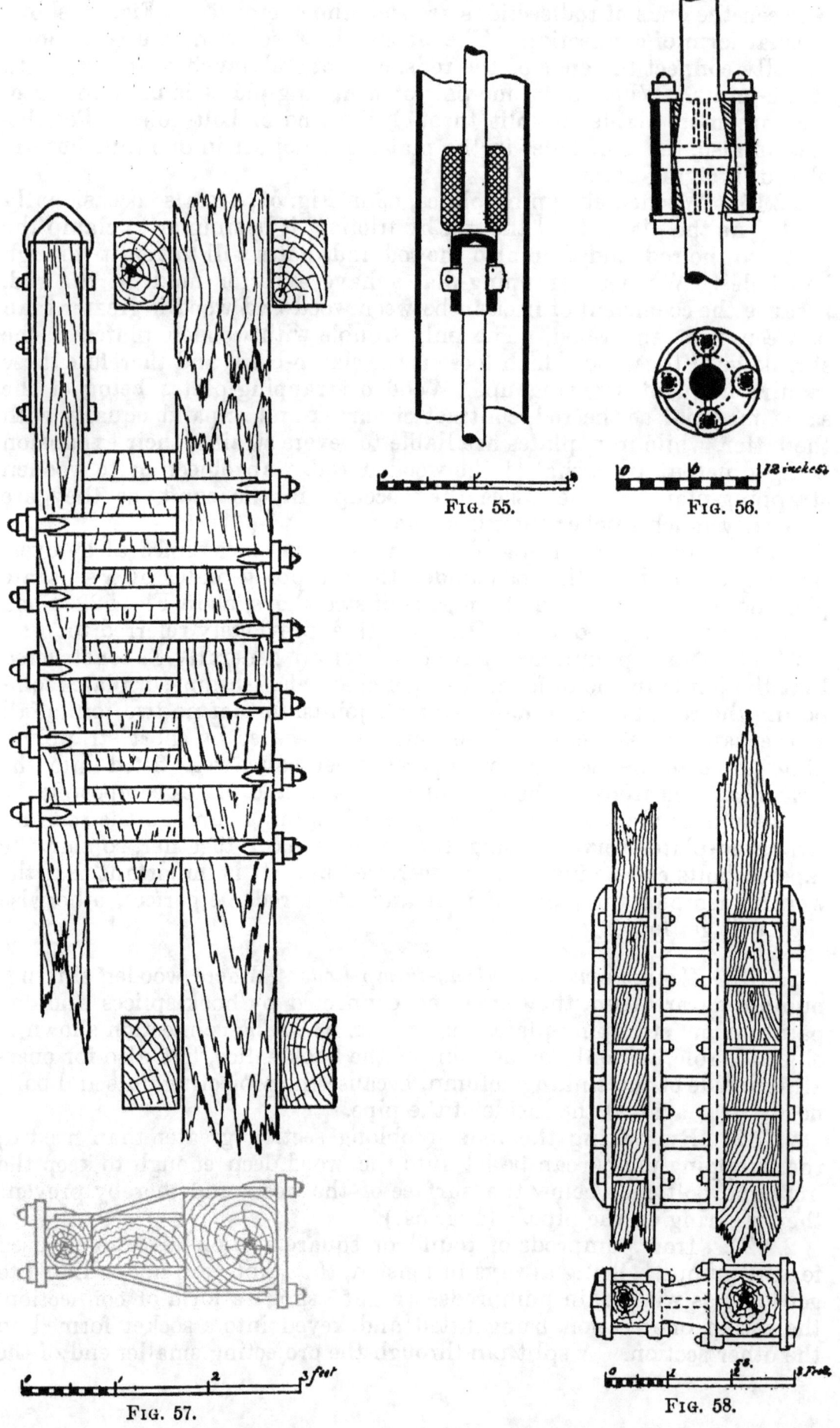

Fig. 55.

Fig. 56.

Fig. 57.

Fig. 58.

key prevents it getting loose. Fig. 56 illustrates a sinking-rod joint much used in Germany. The ends of both sections are upset, as shown, and fit into the cast-iron sleeve a, which is halved on a plane through its axis so that it can be put over the upset ends. A wrought-iron ring b holds the halves of a together, and bolts $c\ c$ keep b in place. The swelled or upset ends of the rods are shown tapering, but they can also be made cylindrical in form to fit into a corresponding recess in the halved cast sleeve.

2.2.21. In order to prevent the projecting parts of iron sinking-rods from wearing against the inside of the column-pipe, rubber hubs or bosses (Fig. 55), extending in diameter beyond any of the projections on the rod, are often used. They are mounted loosely on the rod which they surround, so that they can lag behind the rod in its motion, and thereby distribute the fluid resistance of the passing liquid over the up- and down-strokes. Their cross-section can be as large as half the clear area of the sinking-column.

2.2.22. *Connections of Sinking-Rods to Main Rods.* The usual disposition of sinking-rods in relation to main rods was described in 2.2.04. The manner of offsetting and clamping the sinking-rod can be carried out in various ways. Wooden sinking-rods are often clamped to a wooden block or distance-piece, which is also secured by clamps or bolts to the main rod, as shown by Fig. 57. Clamping the two rods by one

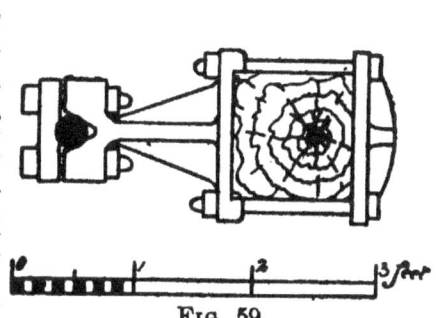

FIG. 59.

set of clamps, with the block between the rods, is a bad plan, as the shrinkage of so thick a body of wood is more liable to loosen the clamps. For this reason it is still better to use a cast-iron distance-piece, like Fig. 58, between the rods. Fig. 59 shows a distance-piece for a round iron sinking-rod. Where a sinking-rod is connected to the main rod, guides should be placed as near as possible above and below the connection.

2.2.23. *Preservation of Pumprods.* The iron-work of pumprods must be protected against rust, particularly the joints and the inside of hollow iron pumprods. Hauer recommends pickling in acid to remove rust, then coating with warm oil, and finally painting with red lead.

2.2.24. The rusting of iron strapping-plates and bolts has a tendency to rot the wood in contact with them.

2.2.25. Wooden rods last better if planed and painted, as the water runs off more readily. The abutting ends of wooden rods should always be well painted with thick paint, as this is where rotting usually first commences.

2.2.26. *Connections to Motive Power.* Wooden pumprods, where operated, as is usually the case, from a beam or bob, are generally coupled directly to the pin in the bob-nose, without an intermediate connecting-rod or link. The upper end of a rod so coupled necessarily sways back and forth during each stroke by an amount equal to half

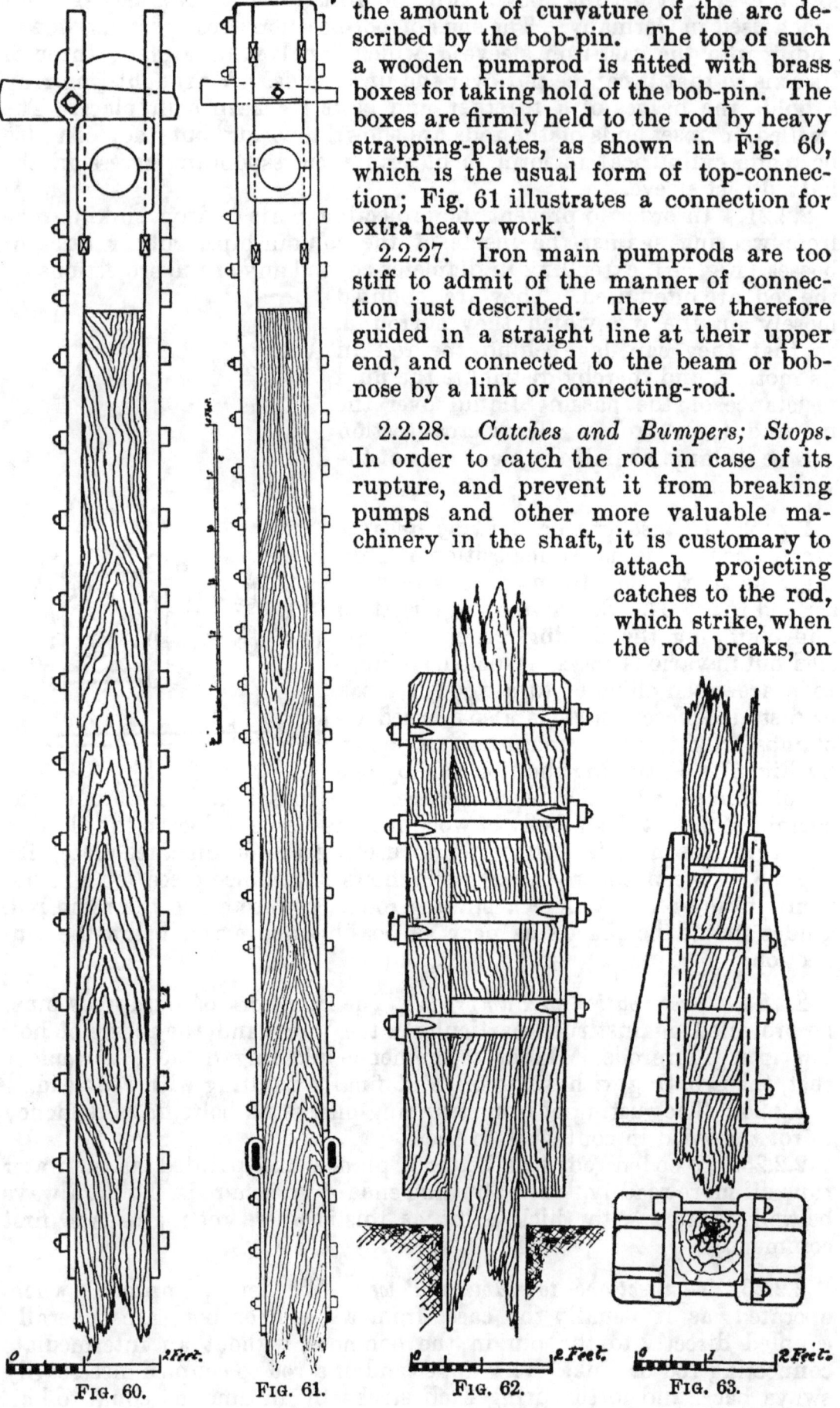

FIG. 60.   FIG. 61.   FIG. 62.   FIG. 63.

the amount of curvature of the arc described by the bob-pin. The top of such a wooden pumprod is fitted with brass boxes for taking hold of the bob-pin. The boxes are firmly held to the rod by heavy strapping-plates, as shown in Fig. 60, which is the usual form of top-connection; Fig. 61 illustrates a connection for extra heavy work.

2.2.27. Iron main pumprods are too stiff to admit of the manner of connection just described. They are therefore guided in a straight line at their upper end, and connected to the beam or bob-nose by a link or connecting-rod.

2.2.28. *Catches and Bumpers; Stops.* In order to catch the rod in case of its rupture, and prevent it from breaking pumps and other more valuable machinery in the shaft, it is customary to attach projecting catches to the rod, which strike, when the rod breaks, on to supports or bumpers fixed in the shaft. Either the catches or the bumpers must be armed with elastic cushions to break the force of the blow.

2.2.29. In order to reduce the stresses due to arresting the falling rod, the energy of the fall can be consumed gradually by causing the rod in its fall to perform some work of deformation or friction, such as breaking successively the individual boards of a pile, or causing a tightly gripping clamp to slip a short distance on the rod, which will bring it gradually to rest.

2.2.30. Catches and stops or bumpers are particularly needed with engines of the non-rotative type, because with these there exists also the danger of the stroke becoming greater than its intended limits, through variations in steam pressure or neglect in regulation of the pumps. Such engines also re-

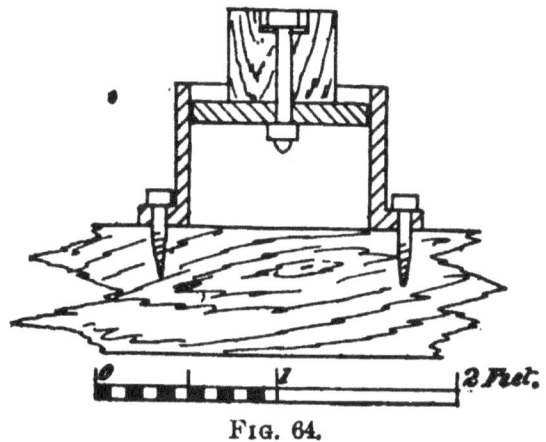

FIG. 64.

quire catches to limit the up-stroke, though only one or two are required close to the engine; generally at the beam. These engine-catches and stops require to be only moderately elastic, as they merely operate to prevent the engine from exceeding the proper limits of its stroke during its regular work, while the rod-bumpers have to consume the energy of a weight falling possibly a distance equal to the stroke of the pumprod.

2.2.31. Fig. 62 shows wooden catches, and Fig. 63 catches of cast-iron clamped to a wooden rod. It is best to secure the catches to the rod with bolts not passing through the latter, as shown, because then the bumpers can slip a little under a heavy fall, while the blow will be less severe, as the friction work produced by the slipping will help to gradually consume the energy contained in the falling rod. Clamping on the catches also secures their adjustability.

2.2.32. The elastic cushion or bumper proper is usually most conveniently attached to the fixed bumper-frame. The one shown in Fig. 64 consists of cork or old rope confined in an iron box or cylinder, and covered by an iron plate with a wooden block, on top of which the catches strike.

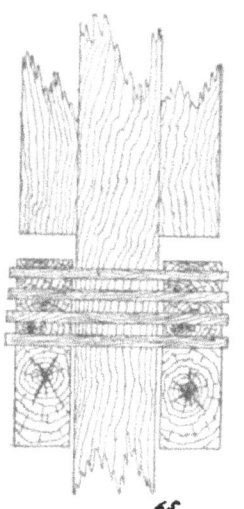

FIG. 65.

2.2.33. Fig. 65 shows a bumper constructed of boards with intervening spaces. The boards are successively broken by the catches on the falling rod, and by their resistance to breakage gradually lessen, if not entirely consume, the destructive energy of the falling rod before the last board is broken.

2.2.34. In constructing rod-bumpers the distance passed through by the rod in overcoming resistance must not be as great as the space in the pump-barrel below the plungers when in their lowest position; otherwise, the plungers will strike the bottom, and parts of the pump may be broken. Similarly other projecting attachments of the rod or balance-bobs must not come closer in operation to fixed parts than permitted by the range of the bumper. It is a good plan to place a bumper above every set of pumps.

2.2.35. *Guides or Stays.* Pump-rod guides, also called stays, should be placed sufficiently close together to render the rod safe from buckling under compressive strains, and they should be kept carefully in line

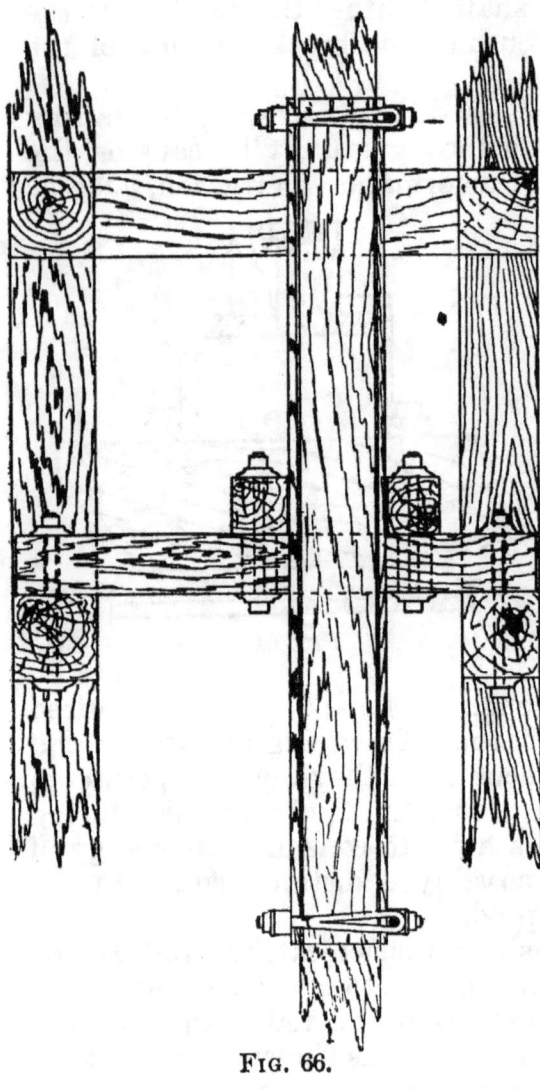

FIG. 66.

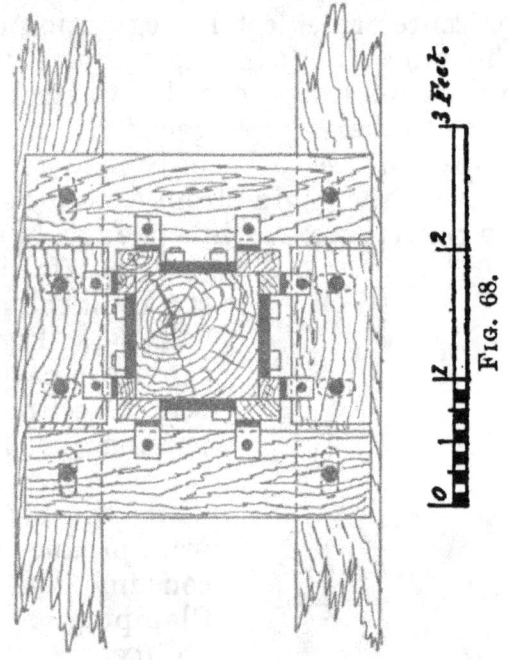

FIG. 68.

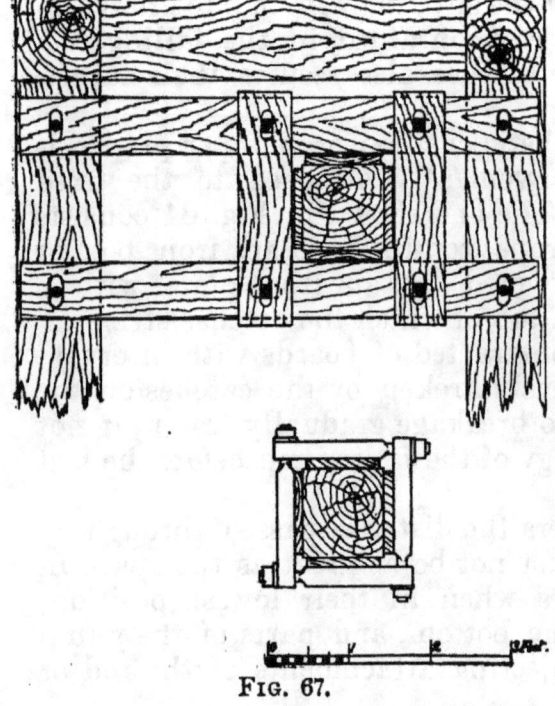

FIG. 67.

where these strains are great. They should always be located as near as possible to points where lateral strains might cause deflection, as above and below balance-bob and pump connections. The guide frames are fixed to the shaft timbers, and are provided with wearing-blocks made adjustable for taking up wear and keeping the rod in line, and the rod is armed with interchangeable wearing-strips at the guides. Fig. 66 shows a guide with wooden wearing-strips on the rod, the strips consisting of pine boards, which should not be nailed to the rod, but clamped to it at the ends by a frame of eye-bolts, as in Fig. 67, called "lamb's legs" by the miner. Where strapping-plates occur on the rod, a construction like that shown in Fig. 68 must be used. On account of the reduced surface, the wearing-strips and wearing-blocks are faced with flat iron. For the sake of uniformity, all the guides are often made the same as those at the strapping-plates.

2.2.36. The lubricants mostly used with guides are tallow for wood, and a mixture of cheap mineral oil and tallow for iron. Albany compound and axle-grease are too expensive for general use.

2.2.37. It was mentioned in 2.2.26 that wooden main pump-rods are generally connected directly to the pin in the nose of the beam or bob operating the rod, and that thereby the upper end of the rod is deflected alternately in opposite directions, while its lower end follows the arc described by the bob-nose. The aim should be to distribute this deflection of the rod uniformly over a considerable length so as to minimize the deflection strains.

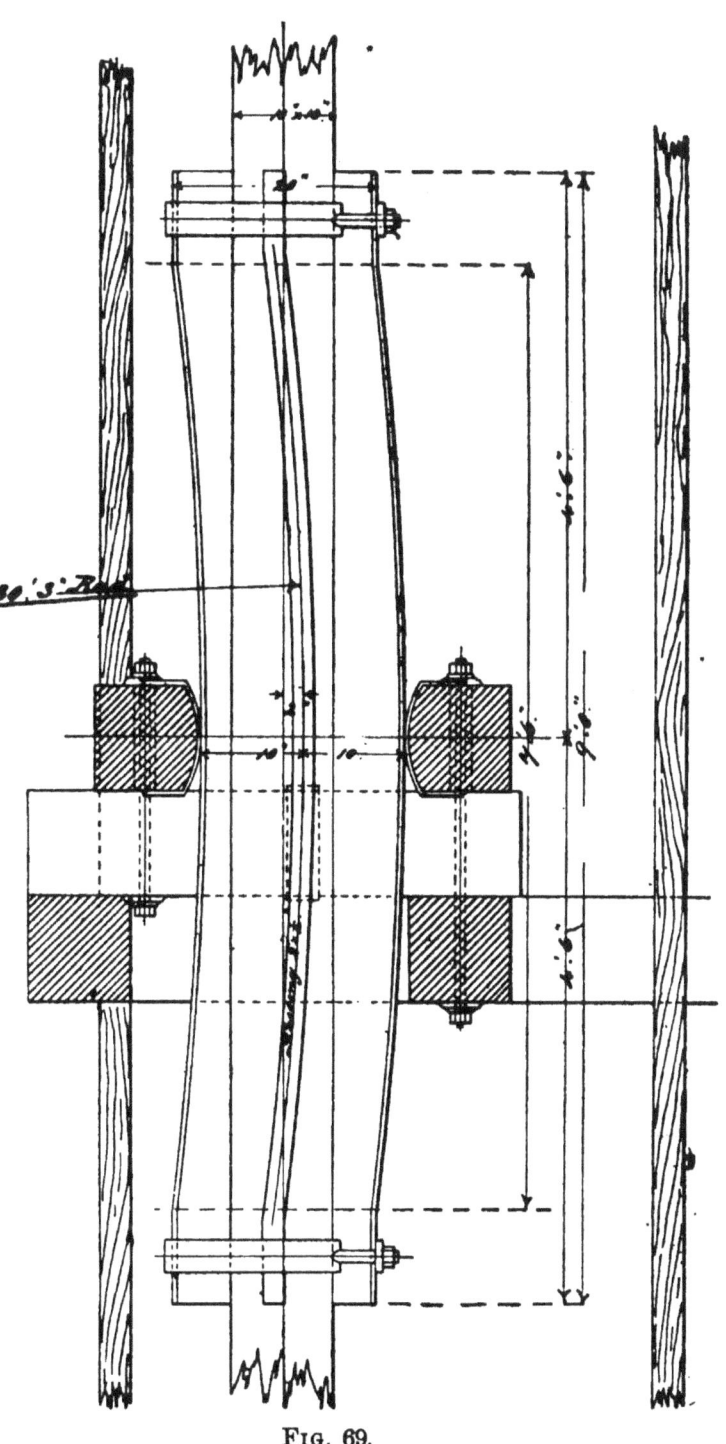

FIG. 69.

The greater the deflection and the larger the rod, the longer must be the part over which the deflection must be distributed. In order to obtain the least and at the same time the most uniform deflection strains, the rod must be curved to the arc of a circle. It is therefore necessary to stay and guide each part in its proper path. The part of the rod working in the guides which fall within the range of deflection has therefore iron curved wearing-strips, as shown in Fig. 69, the amount of curvature being greater the nearer the guide is to the upper end of the rod. Such an arrangement of guide and wearing-strips is generally called a "sweep-stay."

2.2.38. The proper distance apart of guides depends upon the load on the rod and upon its cross-section, and can be determined properly only by calculation.

2.2.39. Where lateral strains are introduced, guides should be made of extra strength, and the wearing-surfaces increased correspondingly with the greater pressure.

2.2.40. Pumprods in inclines are supported and guided by rollers. The guide-rollers are generally stationary and supported in frames fixed to the shaft timbering. Sometimes, however, rollers are attached to the rod, and travel with it on tracks and under fixed top guard-rails. The points of support should be numerous, in order to prevent sagging of the rod. The rollers should be adjustable to enable keeping the rod in line.

2.2.41. The application of the Cornish system in inclines is attended with many drawbacks. The rods are, on account of sag, less able to bear great compressive strains; the friction is great where the inclination is considerable, and the plungers wear on one side only and are hard to keep tight. Less speed is also admissible on account of greater masses than for the same lift in a vertical shaft.

2.2.42. Sinking-rods inside of a sinking-column cannot be used in inclines. It is therefore necessary to use a jackhead pump with outside rod guided like the main rod, unless direct-acting pumps, driven by steam or compressed air, are used for sinking.

2.2.43. *Adjustment of Weight of Pumprods; Balancing Appliances.* In designing a pumprod for operating a series of single-acting plungers, with or without a sinking lift-pump, the aim should be to get the rod of such weight that the work on the up-stroke shall be equal to that on the down-stroke without resorting to the use of balancing mechanism. It was shown in 2.1.04 that in order to secure this result the weight of the pumprod and attachments must be equal to one half the aggregate pressure on the plungers, plus one half the upward thrust due to the buoyancy of the sinking-rod (where this operates inside the sinking-column), minus the total pressure on the lift-pump-bucket. Balancing appliances in general can only be avoided by the use of iron rods. Wooden rods, in nearly every case of deeper mines, require counterweights to equalize the work of the two strokes.

2.2.44. The overweight of rods may sometimes be balanced without the use of other appliances, by placing the plungers a considerable distance below the supply-tank, thereby increasing the height of both the suction- and discharge-columns, so that the work on the suction-stroke is decreased by the lifting effect of the downwardly extending suction-column, and that on the forcing-stroke increased by an equal amount due to the increased height of discharge column. This plan is, however, only applicable where a moderate amount of balancing is required, as it subjects the plungers to higher pressure and greater friction, and is liable to cause heavier shocks, while it also reduces the admissible speed at which the pumps can be safely operated.

2.2.45. The use of larger pumps operated at lower speeds may also sometimes serve to overcome a moderate difference in the work of the two strokes. It is true that the rod would, by the arrangements described in this and in the preceding paragraph, on account of the greater compressive strains, also need to be increased in size, and consequently in weight, but this increase will be less than that of the increase in strength required by the greater resistance of the larger pumps.

2.2.46. In most cases the amount of counterbalance required, necessitates the use of special appliances, such as balance-bobs, or hydraulic or pneumatic counterbalances.

2.2.47. The main bob or beam generally employed to work the pumprod from the surface is usually arranged with a balance-weight. As the depth increases and additional counterbalance becomes necessary, other balance-bobs are connected to the rod below ground at intervals. These balance-bobs consist of braced beams, as in Fig. 70, which shows a balance-bob constructed of wood with wrought-iron tension members and cast-iron bishop-head and nosepiece. The bob is shown in midstroke, the inclined position of the frame being necessary for this point of the stroke in order to bring the bob-nosepiece on a level with the bob-center. The counterweight should also have its center of gravity on the same level. In this manner only will the leverages be equal for the ends of the stroke. Rubber cushions mounted on blocking below the beam are also usually put in to act as bob-bumpers in case the limits of the stroke are exceeded, or as catches when the bob-connection is ruptured. Fig. 71 gives an example of a bob with cast-iron arms.

2.2.48. Where hoisting-compartments are connected with the pump-shaft, balance-bobs should be located at the side opposite to the hoisting-compartment, because in this way the ground around the shaft at the bob-stations remains in the best supported condition.

2.2.49. The balance-weight consists usually of a large number of short sections of old rails placed in a strong wooden box, the construction of which is shown in Fig. 70. The number of small pieces enables adjusting the amount of counterweight easily. The latter sometimes amounts to as much as 30 tons. As sinking proceeds and the main or the sinking-pump rod is lengthened, increase or decrease of the balance-weights becomes necessary. The difference in speed between the up- and down-stroke will indicate in what sense the balance-weight must be changed. In non-rotative pump-engines particularly, the work of the up- and down-strokes of the engine can be adjusted to a limited extent to suit the unbalanced condition, instead of adjusting the balance-weight to obtain equal work on both strokes. The inequality will then be shown by indicator-cards taken from the engine.

2.2.50. The connection to the pumprod, by means of a single link coupled to one side of the pumprod, as in Fig. 72, is now seldom used, as it introduces objectionable bending-strains.

2.2.51. Double links at two opposite sides of the rod are now generally used, by which central strains are obtained on the rod. Such connections require a forked nose on the bob. The construction shown in Fig. 70 is, notwithstanding its faulty design, the most used. The coupling-plates on the rod for connecting to the links should be clamped to the rod by bolts not passing through the rod, so that they can be adjusted in their proper position. A difficulty with such double links is to keep them adjusted so that each link will receive its proportion of the load.

2.2.52. The increase of mass due to the use of balance-weights reduces the speed at which the pumping-system may be safely permitted to operate, but it also permits higher degrees of expansion to be used in the case of operating by non-rotative engines.

2.2.53. Hydraulic counterbalances, consisting of a plunger operating in a barrel like that of a pump, against a column of water, which constitutes the counterweight, are more objectionable than the rigid balance-bob.

2.2.54. By using compressed air, instead of a column of water, in a similar manner, a counterbalance is obtained without the evil of increased mass. Such counterbalances require no excavations like balance-bobs.

2.2.55. The distance apart at which counterbalances are required depends greatly upon conditions. The closer they are together, the smaller will be the units and the less will be the maximum compression strains on the rods, but the greater will also be their cost, particularly if the balancing appliances require the excavation and timbering of stations. Counterbalances decrease the tensile strain and increase the compression strains in the rod. Their distribution and amount should be determined by careful calculations.

2.2.56. The rod can sometimes be divided, and the two parts connected to opposite ends of a beam, as shown in Fig. 73, thus obtaining a balance without the use of extra counterweights. The plan illustrated was carried out at a mine in Belgium, but can only have application under special conditions, as the offset in the shaft is objectionable. The offset could in some cases, however, be avoided by such a construction as shown in Fig. 74 or Fig. 75.

2.2.57. The principle just illustrated can often be applied with advantage where a change of direction necessitates a bob or bell-crank to connect pumprods at an angle; the bob can be designed and located so that the rods take hold of opposite arms of the bob and balance each other more or less completely. Such an arrangement, shown in Fig. 76, was introduced in the Lady Bryan Mine, near Virginia City, Nev. The double pumprod arrangement used at the Alta Mine, Virginia City, and illustrated in Fig. 77, also affords a perfect balance, a moderate counterbalance being only required when the lift-pump work is done entirely by one of the rods. The double rod arrangement, however, takes up considerable room in the shaft.

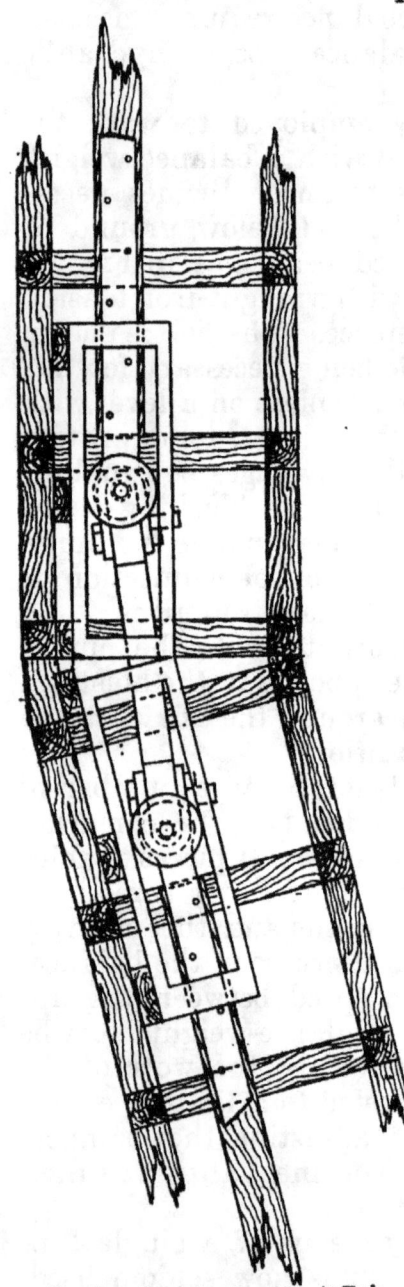

Fig. 80.

2.2.58. *Changes in Direction of Pumprods; Angle-Bobs.* One very good arrangement has already been described in the preceding paragraph. It is, however, not always possible to place the bob as shown in Fig. 76, and other less advantageous constructions have to be used. Figs. 78 and 79 illustrate forms of bobs and bell-cranks that are common. In Fig. 79 the bell-crank serves also as balance-bob.

2.2.59. Slight changes in direction are often made without the use of bobs or bell-cranks, by having the ends of the rods fitted with rollers guided in straight lines, and simply coupling them by a link, as shown in Fig. 80.

2.2.60. *Strains in Pumprods.* During the up-stroke the main pump-rod is in tension, due to the weight of the rod and attachments plus the pressure of the column of water on the lift-pump-bucket. On the down-stroke the strains change to compression, and these are due to the excess of the resistance overcome by the plungers over and above the weight of the rod. The counterbalances reduce the tension strains above the sections where applied by an amount equal to the upward force exerted by them, and increase the compression strains by an equal amount. The resultants of these strains are modified by those due to the inertia of the rod attachments and counterweights; that is, by the force required to give the rod the required velocity in a given time during the early part of the stroke, and that required to be subtracted from the motive force during the final part of the stroke, so that the rod may come to rest quietly within the limits of its travel. In addition, bending strains are often introduced by lateral disposition of single plunger pumps, or sinking-rods. Those due to the latter are of little moment, because they occur near the lower end of the rod, where the other strains are light. The upper end of the rod is generally strained the most in tension, while the point at which the greatest compression strain occurs depends upon the distribution of pumps and balancing appliances.

2.2.61. Owing to their elasticity, long pumprods extend considerably under tension on the up-stroke, and shorten under compression on the down-stroke. The result of this is, that the lower pumps do not operate at their full stroke. Another result which follows, and is intensified by the inertia of the rod, particularly at higher speed, is, that the upper part of the rod will be already in motion and will have performed a portion of its stroke when the lower end of the rod begins its motion. It is therefore evident that the ends of the strokes of the successive pumps attached to the rod at different levels cannot occur at exactly the same time. Attention was first called to the logical necessity of these results by Hraback and Bochkolz. (See Hauer, Wasserhaltings-machinen.) Experiments made by W. R. Eckart, in 1880, on pumps of Comstock mines, proved the correctness of this reasoning.

2.2.62. Wooden rods are more elastic than iron ones. Their weight is also greater for the same strength, and therefore the variation in extent and coincidence of stroke of pumps must be greatest with wooden rods.

2.2.63. *Wire Rope.* Single wire ropes are occasionally used instead of rods for operating draw-lift pumps, in which case there must be a heavy weight connected with the bucket to effect its down-stroke. A recent arrangement of this kind which has been successfully used for sinking-pump work in a deep mine in Bohemia is described in 2.3.33 and 2.3.34.

2.2.64. Two wire ropes connected to opposite ends of double-armed levers at the surface and bottom, so as to act like a rigid rod, have been also used for double-acting pumps at moderate depths. This system has, however, only limited application.

CHAPTER III.

Sinking-Pumps.

2.3.01. *Types of Sinking-Pumps.* It has already been stated in 1.1.03 and 2.1.02 that where pumps are operated by rods, as in the Cornish system, the lowest or sinking-pump, unless it be of the direct-acting, steam-pump type, is nearly always a lift pump, because this type can be more readily operated and repaired when obliged to work under water. The lift pumprod, where the total lift is too great for one pump and one or more plungers are needed above the sinking-pump, is usually coupled to an offset bracket bolted on the main pumprod, as described in the preceding chapter (2.2.04 and 2.2.22). Where the total lift is within the range allowable for the sinking-pump, it is worked directly from a bob driven by a steam engine or other motor.

2.3.02. Sinking-pumps in deep shafts have in some recent cases been operated by a wire rope, worked by a bob. (See 2.2.63.)

2.3.03. Sinking-pumps are subject to much greater wear and tear in the shaft than the other pumps, because in their case it is not practicable to settle the sand or mud from the water before it enters the pump, which can very easily be done with the other pumps.

2.3.04. In designing a sinking-pump the aim should therefore be more to secure uninterrupted operation, or facility for rapid repairing, than economical pumping. Economy is not of great importance here, particularly in deep mines, where the sinking work usually constitutes but a small proportion of the entire pumping work.

2.3.05. The ordinary or English form of lift pump, generally used in vertical shafts, has its working-barrel in line with its discharge-column, and the sinking-rod works inside the latter. Fig. 81 illustrates a common type of Cornish sinking-pump. The suction-pipe is either a rigid casting, as shown, with which the pump rests on the bottom of the shaft, or it is made with a slip-joint so that it can be raised or lowered while the pump is temporarily secured, or it is simply a suction-hose with strainer, as in Fig. 82. The pumps are usually attached to a sinking-frame guided in the shaft, as in Fig. 83, the frame with pump being raised or lowered, as required, by chain-blocks, winches, or a special pump-hoist at the surface. In most instances, the sinking-frame is only as long as the pump, and the column-pipe is guided independently. The pumps with rigid suction-pipes are often not guided in the shaft, so that the lower end of the suction-pipe may be swung around to a limited extent. The chamber for the suction-valve below the working-barrel is fitted with a door for gaining access to the valve. A casting with a door is placed on top of the working-barrel to get at the bucket without having to draw it up through the entire length of the column-pipe, which need only be done when the pump is submerged. This type of pump being all in one line with the column-pipe, occupies very little room in a shaft, and is therefore most generally used with the Cornish system in vertical shafts.

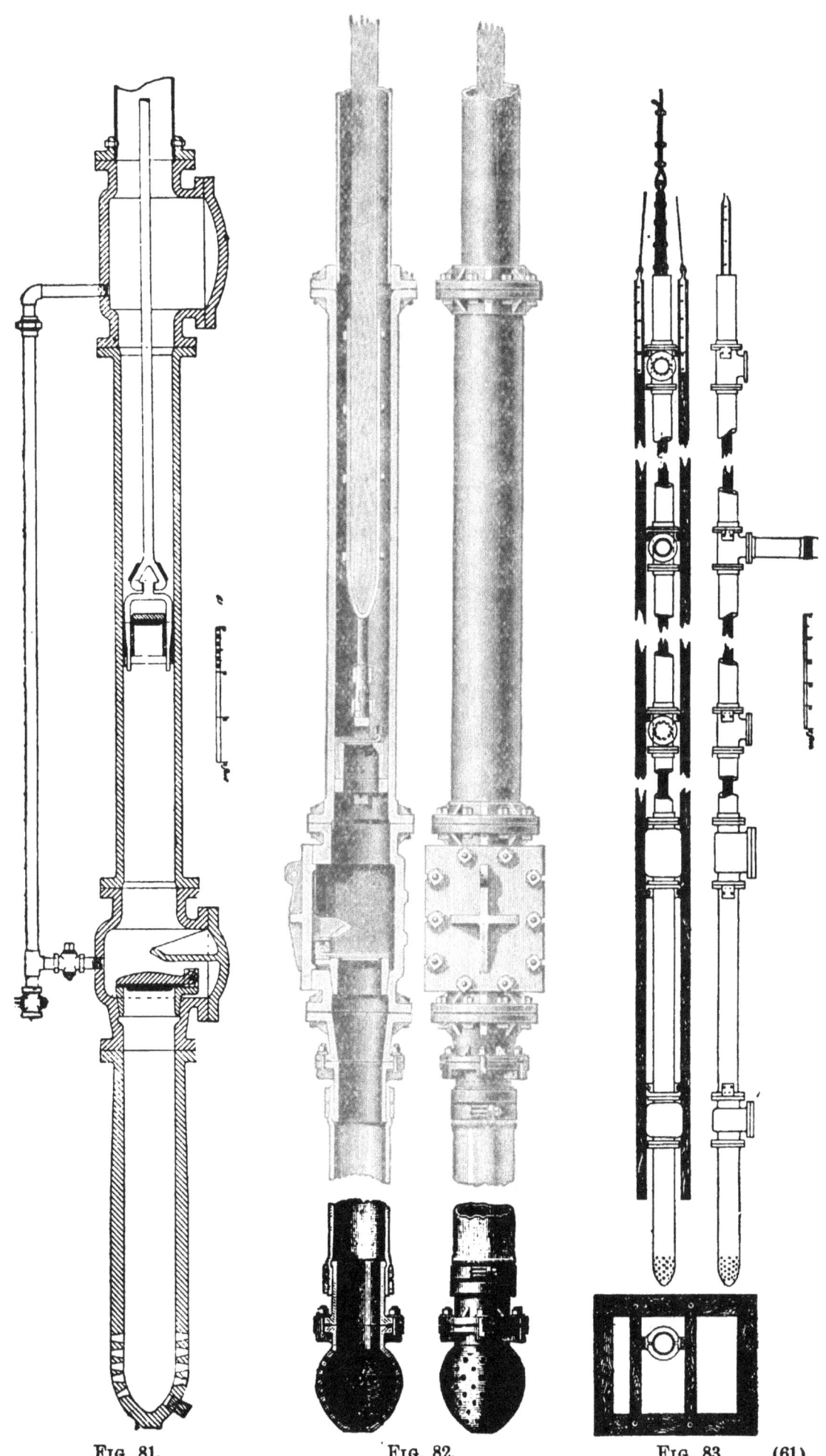

Fig. 81. Fig. 82. Fig. 83. (61)

2.3.06. In inclines it is not possible to use the kind of pump just described, because, as stated in the preceding chapter, the rod would rub heavily against the inside of the column-pipe and soon wear it through. It is, therefore, necessary to use a pump having the rod outside of the column-pipe. The jackhead lift-pump (Fig. 84) is of this kind. The column-pipe is here laterally connected at the top of the working-barrel by a gooseneck, while the bucket rod passes out through a stuffing-box in a cover, bolted to the top of the pump-barrel, and is connected to the sinking-rod outside. In inclines particularly, the latter must be guided like the main pumprod.

2.3.07. The arrangement of jackhead pumps varies with reference to relative position of valves. Some have the suction-valve below the working-barrel, like the English sinking-pump. In others, the valve in the bucket constitutes the suction-valve, and the discharge-valve is placed above the gooseneck, which arrangement admits of taking out the bucket without letting the water out of the column-pipe. Another form has a valve below the barrel and one above the gooseneck, as in Fig. 85. It is evident that where the usual hinged valve or clack is used with jackheads in inclines, the hinge must be at the upper edge of the valve.

2.3.08. Jackhead pumps cannot be operated under water for any length of time, because the vital part (the bucket) which requires frequent repairs cannot be hauled up through the column-pipe. They are, therefore, usually attached to long frames, which are sometimes sufficiently long to carry also the column-pipe, the whole frame being mounted on rollers, or otherwise guided, and arranged for hoisting, by means of tackle or engines, when the pump is submerged and requires repairs.

2.3.09. Although the bucket lift-pump is generally used for operation by rods in sinking, specially designed sinking plunger-pumps are also occasionally employed where there is no danger of being drowned out, or where the pump can be arranged to be hauled up for repairs. The plunger will remain tight much longer than the bucket, as it is not exposed to wear from sand, but it cannot be arranged to be packed when the pump is under water.

2.3.10. In many applications of the rod-pumping system in the present practice on this coast, particularly in inclines, and in places where lift pumps would be subject to excessive wear, direct-acting sinking-pumps, operated by steam or compressed air, are used. Such pumps are described in the chapter on "Direct-Acting Pumps."

2.3.11. *Sinking-Rod.* In the regular Cornish system the sinking-pump is operated from a bracket bolted to the main pumprod, as described in 2.2.04 and 2.2.22, and illustrated by Figs. 57, 58, and 59. The sinking-rod is clamped to the bracket in such a manner that it can be quickly loosened, lowered, and secured again, as sinking proceeds. The sections of the pumprod, except the topmost one, should be short, so that they can be easily handled, and all should be of equal length; then, knowing the number of sections, the distance to the bucket can be quickly figured out and its position in the pump-barrel determined. (See also 2.3.16.) The rod sections not in use are generally brought to the surface. The top of the upper section of the rod should carry a bail or ring for attaching the cable to raise the rod; the length of this sec-

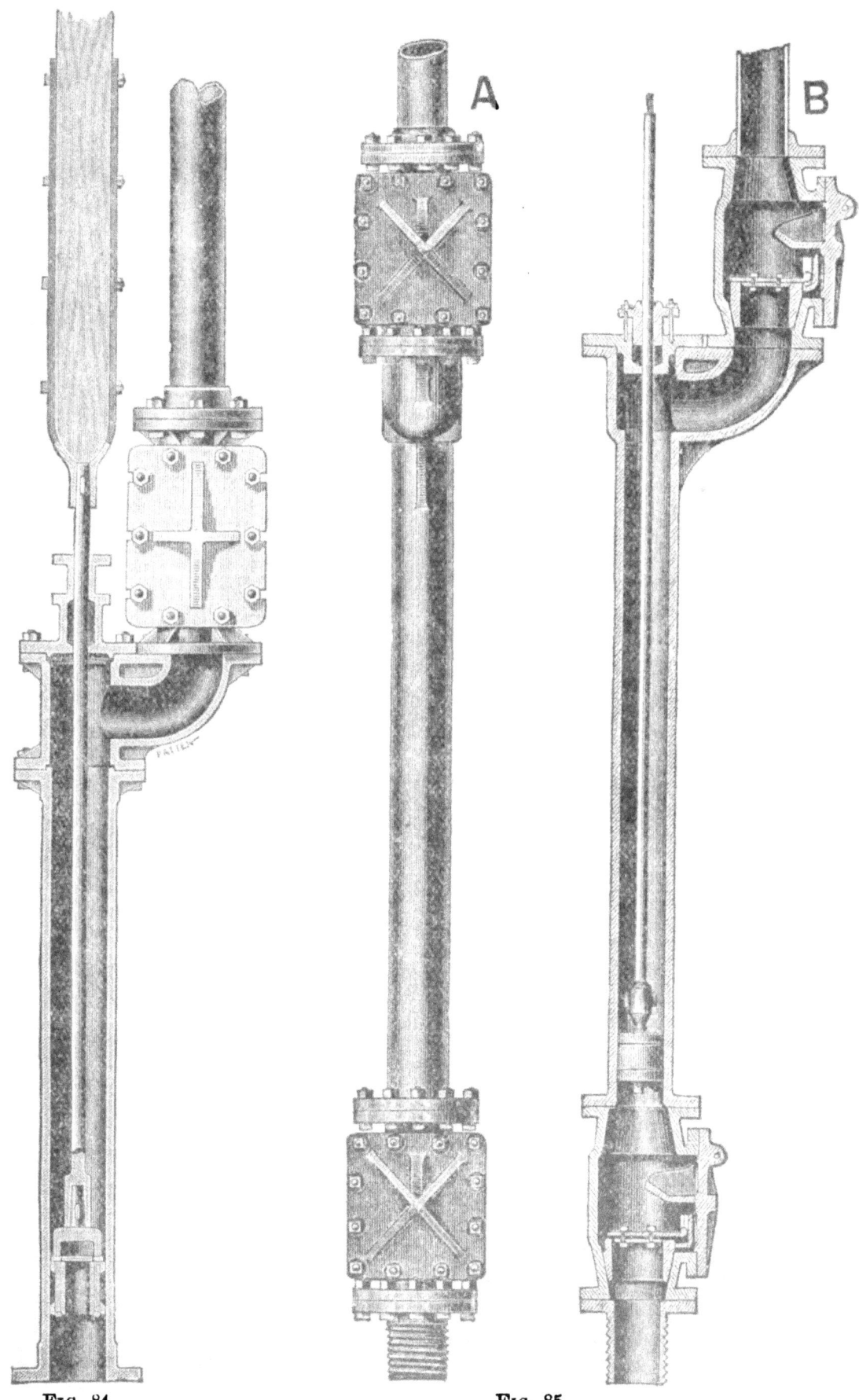

Fig. 84. Fig. 85.

tion should be sufficiently greater than that of the other sections, so that one of these with its joint or strapping-plates can be inserted below the clamps, when the top of the upper section has been lowered as far as the clamp, and then raised so as to admit the new section below it.

2.3.12. *Column-Pipe.* The discharge from the column-pipe must be kept sufficiently high above the top of the station-tank so as to admit of lowering for some distance, before it becomes necessary to extend the column-pipe. The extension of the column is made, as with the rod, by inserting a section of pipe below the discharge top; the latter is made usually of galvanized iron, for the sake of lightness, with a lateral branch, generally carrying a canvas hose, leading into the station-tank or into a launder connected with it. (See Fig. 24.) In order to save time, the column-pipe is usually lengthened whenever a new rod section is inserted. The amount of column extension to be made at one time is generally much less than the length of a full section of pipe; it is, therefore, necessary to insert, for the first section, a shorter piece, which is to be taken out and replaced by a longer one, or by a full section at the next extension, and so on, the short piece being put in and taken out alternately. In order to avoid keeping such sections on hand, and also to reduce the number of times that extensions must be made, the discharge top can be made with several discharge branches, as shown in Fig. 86. These branches have flanges for bolting on either a blind flange or a thimble carrying the discharge-hose. At first the discharge takes place at the lowest branch; and when, in lowering, this has reached the level of the top of the station-tank, the discharge-hose is taken off and fixed to the next higher branch, while the lower branch is closed by a blank flange. By making the discharge top sufficiently long, the column can be lowered a considerable distance, before it becomes necessary to extend it.

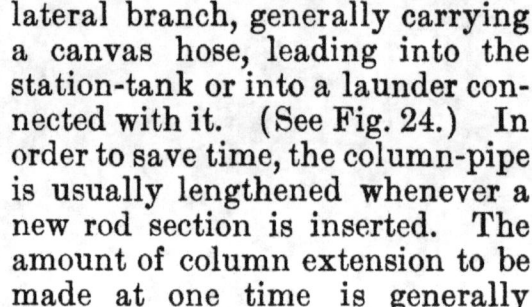

2.3.13. In some recent lift sinking-pumps for deep vertical shafts, the sinking-rod in the column-pipe is not operated from the main rod, but by means of a wire cable from a geared engine at the surface, which also serves as pump-hoist. (See 2.2.63 and 2.3.02.) In one arrangement referred to, the entire length of both the pumprod and column-pipe is never reduced, the pump and column being lowered as a whole. Each section of the column-pipe has a lateral discharge branch, as described in the preceding paragraph. All of the branches below the one discharging into the station-tank of the next higher pump are necessarily closed by blank flanges. As the column-pipe is lowered, the higher branches are successively used as discharge-pipes. The arrangement described admits of rapid manipulation in sinking, and is well suited for this purpose. Wire ropes are,

FIG. 86.

however, on account of their elasticity, not to be recommended for operating fixed pumps in permanent installations.

2.3.14. Pumps operated by rods or ropes are sometimes used for sinking, when the other shaft pumps are direct-acting steam or compressed-air pumps.

2.3.15. The column-pipe, or at least the upper part of it, must be guided in its descent in the shaft. The guides are usually wooden pieces, cut out to fit the pipe, and bolted to or wedged against the shaft timbering in such a manner that they can be separated in order to allow flanges to pass.

2.3.16. *Pump-Barrel.* The pump-barrel is generally considerably longer than required for the stroke of the bucket, so that the latter need not always be lowered whenever the pump is lowered a moderate amount. Some lift pumps have a stop at the lower end of the barrel, formed by the reduced opening of the suction clack-chamber, which prevents the bucket from dropping into the chamber. In lowering the rod this stop can serve to indicate the position of the bucket. (See 2.3.11.) The barrel is usually made of cast-iron, and being bored its inner surface is liable to rapid destruction in case of acid water. Brass and copper linings are sometimes used where the water is very bad. The barrels of sinking lift-pumps are also subject to great wear from grit in the water.

2.3.17. *Suction-Pipe and Strainer.* As described in 2.3.05, the suction-pipe is either a rigid, heavy casting bolted to the pump, which is supported by it on the bottom of the shaft, or it is made extensible after the manner of a telescope, or it consists of a flexible suction-hose, in both of which last-named cases the pump and the column-pipe must be supported, when not being lowered, by timbers on the shaft sets.

2.3.18. Fig. 81 illustrates a pump with a rigid suction-pipe. It is made very heavy, so as not to fracture under the blows from flying rock when blasting. It is generally further protected by planks, which cushion the blows of large rocks. Often for this purpose the suction-pipe or -hose is permanently wound with old rope, canvas, or similar material. When the suction-pipe is heavy, it is even admissible to blast right under it. The bottom is shaped to a rounded point, so that drills can be operated close under the suction. The strainer-holes should be conical, with the larger diameter inside, so that small pieces of rock will not jam into the openings. It is well to have two or three larger openings, ordinarily closed by wooden plugs, for getting out any small pieces of rock which may have been drawn into the strainer.

2.3.19. The suction should always be at the lowest part of the shaft, so that the men will not have to work in any deeper water than necessary. If the pumprod is long, the pump with column and rigid suction can be swung out of line to some extent, so as to allow placing the suction in the most advantageous position within the reach of the deflection. (See 2.3.05.)

2.3.20. The flexible suction has the advantage that the suction end, which is fitted with a strainer, can be moved to any part of the shaft bottom, so as to reach the lowest point, wherever that may be located. It also permits placing a foot-valve above the strainer, which is an advantage in many cases. (See 3.2.05.) Suction-hose is made of rubber, with layers of canvas between, and has a steel or iron stiffening-spiral on the inside, to prevent it from collapsing. In some hose the spiral is again covered inside with rubber, to prevent corrosion.

2.3.21. The suction height is the vertical distance from the bucket to the water-level in the sump. Its admissible maximum is much less in high altitudes than at sea-level. High speed of pumps, narrow and long suction-pipes, and great resistance of suction-valves also tend to reduce it.

2.3.22. If the water carries much sand, it is well to make the suction-pipe large, for then the velocity of flow will be reduced, and less sand will be drawn into the pump.

2.3.23. If the water in the sump is lowered, so that the upper strainer-holes are exposed, these are plugged up by the men working in the bottom of the shaft. Small quantities of air entering in this manner find their way through the valves into the column-pipe. If too much air has been drawn in, so that the pump loses its suction, it must be primed, or the sump-water must be allowed to accumulate to such a depth that the pump will prime itself. This it will do the more readily, the less the volume of the space between the suction-valve and the bucket in its lowest position as compared with the volume of the pump displacement, because the air will be the more rarified the greater the ratio of these two volumes. Self-priming will also be the more readily accomplished the smaller the suction lift as compared with the barometric head.

2.3.24. *Suction-Valves.* The suction-valves of sinking-pumps are often at a considerable height above the water-level in the sump; it is therefore important that they should open easily, as the available amount of overpressure beneath, tending to open them, may be only slight. Light valves naturally open more readily than heavy ones, and small valves can be made more than proportionally lighter than large ones, so that the use of multiple valves would be of advantage in this respect. (See 1.3.20 and 1.3.21.) Multiple valves cannot well be designed to admit of hauling up through the column-pipe, but this is not generally provided for in modern plants, as direct-acting steam sinking-pumps or large bailing-tanks can generally be used in case of emergency.

2.3.25. Suction-valves of sinking-pumps should be so constructed that they may last and remain tight for as long a time as possible. They should, however, be readily accessible to facilitate repairing when needed. The suction-valves are generally single or double clack-valves. The valve-chambers are made as described in 1.3.02 and illustrated in Fig. 48. They are often made extra heavy, to admit of their being brought down closer to the sump, where they are more subject to the effects of blasting. Extra clack-chambers, with valves in place, and also suction-pipes, should be on hand and in readiness to replace broken ones with the least possible delay. Steel cast valve-chambers have recently come into use. They can be made much lighter, and at the same time are less liable to breakage, than those made of cast-iron. The greater lightness of steel doors facilitates their handling when changing valves.

2.3.26. *Buckets.* Lift-pump-buckets should be so arranged that they give the greatest possible area for the passage of water through them on the down-stroke, and they should fit the barrel as closely as possible. When much sand is carried into the pump-barrel, the buckets have sometimes to be taken out and fitted with new packing every few hours. The valve in the bucket is generally either a clack or a straight-lift valve. Conical flexible valves have been used, but are suitable for only low lifts; they are extensively used in hand pumps. Leather is the most common material for packing the body of the bucket against the pump-barrel. It is so arranged that the pressure of the water will force the leather against the bore of the barrel during the up-stroke. Fig. 87 illustrates a common form of lift-pump-bucket. The ends of the leather forming the ring are beveled off and riveted together by

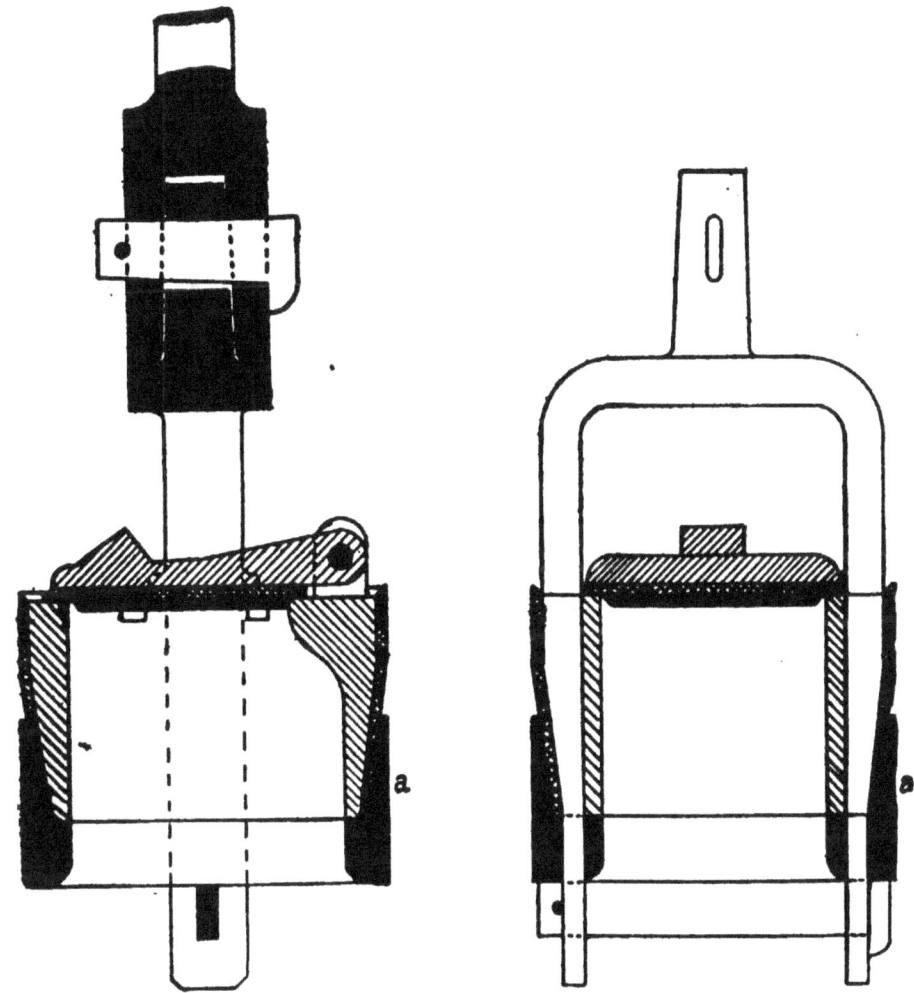

Fig. 87.

copper rivets. The leather should be of the best quality, and should present the flesh side as wearing surface, so that the more compact, hair side, which holds the leather together, will remain intact. It is best to soak the leather in tallow for some time before using. The packing is held in place by the taper-bored ring a, secured by a follower and key. The body of the bucket is made either of cast-iron or brass. The yoke by which it is connected with the rod, and the bevel ring and follower, and generally the valve also, are of wrought-iron.

2.3.27. The bucket should be quickly detachable from the rod, and it must be possible to immediately replace it by another, so that the pump need not long remain idle. The bucket taken out should be repaired and kept in readiness for going into the pump when in turn the one in place requires repairs. This can always be seen by the decreased quantity of water delivered by the pump, and the sinking of the water-level in the tank supplied by the sinking-pump. Key connections to the end of the rod, as shown in Fig. 87, are inconvenient to get at when taking off the bucket at the pump, and often require much time to loosen. The connection shown in Fig. 88 is a very convenient form for this purpose. The tapering sleeve, which surrounds the spearhead and claw, remains in place simply by its weight.

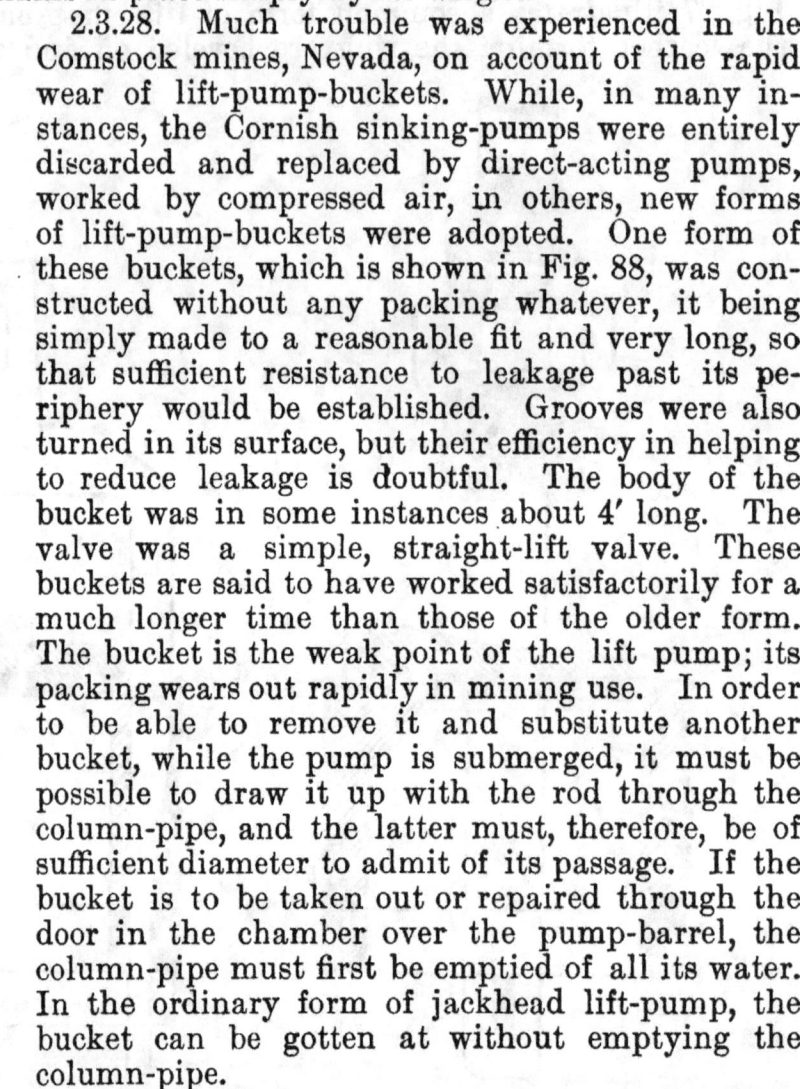

Fig. 88.

2.3.28. Much trouble was experienced in the Comstock mines, Nevada, on account of the rapid wear of lift-pump-buckets. While, in many instances, the Cornish sinking-pumps were entirely discarded and replaced by direct-acting pumps, worked by compressed air, in others, new forms of lift-pump-buckets were adopted. One form of these buckets, which is shown in Fig. 88, was constructed without any packing whatever, it being simply made to a reasonable fit and very long, so that sufficient resistance to leakage past its periphery would be established. Grooves were also turned in its surface, but their efficiency in helping to reduce leakage is doubtful. The body of the bucket was in some instances about 4′ long. The valve was a simple, straight-lift valve. These buckets are said to have worked satisfactorily for a much longer time than those of the older form. The bucket is the weak point of the lift pump; its packing wears out rapidly in mining use. In order to be able to remove it and substitute another bucket, while the pump is submerged, it must be possible to draw it up with the rod through the column-pipe, and the latter must, therefore, be of sufficient diameter to admit of its passage. If the bucket is to be taken out or repaired through the door in the chamber over the pump-barrel, the column-pipe must first be emptied of all its water. In the ordinary form of jackhead lift-pump, the bucket can be gotten at without emptying the column-pipe.

2.3.29. A small pipe is usually placed by the side of the pump, as in Fig. 81, which, on turning the cock a, and thereby opening communication between the column-pipe and the space between the bucket and suction-valve, permits the charging of the pump. This cock should be placed down at the lowest part of that space, so that any sand carried into the pump can be periodically blown out. The other cock is for letting the water out of either the pump-barrel or the column-pipe. By connecting it with a float in the sump in such a manner that the cock is opened and lets water out of the column when the sump water-level falls below a certain point, a means could, if desirable, be obtained for keeping the pump charged and working, even when it runs faster than necessary to keep down the water.

2.3.30. *Piston Sinking-Pumps.* Mr. S. N. Knight, of Sutter Creek, Cal., has built sinking-pumps with solid pistons, in which the work is done on the down-stroke, the pump really operating like a plunger pump. Its construction appears from Fig. 89. The pumprod is here subjected to compression instead of tension, and must, therefore, be very well guided. It requires comparatively little repairing for a sinking-pump, as the course of the water is not through the piston. This pump, like the jackhead, is more difficult to support in a vertical shaft, and it requires more room than the ordinary single-axis lift pump. This construction has many features to recommend its use in inclines, and ought to be preferable, in most cases, to the jackhead. The work being done on the down-stroke has also the advantage that less counterbalance is required for the main rod. The piston must not quite reach to the top of the barrel at the upper end of the stroke, so that there may always be a quantity of water on top of the piston, which will seal it if leaky, and prevent the influx of air on the suction-stroke, while the escape of air past the piston on the working-stroke would not be obstructed by the water. The pump illustrated was constructed with its valve-chambers and other large castings of steel. This makes possible a lighter construction and admits of a somewhat more compact arrangement.

2.3.31. *Admissible Lift of Sinking-Pumps.* The lift of a sinking-pump increases as the shaft goes down, until it becomes necessary to relieve it by placing a fixed plunger pump with tank-station in the shaft. When this plunger is ready for operation, but not before, the sinking-rod is detached from its connection to the main rod above the next higher plunger pump, the rod and column-pipe shortened by taking out the sections between the top and bottom pieces, and the rod clamped to the main rod above the new plunger pump, while the discharge of the column-pipe is diverted into the tank of that pump. The lift pump must therefore be capable of working, at least for a part of the time, against a head a little greater than the highest head under which any of the plungers in the shaft are working. In exceptional cases lift pumps have worked against a head of over 300'. Unless absolutely necessary, however, a head of 200' should not be greatly exceeded.

2.3.32. The extreme lift of sinking-pumps is quite often kept within moderate limits by dividing the total sinking-lift between two pumps working in a series, so that the lower pump raises to a small tank fixed around the suction-pipe of the upper pump. In this arrangement the upper pump is not put into operation until the limit allowed for the lower, or sinking-pump proper, is reached. The lower pump usually advances as the shaft goes down, while the upper one is temporarily fixed and lowered only at intervals. The latter is usually also a lift pump of the same pattern as the lower one, so that either pump can, in case of emergency, be used for sinking. If anything happens to the lower pump and it is drowned out, the upper one can be lowered and used under increased lift until the lower pump can be drawn up and repaired. In the same way, if the upper pump is disabled, the lower one can have its column-pipe extended to increase its lift, so as to include that of the upper.

2.3.33. An interesting sinking operation described by Professor Riedler, in the Zeitschrift des Vereins Deutscher Ingenieure, Vol. XXXVI, No. 16, 1892, was carried out in 1889–90, in bringing down the Max shaft of the "Prague Iron Industrial Company," in Kladno,

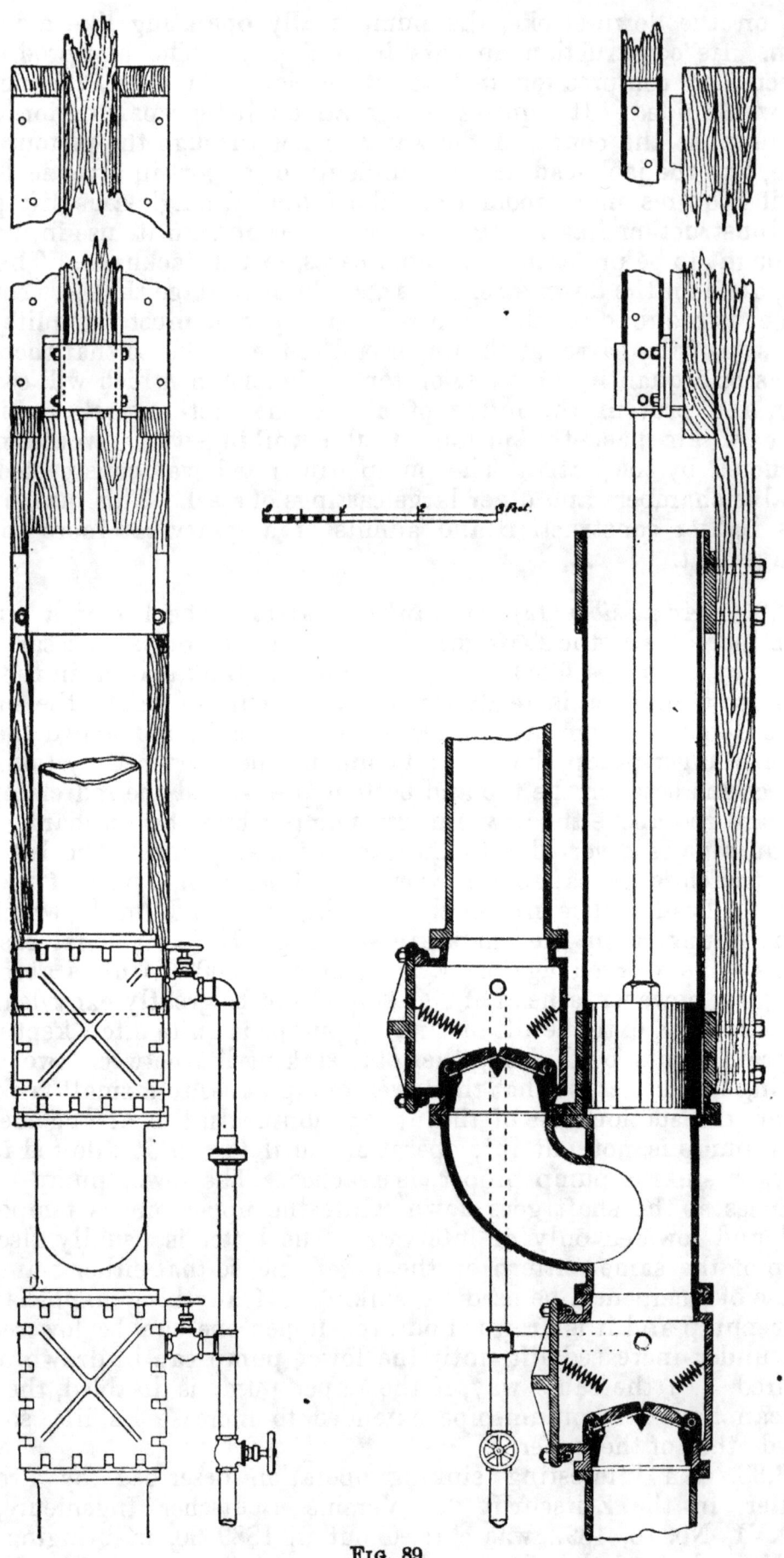

Fig. 89.

Bohemia. Riedler steam pumps were used for the permanent installation, and the extreme sinking-lift came to about 475'. Two sinking-lift pumps of Karlick's patent (which will be described presently) were used in series, after the manner described in the preceding paragraph. The extreme lift allowed for each of these pumps was 200', so that a depth of 400' could be sunk by their combination. The remaining 75' was overcome by the use of a Hall pulsometer. By this arrangement, the most wasteful machine in the use of steam—the pulsometer—was detailed to do the lesser part of the work, while at the same time its heating effect was removed far from where the men worked in the bottom of the shaft. A steam sinking-pump would doubtless have been better than a pulsometer.

2.3.34. The Karlick sinking-pump, illustrated, with its sinking-frame, in Fig. 83, consists of an ordinary English lift pump with the pumprod inside of the column-pipe. The latter is constructed as described in 2.3.13, the sections each having a nozzle, which can be used as discharge or closed by a blind flange bolted on. In sinking, except when doing so from the surface, the sections of the column-pipe always remain connected with the pumps, the water being discharged first at the lowest nozzle, which is opened for the purpose. As the pump goes down, the next higher nozzle is connected with the discharge-hose, and the lower nozzle is closed. In this manner very little time was lost through stoppages. The bucket-rods of the pumps extended only a little beyond the top of the column-pipe, and were there operated each by a wire rope from bobs at the surface. The weight of the pumprod and bucket kept the rope taut on the down-stroke. The ropes were clamped to links hinged to the bob-nose, so that they could be quickly loosened, lowered, and secured again as the sinking went on. The pumprod sections were never disconnected, except for repairs. Breakages of ropes, when they did occur, were quickly repaired by means of clamps. The pumps were made of steel castings in order to secure lightness.

2.3.35. By arranging the pumps with guides for a considerable distance above them, they could be raised when drowned out, and an additional safeguard against the entire flooding of the mine was thus obtained.

2.3.36. The system of sinking just described deserves a wider application, on account of the rapidity with which the different manipulations may be carried out.

2.3.37. *Volumetric Effect of Lift Pumps.* Owing to the wear of pump-barrel and bucket-packing, and the consequent leakage, lift pumps at low speeds raise a smaller quantity of water than that due to the volume displacement of the bucket. At high piston speed the leakage is less in proportion to the volume displacement, and the latter is more nearly approached by the quantity of water raised, particularly in the common lift pump, where the energy of motion of the water assists in its own advancement, sometimes to such an extent that the quantity of water actually raised exceeds by several per cent that due to the volume swept through by the bucket. But when ordinary pumps are run in this manner, their operation is generally accompanied by severe shocks, and the pumping is done with less efficiency and with less security against breakdowns.

CHAPTER IV.

Plunger Pumps.

2.4.01. It has already been stated in 1.1.04 that plungers are more easily packed, admit of pumping against higher heads, and remain tight much longer than buckets or pistons, and that they are much less subject to wear, since the rubbing surfaces are located at such a point that very little of the sand usually carried by the water will reach them. The objection that they cannot be packed under water, applies only where they are liable to be drowned out. The use of plungers also admits of equalizing partially or entirely the work on the up- and the down-strokes, so that much less counterbalance will be required than if lift pumps only were used. (See 2.1.04 and 2.2.43 *et seq.*)

2.4.02. For these reasons, the pumps of the Cornish system, with the exception generally of the lowest, or sinking-pump, are designed as plunger pumps.

2.4.03. *Relative Arrangements of Parts.* A usual type of plunger pump for a vertical shaft is shown in Fig. 90. The disposition of valves in relation to the working-barrel is the one which long experience has demonstrated to be the most convenient. This pump can also be used in an incline. The clack-chambers can then either be placed on top or at the side, in which latter case, however, the clacks will have to be turned around 90°. Where straight-lift valves are used in an inclined pump, the valve-chambers should be placed in a vertical position.

2.4.04. *Plungers.* These are generally made of cast-iron, though brass is a better material, as it works through the packing with much less friction than iron. Sometimes, therefore, the plungers are made of or lined with brass. Brass also resists better the action of acid water. Thick grease will protect cast-iron plungers to some extent, if the water

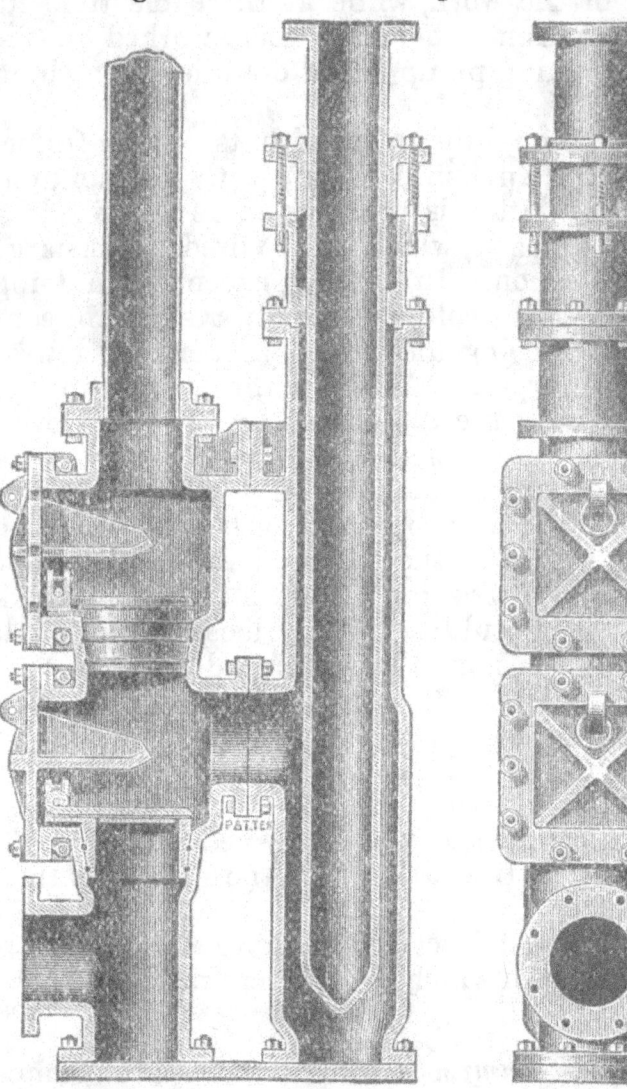

Fig. 90.

is not too warm to melt the grease and float it off. The plungers are properly formed with a rounded, point-shaped bottom, as shown in Fig. 90, so as to reduce shocks on striking the water in the barrel, in case the pump does not quite fill on the suction-stroke. The top is formed with a flange, which is bolted to a bracket that is clamped to the pumprod. Fig. 91 shows such a bracket. Where two pumps are attached at opposite sides of the rod the clamping-plates are dispensed with and two such brackets held in place by the same bolts. Clamping the brackets to the rod insures their adjustability. It may also permit a ruptured rod to slip through the clamp under the severe strain due to

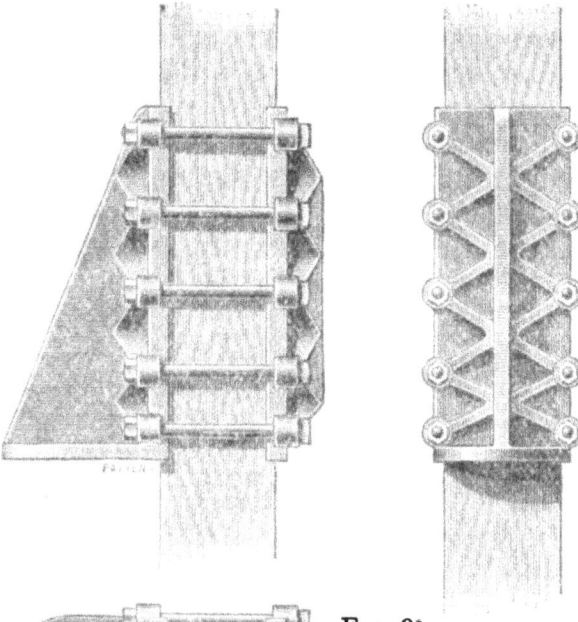

FIG. 91.

the fall, and by slipping prevent injury to the pump. (See 2.2.29–2.2.34.) Where the lift is so small as to only require a single plunger pump, the plunger is attached to the lower end of the rod, and in line with it, by a split socket casting, as shown in Fig. 92. The plungers of inclined pumps are subject to one-sided wear, which makes it hard to keep them tight and hold the leakage down to an allowable amount.

2.4.05. *Stuffing-Boxes.* The stuffing-box for packing the plunger is generally cast separate from and bolted to the top of the pump-barrel, as in Fig. 90. The usual packing consists of square braids of hemp, flax, or cotton, soaked in tallow, Albany compound, or a mixture of tallow with beeswax. For cold water braids of flax, thoroughly impregnated with Albany compound, give good results. The wasting of the compound should be made up by periodically smearing some on the plunger. For hot water this packing is not suitable, as the compound becomes too fluid and is carried off by the water very rapidly. Square braids of cotton impregnated with powdered plumbago work very well in the hot water. The braids should be put in in level layers, not wound around in the form of a spiral. For such and other packing of a fibrous nature, the bottom of the stuffing-box and gland are, with advantage, made in a grooved form, as shown in Fig. 93, for by such construction fibers will be less liable to be dragged along by the

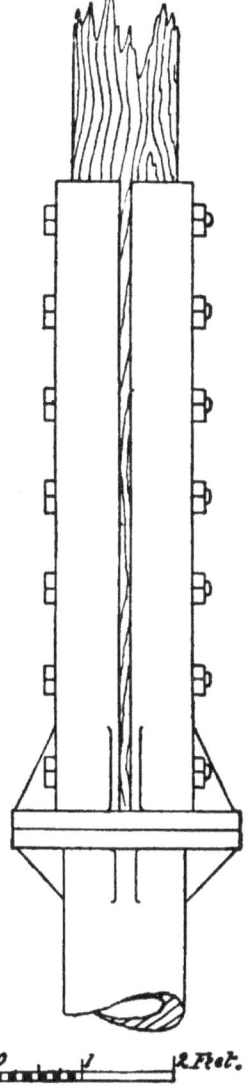

FIG. 92.

plunger and forced between it and the metal of the stuffing-box, thereby causing one-sided wear of the plunger. The gland, for vertical pumps, should be cast with an annular bead, forming a cup to surround the plunger and keep the grease from spreading. This cup, by being filled with grease and water, also prevents air from being drawn in through the stuffing-box on the suction-stroke.

2.4.06. Ordinary stuffing-boxes generally cause considerable friction, because they are drawn up too tight. They should be drawn up just enough to permit a little leakage. In screwing up the gland, care should be taken to keep it true with the plunger. Plungers wear unevenly, and when it is attempted to prevent leakage by screwing up the packing, the friction becomes excessive. When the plungers are so worn, they should be replaced by spare ones kept on hand. Those taken out should then be trued up and kept ready for putting in again. As they are reduced in size by repeated truing-up, the stuffing-boxes become too wide for the plungers, and they, or their linings, must be replaced by new ones.

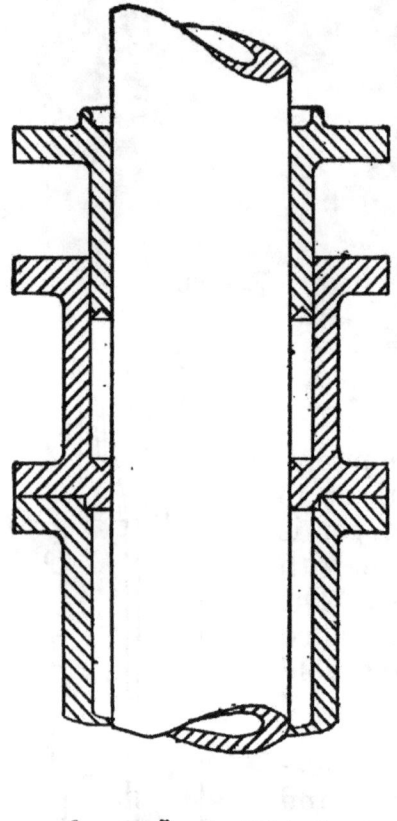

Fig. 93.

2.4.07. With large pumps in inclines, the stuffing-boxes give a great deal of trouble, because the heavy plunger presses on the packing on one side only. (See 2.4.04.)

2.4.08. *Pump-Barrel.* Where, as in the ordinary designs, the connection to the valve-chambers is about mid-height of the barrel, the part below the connection should have a cross-section area equal to about twice that of the plunger, as in Fig. 90, so that the water can flow freely along the plunger during the lower half of its stroke, to fill or empty the space swept through by it.

2.4.09. With the connection to valve-chambers below the top of the barrel, air will accumulate in the upper part. For this reason pumps sometimes have the connection at the upper end of the barrel. But this makes an inconvenient form to support and place in an accessible manner in the shaft. It is, therefore, better to provide a small pipe-connection from the highest part of the pump-barrel to the column-pipe. On the working-stroke, water will be forced through this connection into the column-pipe, while on the suction-stroke some water will flow back into the barrel. The pipe-connection should therefore be provided with a cock to regulate the amount of opening, and to close it in case the suction-valve has to be inspected. A small check-valve would prevent the back-flow of the water, but in order to be operative it should open easier than the main discharge-valve. Some air generally escapes through the leaky stuffing-box, and many pumps are therefore made without the aforesaid connections.

2.4.10. A cock to let out the air on filling the pump must also be fitted to the top of the barrel, as stated in 1.1.12 and 1.1.13, where the manner of starting and priming pumps is described.

2.4.11. Near the bottom of the barrel there should be a hand-hole, or a nipple with valve, to clean out accumulated sediment.

2.4.12. *Valves and Valve-Chambers.* The valves and their chambers are generally superposed as in Fig. 90. Single or double clack valves are most generally used on this coast. The advantages of multiple, light, spring-loaded valves have been pointed out in 1.3.21. A pipe, as in Fig. 94, must be arranged at the side of the chambers, with valve-connections to the space above each valve, and a waste-connection to the station-tank or to the suction-pipe. The connections above the valves should be placed as low as possible, so that they

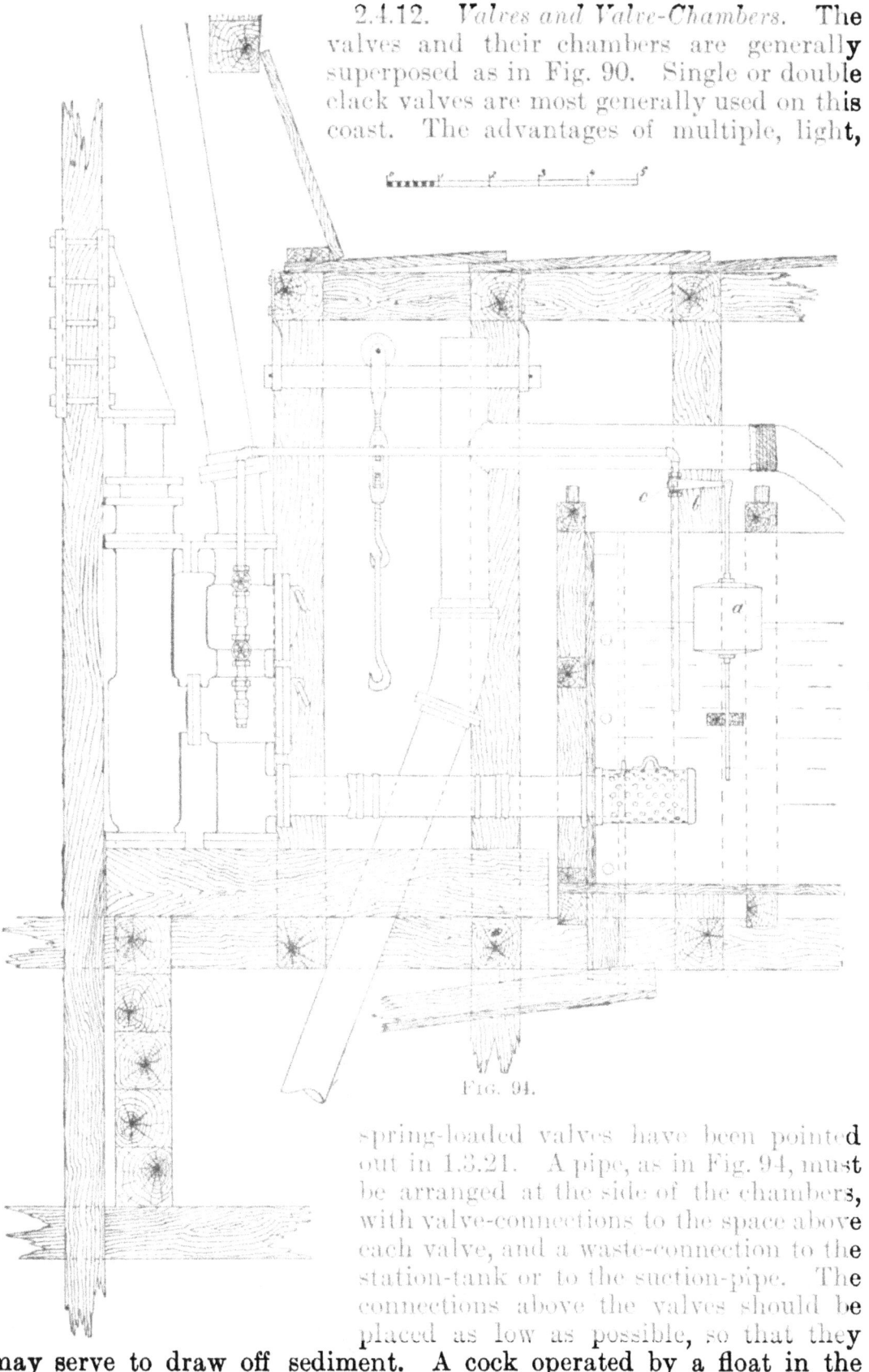

Fig. 94.

may serve to draw off sediment. A cock operated by a float in the station-tank is generally also placed in the pipe connecting the spaces above the valves. The arrangement of pipe and valves serves to regulate the relative capacity of a series of pumps, according to the varying duty at each station, and it also serves for priming the pump or emptying the column-pipe when necessary.

2.4.13. For handling the heavy valve-chamber doors, when access to the valves becomes necessary, a hook, vertically adjustable by a screw-connection, and suspended from a roller traveling on a bar, either fixed or capable of swinging in a horizontal plane, is generally provided. Fig. 94 illustrates an arrangement of this kind. (See 1.3.22 *et seq.*)

2.4.14. *Connection to Supply-Tank.* In most cases the suction-pipe runs only horizontally, and connects directly to the side of the station-tank or reservoir, as in Fig. 94 or 95. Where, like in Fig. 96, the pump is placed at a distance below the tank, the suction-pipe turns upward, and is connected to the bottom of the tank. In either case the horizontal portion of the suction-pipe should have a flexible part inserted, to admit of unequal settling of the pump and tank. This flexible part consists usually of a piece of heavy rubber suction-hose, with internal metal rings, and longitudinal strips to keep them in place, the ends of the hose being held by clamps to thimbles, as in Fig. 97, having flanges for connection to the other parts of the suction-pipe. A couple of layers of canvas coated with pitch are often used inside of the rubber hose, as a protection for the latter. It is also best to wrap the hose with tarred marlin, particularly where it has to withstand considerable pressure, as when the suction is arranged like in Fig. 96. It is evident that the suction-pipe must be air-tight. The end connected to the tank should be a little above the tank floor, to prevent sediment from being drawn into the pump. A strainer of ample area should form an extension of the pipe inside the tank, and should be removable for cleaning. The tank end of the pipe is sometimes flared out to a larger diameter, so that the water will enter with less current, and therefore not sweep in so much sediment. It is

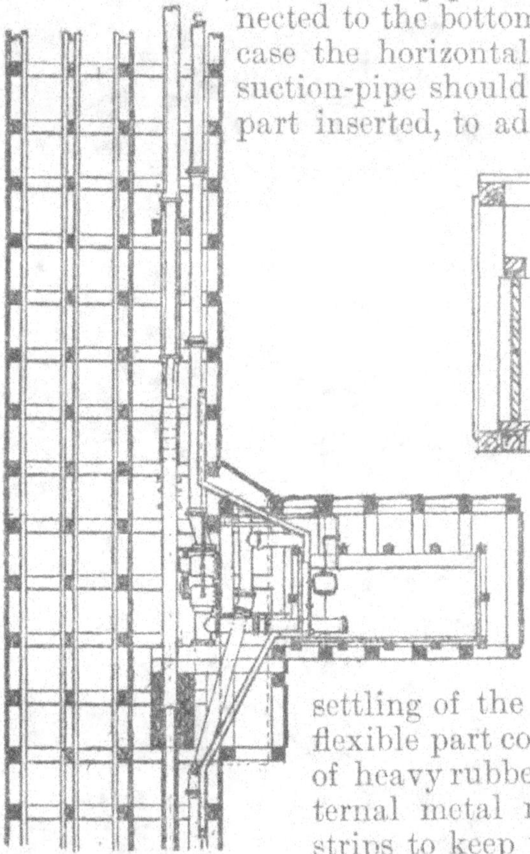

FIG. 95.

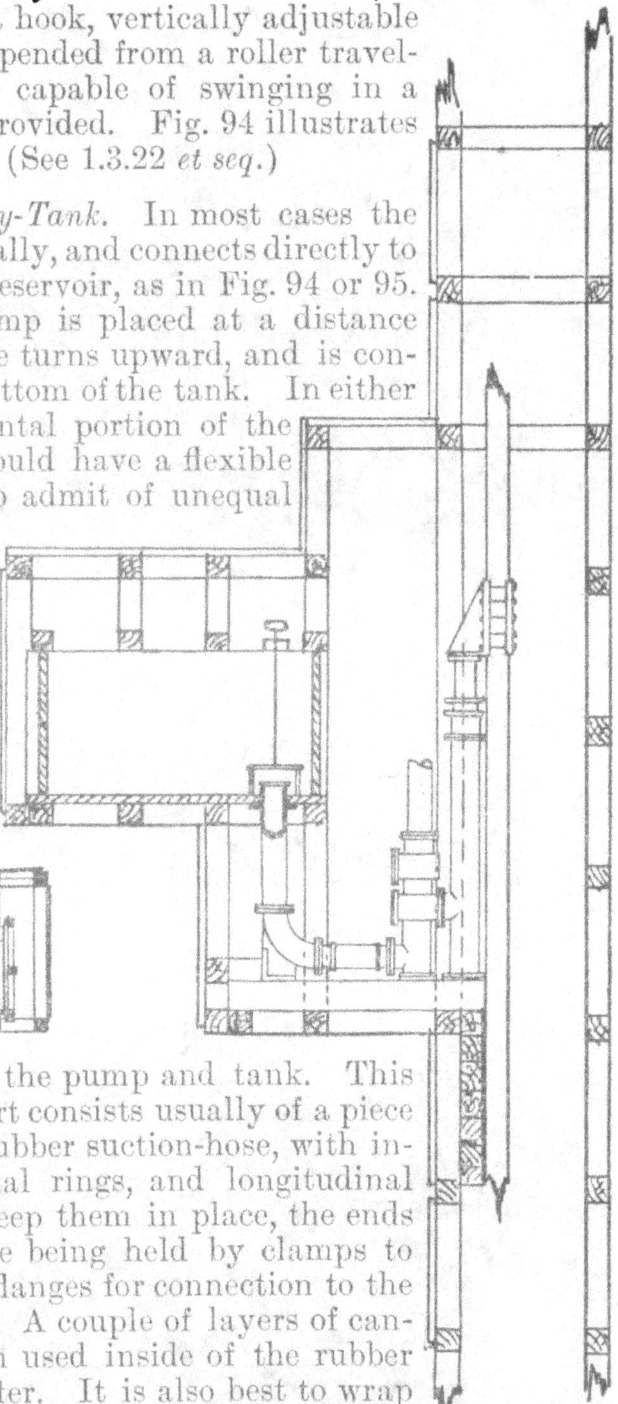

FIG. 96.

also a good plan to arrange the end of the pipe with a tight cover, which, when the pump is working, is left off, but which can be closed when it is desirable to drain the pump for inspection, without also draining the tank.

2.4.15. *Supply- or Station-Tanks.* Where the station is in hard, self-supporting ground, requiring no timbering to support the roof, a reservoir can be made by lining the bottom part of the excavation up to water-level with cement, and throwing up a small masonry dam in front. Generally, however, the stations have to be timbered, and then wooden tanks are set up, as in Figs. 94 and 95. It is advantageous to have the tanks of large capacity, so that they can take up a considerable inflow from levels, or overflow from upper tanks, and prevent it from reaching the sump.

2.4.16. The water from the lower pumps and other sources should flow into the tanks quietly, with the least possible disturbance of their

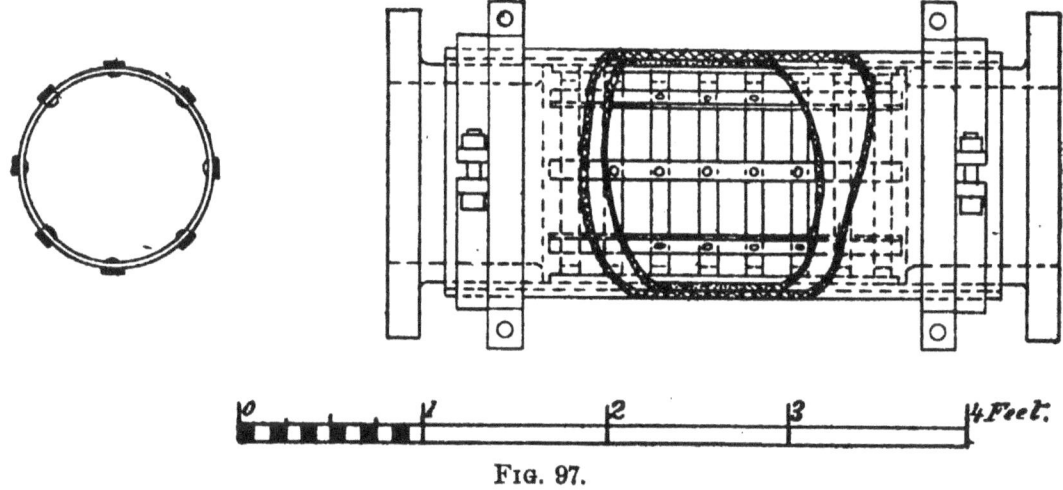

Fig. 97.

contents, in order to give sand and mud a chance to settle. Partitions in the tanks are useful to confine the bulk of the sediment to those parts where the water enters, and keep it from reaching the suction-pipes of the pump. A drain with a pipe or wooden box leading to the next lowest tank serves to draw off the water when repairs or cleaning of tanks is necessary. There should be draw-off plugs at different levels, so that the water can be drawn off without disturbing the settled mud and sand. This can best be removed through a separate plug into a small tank on the hoisting-cage, and brought to the surface. A notch near the top of the tank must also connect with the drain, in order to divert a possible overflow into the next lower tank.

2.4.17. *Pump Supports.* The foundations of Cornish plunger pumps operated by rods consist of beams or arches built in across the shaft. They have to bear, not only the weight of pumps, with column-pipes and water contained in them, but they are also subject to heavy, sudden strains from water-ram, on which account they should possess, besides strength, a certain amount of elasticity, so as to better resist shocks.

2.4.18. Smaller pumps can generally have their foundations supported directly by the shaft timbering, the load being distributed over several sets. Very small pumps are only bolted to the sets themselves,

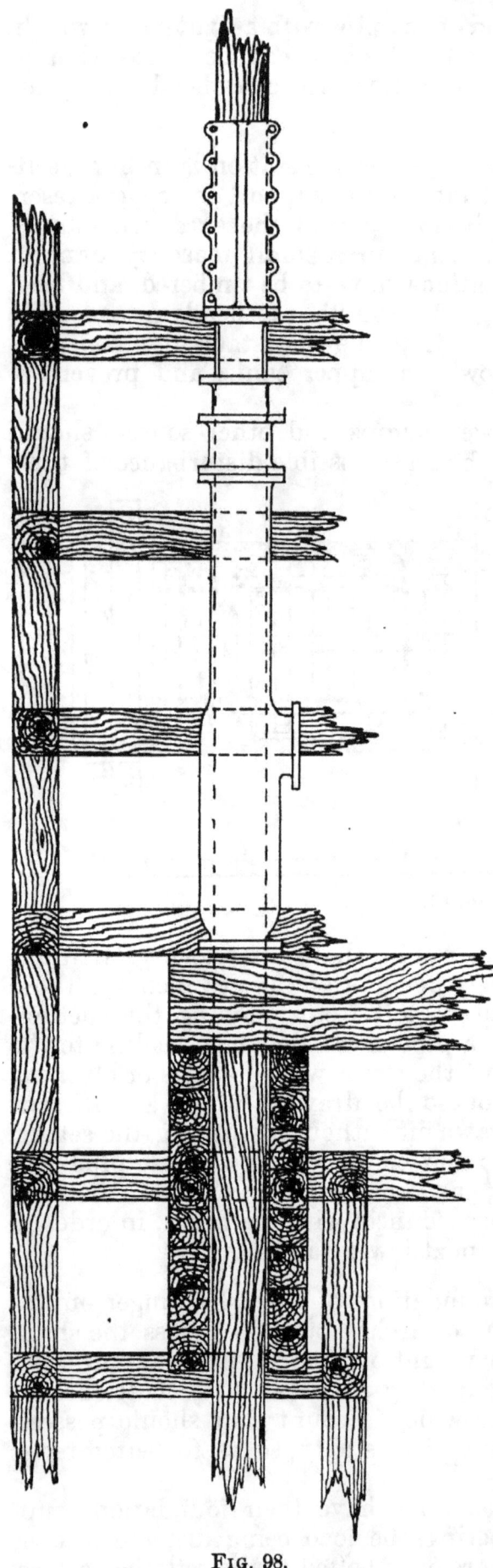

Fig. 98.

being provided with lugs for this purpose. For large pumps, it is always best to rest the supports independently on solid ground, outside of the shaft timbering. In most cases, the supports consist of wooden or wrought-iron beams or trusses. Cast-iron girders or arches are not so good. These may be used, however, if a pedestal of a more elastic material be placed between them and the pump. Where the nature of the ground admits of it, beam supports are preferred, on account of greater simplicity. Masonry, cast-iron, or wooden arches are sometimes used in slaty formations, or where the ground is liable to crumble away under the vertical pressure of beams or trusses, the arch supporting the ground by its principally lateral pressure. Whether beams or arches be used, the excavations for the ends or counterforts should be cut and broken out by hand, not blasted, so as not to shatter the ground too much.

2.4.19. Pump supports of wood are generally used on this coast. The simplest support consists of a pile of beams, like that shown in Fig. 94, built across the shaft. They should be firmly supported and wedged in at the ends, to prevent their displacement. A cross-beam, usually extending into the tank-station, affords ample bearing-surface for both the pump and base of clack-chambers. A simple arrangement like the foregoing can be used where a single line of plungers is offset from the side of the rod, as shown in the plan. Where pumps are placed at opposite sides of the rod, two piles of beams are often used, as in Figs. 98 and 95, to allow the rod to pass between them. Where the space at the side away from the tank-station is scant, the foundation beams are sometimes placed at an angle, as shown in plan in Fig. 99. In some cases, where good supporting-ground cannot be obtained close to the shaft, the main foundation beams have to be of considerable length, and might thereby become too elastic. In such cases, braces can be thrown in as in Fig. 100, or the beam can be constructed as a truss, like Fig. 101.

2.4.20. Sometimes the pumps are only held on the foundation by their own weight and the pressure of the water-columns. This admits

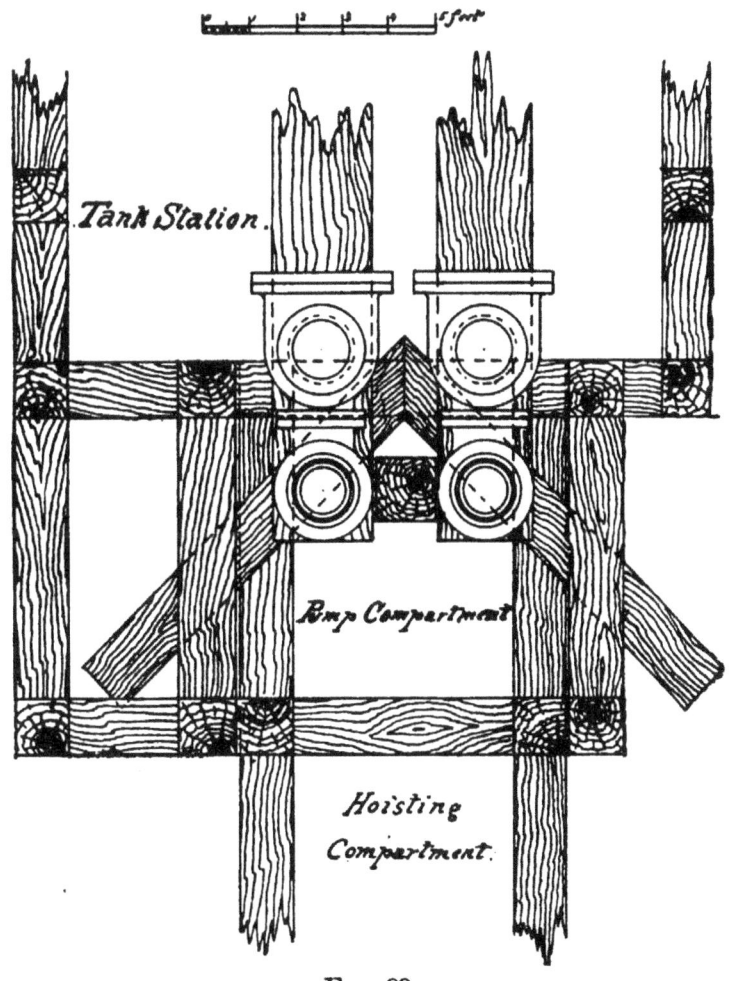

Fig. 99.

of shifting them more readily, to keep them in line if the shaft be in moving ground. It is always best, however, to bolt or clamp the pumps to the foundation. The upper part of the pump-barrel is usually held to the shaft timbering by clamps or strap-bolts, to prevent its lateral displacement.

2.4.21. *Arrangement of Pump-Stations.* Excavations for tanks, like those for balance-bobs, should always extend in a direction opposite to that in which the hoisting-compartments are located. This disposition leaves the ground around the shaft in the best supported condition.

2.4.22. Pump-compartments of timbered shafts, particularly for a double line of pumps, are rarely large enough, without increasing their size at the stations, to admit of such an installation of pumps as to give accessibility to every part, and also leave room in the shaft for lowering or raising parts of the underground machinery, or for running a cage. The pump-shafts are, therefore, generally enlarged at the pump-station, as in Fig. 99.

2.4.23. *Admissible Lift and Plunger Speed.* The Cornish type of plunger pump is rarely used for lifts above 250'. About 200' is the usual lift allowed. Lifts of 400' and 500' occasionally occur, but with

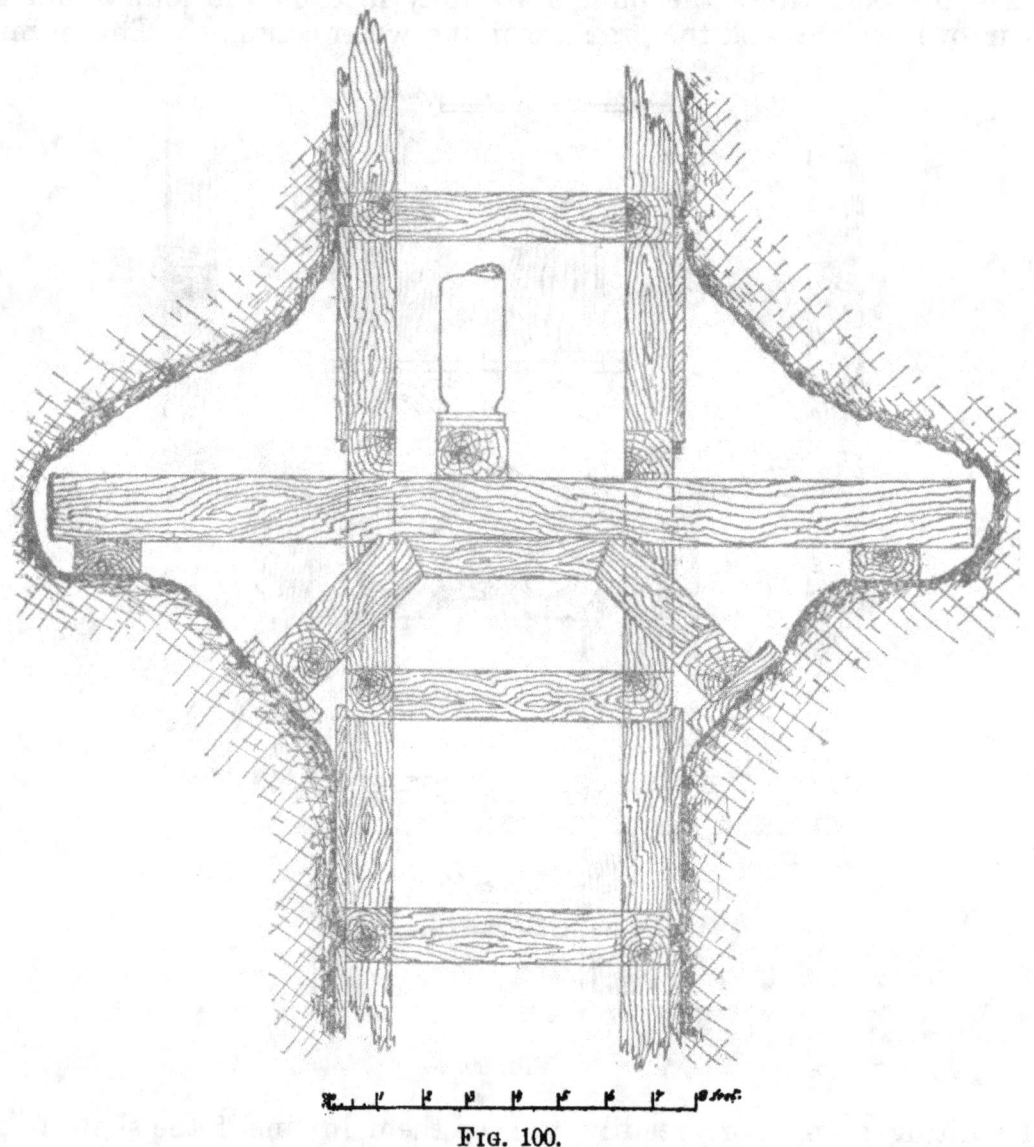

Fig. 100.

such the pumps can operate only at a very much reduced speed, and require, therefore, to be of larger size to handle a given quantity of water. The greater the lift, the slower must the pumps run to avoid too severe shocks. Great length of the pumprod and column-pipes also reduces the admissible number of strokes. (See 2.2.01.) For this reason, inclined pumps cannot be run as fast against the same head as pumps of the same size in a vertical shaft. (See 1.2.38.) The longer the stroke, on the other hand, the greater is the admissible plunger speed. Cornish plunger pumps are usually so placed that the water will run into and almost fill them by gravity. The height above sea-level has, therefore, less influence on the working of Cornish plungers than on that of lift pumps.

2.4.24. *Relative Size of a Series of Plungers.* If the water must all be lifted from the sump of a deep mine, requiring several superposed sets of pumps, the plungers should properly increase in size as they are nearer the bottom, to make up for the decreased length of stroke resulting from the elasticity of the pumprod. Generally, however, the pumps are all made of the same size, so that the parts will be interchangeable.

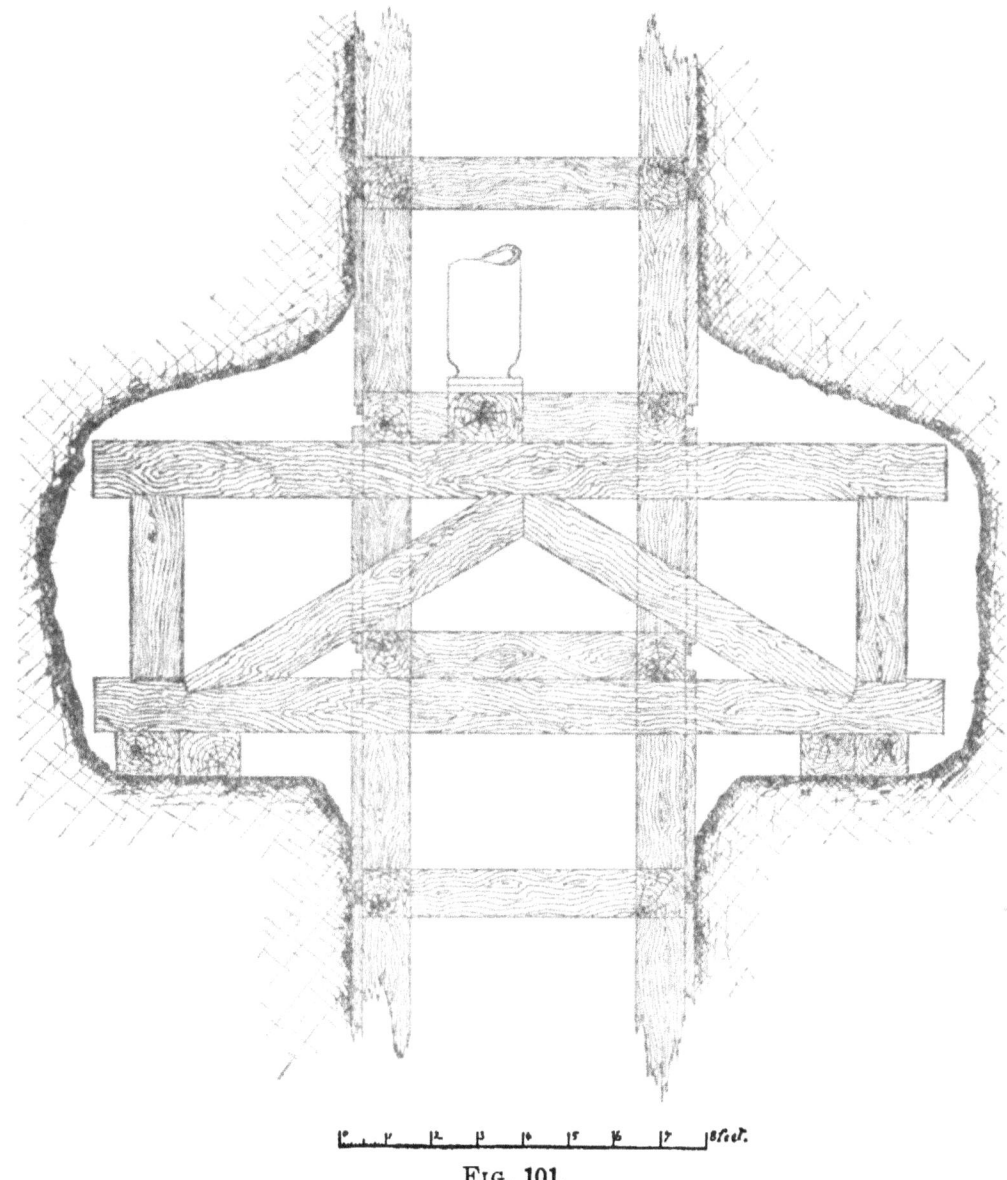

Fig. 101.

But, since plungers and stuffing-boxes are not very liable to breakage, these could be made of the proper sizes, and the valves, valve-chambers, and barrels of all the pumps interchangeable. Where water issues at different levels, the aim should be to adapt the sizes of the corresponding pumps to the water to be handled by them.

2.4.25. Plunger pumps are much more subject to breakage than lift pumps, and extra parts liable to be broken, such as clack-chambers and pump-barrels, should be kept on hand where severe service is required of the pumps. Such parts, as was stated before, are now often made of steel.

2.4.26. Before placing pumps in a shaft, a careful survey of it should be made, in order to determine if it is crooked or twisted out of line, and the relative position in plan of the different sections should then be drawn on paper and compared with the desired arrangement of pumps, in order to see if sufficient space is available for installing them. After pumps, rods, and pipes are in place, they should be kept carefully in line.

2.4.27. It is desirable, particularly in deep mines, to have space in the pump-compartment for running a small cage, in order to enable the pump-men to reach rapidly any point of the shaft. The ladders which are required in every mine are also placed in the pump-compartment. The space allowed for the cage should be large enough to admit of lowering the largest parts of the underground machinery. As the rods and column-pipes, with their guides and stays, take up a considerable portion of the shaft area, the compartments intended for Cornish pumps require to be of large size. In pumping-plants for moderate depth and capacity, the heavy parts are generally lowered by chain-blocks, or by winches, operated by hand. These winches are also often used for raising and lowering the sinking-pumps, and must be of ample strength for the purpose. Sometimes, however, hand winches for the sinking-pumps are located in the shaft near the top of the discharge-column.

2.4.28. Hand winches are too slow for the largest Cornish plants, and with these regular pump-hoists, geared in a large ratio, so as to be able to lift or lower heavy loads, are installed at the surface. These are then generally used also for running a cage in the pump-compartment.

CHAPTER V.

Power-Plants for Operating Pumps Through Rods.

STEAM ENGINES.

2.5.01. The steam engines for operating mining pumps by means of rods will, for want of a better generic name, be simply called rod-pumping engines in the course of this article. They may be rotative, non-rotative, or geared. The non-rotative types can be either direct- or indirect-acting. Direct-acting engines are those in which the piston-rod is in line with and forms an extension of the pumprod. Indirect-acting engines are those in which the piston-rod moves the pumprod through the medium of a beam or bob, as in Fig. 102 or 103.

2.5.02. Rod-pumping engines are now rarely made direct-acting, because the cylinder then obstructs the mouth of the shaft. Where a beam is used in modern mine pump engines it is usually placed below the cylinders, because in this manner the top of the shaft can be kept clear.

2.5.03. Large cylinders of pump engines are best placed in a vertical position, because then the heavy piston will not wear the cylinder on one side, as in the horizontal engines.

2.5.04. *Non-Rotative Engines.* It seems proper to consider the non-rotative engines first, as they are the oldest type.

2.5.05. During the sinking of a shaft in water-bearing ground the work to be done by the pump engines changes, not only according to increase of depth, but also by the opening-up of new bodies of water.

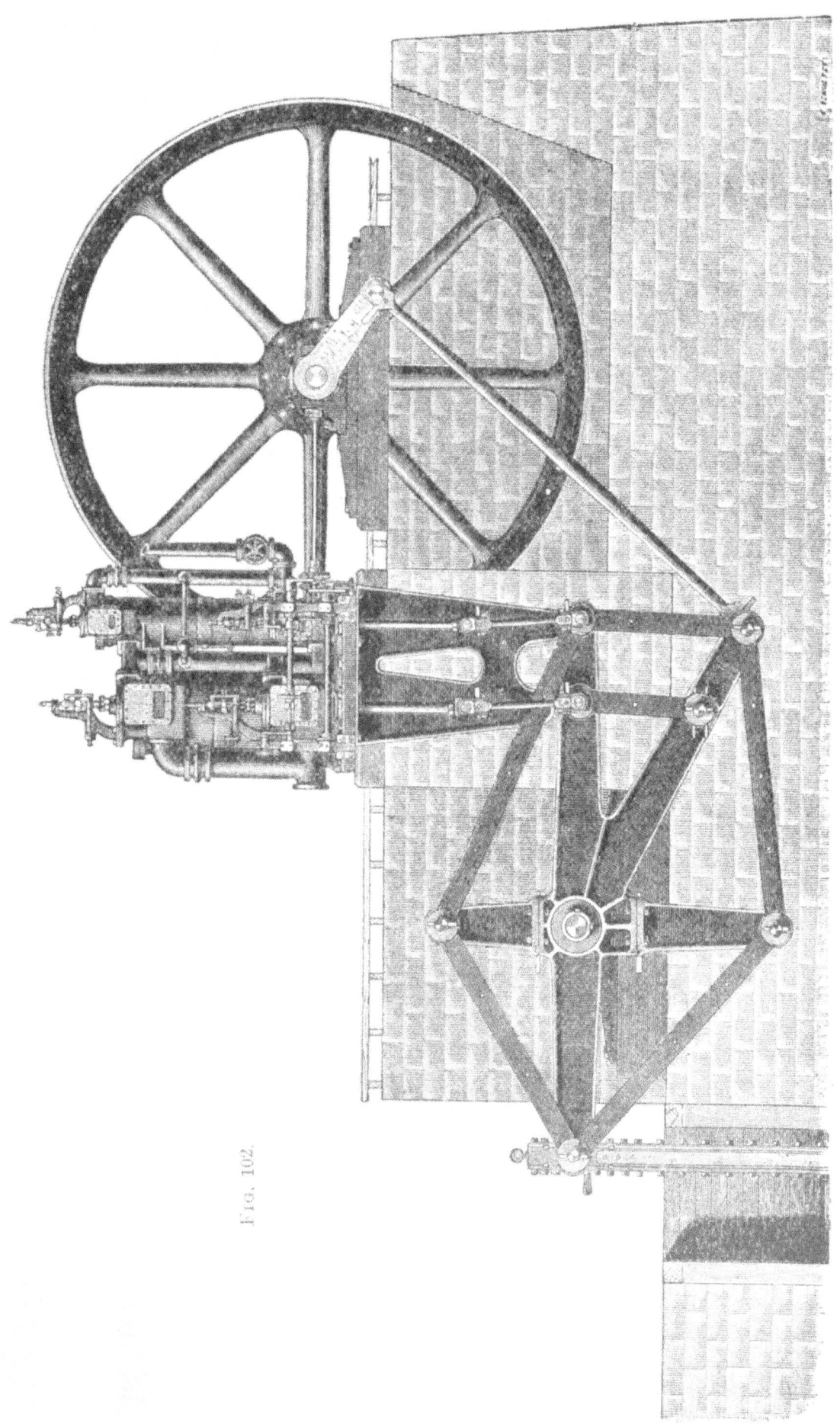

Fig. 102.

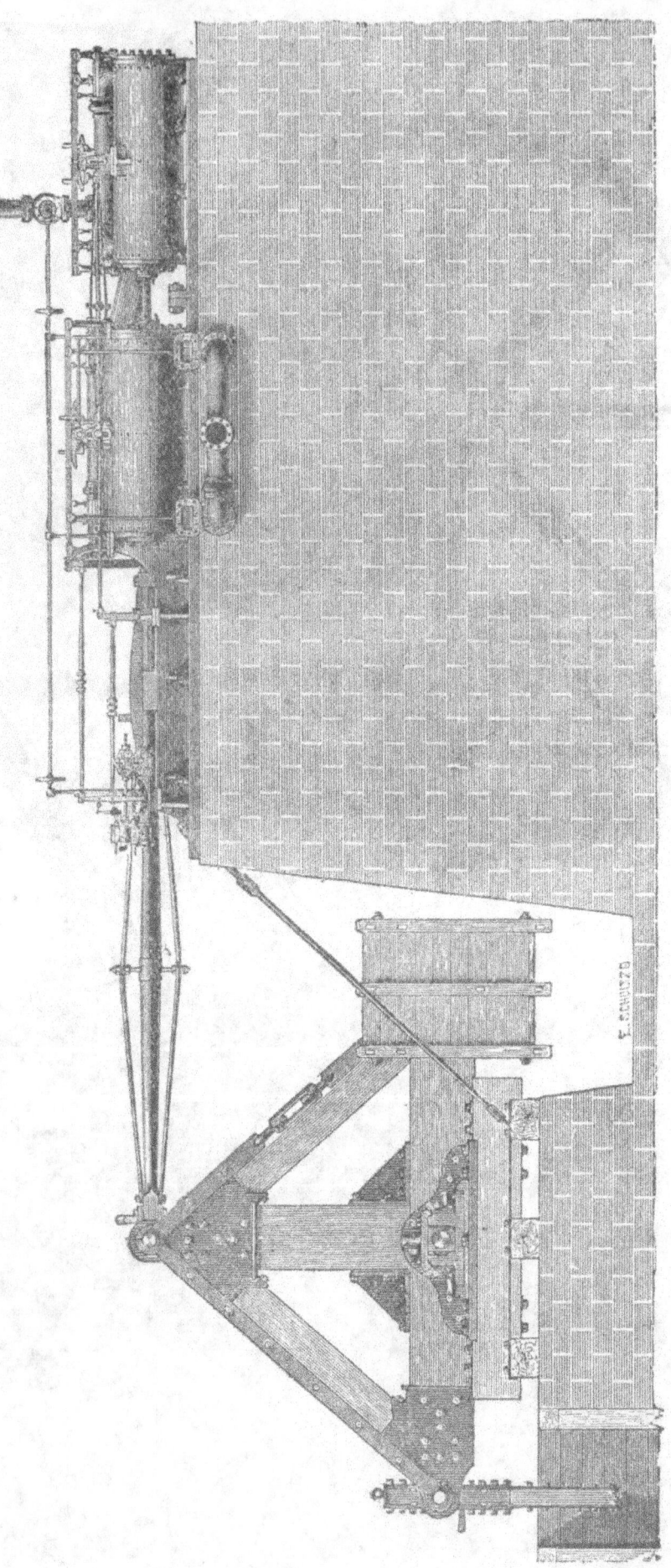

Fig. 103.

The old single-acting Cornish engine and the double-acting engines of the Ehrhardt type, both of which work with variable pauses at the end of the stroke, admit of considerable variation in quantity of water pumped by changing the duration of the pauses between the strokes. Such pauses are useful also in affording time for the pump-valves to seat before the return-stroke is started. These engines are, however,

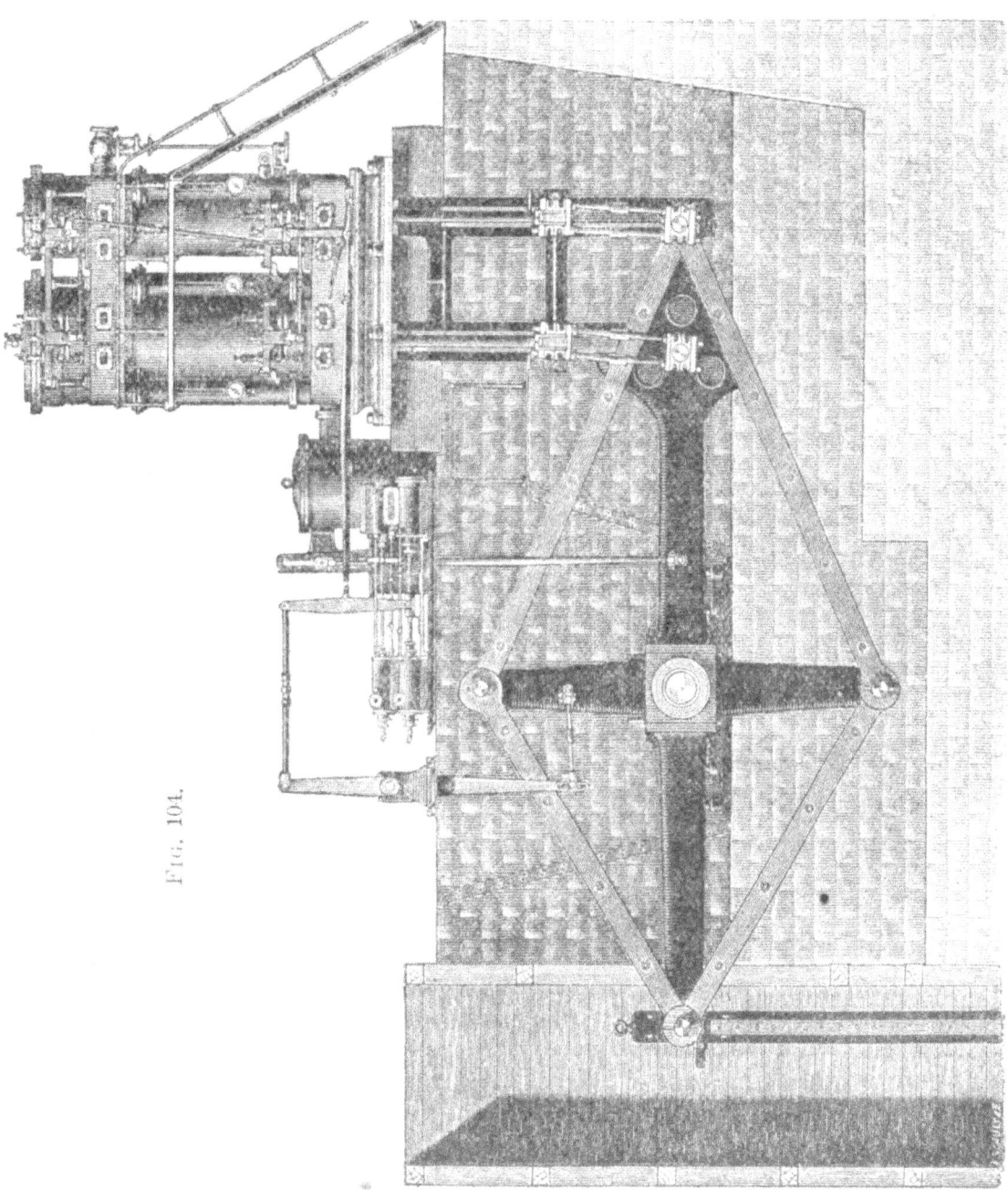

Fig. 104.

even when compounded, not well adapted for sinking, because the engine work during each single stroke can be varied only within comparatively narrow limits.

2.5.06. The only engines of the non-rotative class which have been applied to the Cornish or rod-pumping system in our mines are engines with Davie valve-gear, many examples of which are to be found on the Comstock. Fig. 103 illustrates the Davie engine with bob, examples of which once operated at the C. & C. shaft, the Gould & Curry, and Hale & Norcross at Virginia City, Nev. The Belcher and Overman vertical-beam engines are shown in Fig. 104. The Lady Washington, Lady Bryan, and Alta mines have also had such engines in operation. The steam distribution in these engines is either by slide- or by puppet-valves, controlled by the combined action of a small steam cylinder and that of the main engine itself. The steam is, however, not permitted to act by simple expansion, but is wire-drawn to a great extent. In order to secure the requisite degree of uniformity of pressure during the stroke, and at the same time some of the benefits of expansion, the Davie engines were usually constructed as compound engines.

2.5.07. These engines admit of a somewhat wider range of variation of work per stroke than the older non-rotative engines, but their degree of economy in the use of fuel is considerably behind what might to-day be expected of a first-class pumping engine of the rotative type.

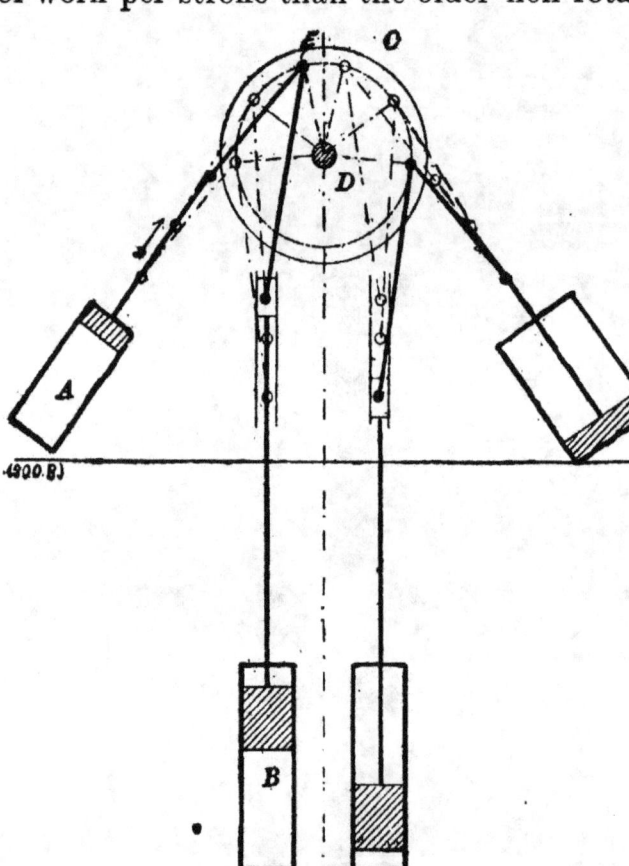

Fig. 105.

2.5.08. In all non-rotative engines, the point at which the stroke is completed is uncertain, on which account they have to be operated with a very large amount of clearance, entailing a considerable waste of steam.

2.5.09. In case the pump-rod breaks near the surface the load will be suddenly removed and the heavy masses disconnected from the engine, so that the latter will immediately attain a higher speed, and strike the bumpers. Many breakages have occurred in this way. It was claimed for the Davie engines that they would automatically shut off their own steam whenever the speed became greater than a given rate; but experience has shown that the Davie valve-gear was no safeguard against accidents of this kind.

2.5.10. Non-rotative engines require to operate with a more uniform pressure during the stroke when the masses to be moved are moderate, as will be the case when a shaft has not been sunk very far, than when the masses are great, as in a deep shaft, where a greater initial pressure can be allowed to accelerate the heavy masses. In other words, the steam should be cut off latest when the work is least, and vice versa. The only way to reconcile these contradictory conditions is by using a very low boiler pressure at first, and greatly increasing it as the shaft goes down. This, however, would be impracticable beyond very narrow limits of pressure. In the Davie engines these defects are corrected to a certain degree by wire-drawing the steam.

2.5.11. In order to enable non-rotative engines to utilize expansion more perfectly, Davie has more recently arranged some of his engines with beams that work the pumps with variable leverage, as shown in Fig. 105. The lever arm stands about normal to the line of motion at the beginning of the pump-stroke, so that, as it swings through a considerable arc, the effective or projected leverage is much reduced at the end of the stroke, thereby causing the moment of the pump-resistance to correspond at each point more nearly with the change of pressure on the engine-piston. As the engine is double-acting, two single-acting, oppositely reciprocating pumps are necessarily coupled up in this

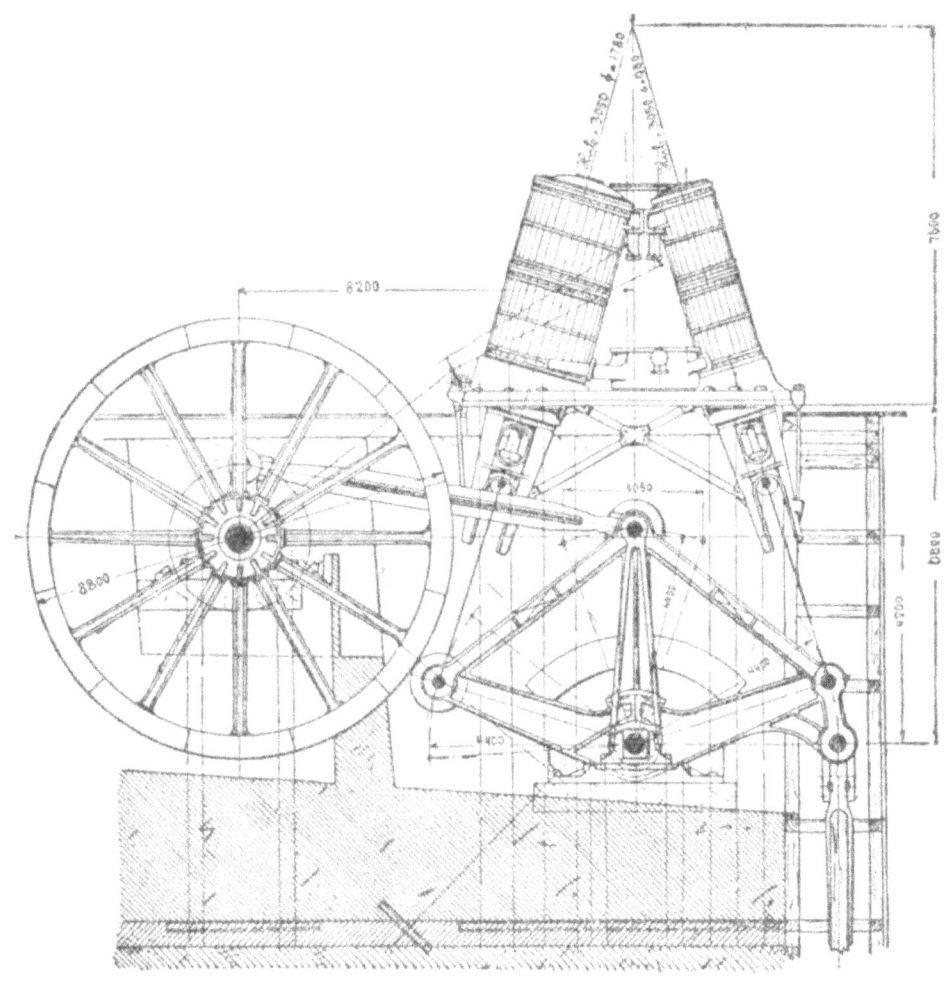

FIG. 106.

manner. The non-rotative principle is not being applied much in recent rod steam-pumping-plants, and it does not seem probable that many more engines of this class will be constructed for mines on this coast.

2.5.12. *Rotative Engines.* The modern forms of these are the most perfect types of rod-pumping engines. They admit of economical steam distribution by comparatively simple valve-gear. They can be operated at higher speeds, on which account they can be made of smaller size, and the work per single stroke can be varied in much wider limits than can be done with the non-rotative engines. The last-named quality makes them well adapted for sinking purposes. On account of these advantages, most of the recently built rod-pumping engines, both in America and in Europe, have, notwithstanding their greater cost, been constructed on the rotative principle. An incidental advantage of the rotative engine is that, in case the pumprod breaks near the upper end, and the load is thereby suddenly removed from the engine, the latter will require time to accelerate the mass of the flywheel and run away,

Fig. 107.

so that the attendant has a chance to close the throttle. Governors which automatically close the throttle or throw a brake onto the flywheel as soon as the speed exceeds a certain limit, can also be easily applied.

2.5.13. Fig. 102 illustrates the beam pump engine at the New Almaden Quicksilver Mine, near San José. Fig. 106 gives the arrange-

MINE DRAINAGE, PUMPS, ETC.

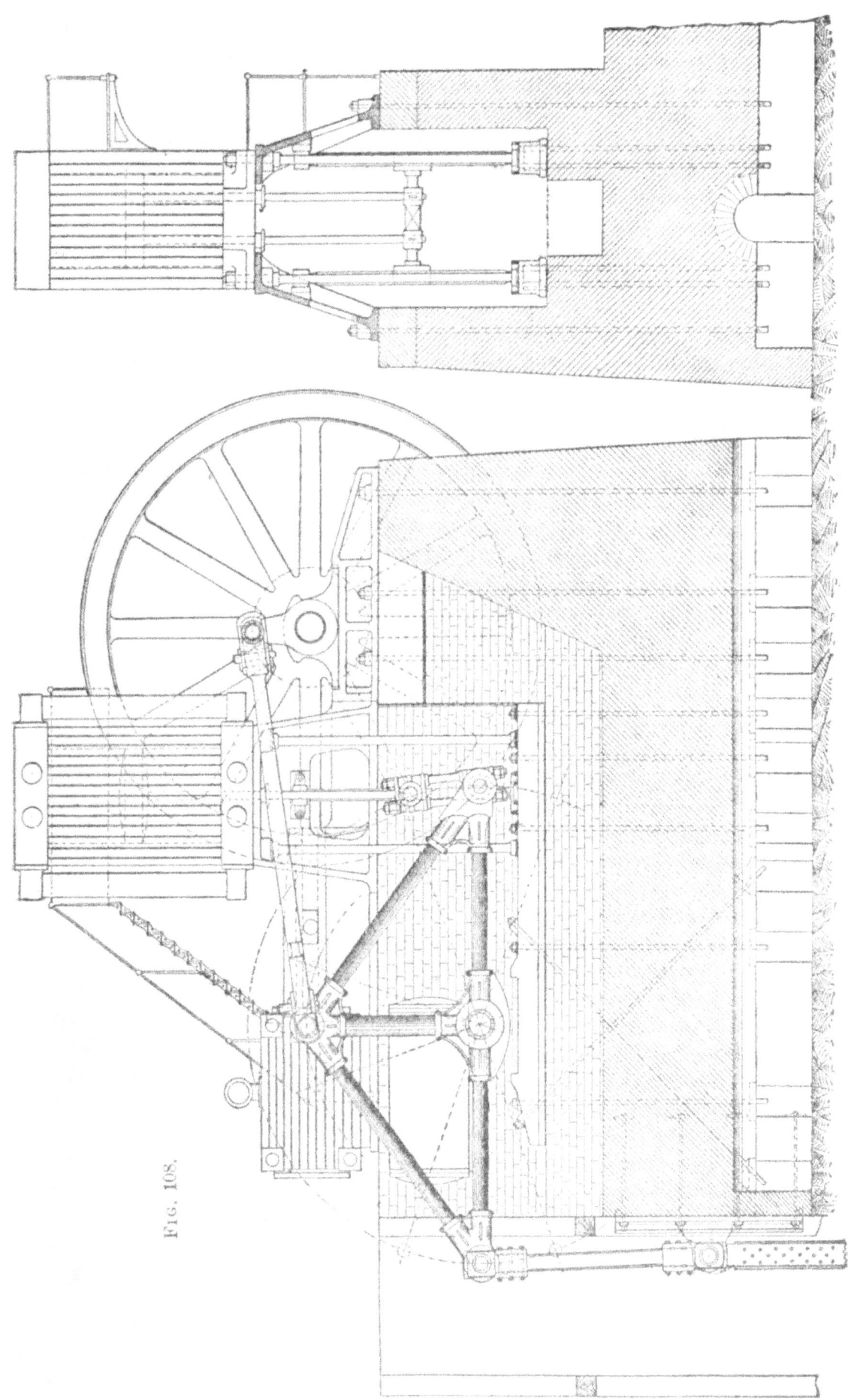

Fig. 108.

Fig. 109.

MINE DRAINAGE, PUMPS, ETC.

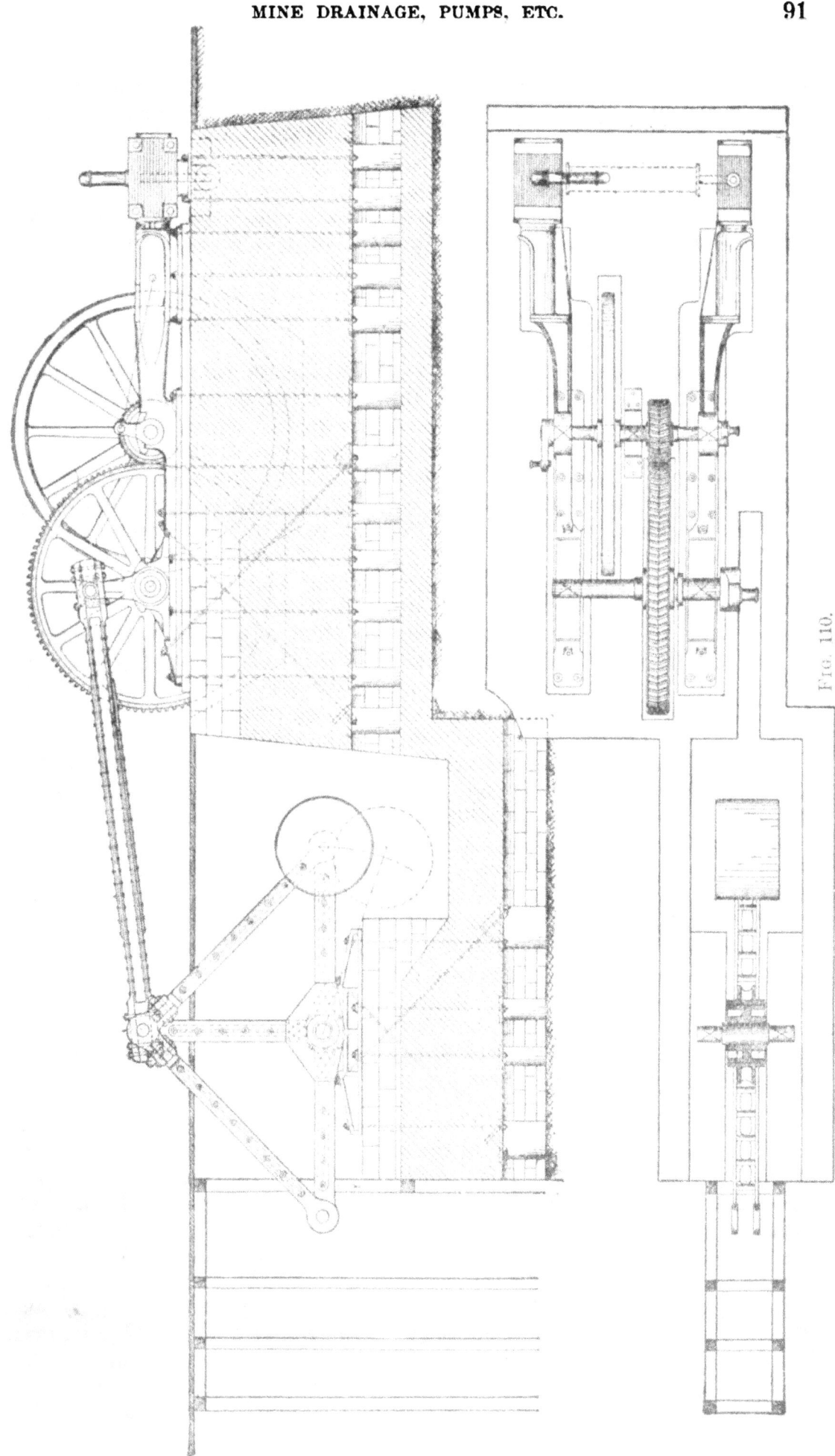

Fig. 110.

ment of the pump engine at the Ontario Mine, Utah, designed by W. R. Eckart, M.E. The inclined position of the cylinders was chosen to take some of the load off the beam-center by letting one cylinder take hold directly above the rod. If vertical, this cylinder would have to come right over the shaft, which is objectionable. Inclined cylinders were first used in this manner by Leavitt.

2.5.14. There are, however, some defects connected with the ordinary rotative engines, like those at the Ontario Mine, New Almaden Mine, and others. This class of engines cannot be run at very low speed, as they will then stop on the center. If very heavy flywheels are used in order to gain some reduction in speed, the engines become too expensive.

2.5.15. In order to combine the advantages of the rotative engines with some of those pertaining to the non-rotative principle, Kley has constructed engines like those shown in Fig. 107, which are connected with a crank and flywheel, like the ordinary rotative engines, but differing from these in that the valve-gear, instead of being operated from the crank-shaft, is a latch-gear of the same type as those in engines of the non-rotative class, and is worked from a tappet-rod a, having a reduced motion coincident with that of the piston. The effect is that the crankshaft may revolve in either direction without changing the function of the valve-gear. This quality is utilized when running at very low speed, down to one stroke per minute, by then adjusting the valve-gear so that the flywheel will come to rest just short of the dead points, and start on the return-stroke in the opposite direction. For higher speeds the valve-gear is adjusted so that the flywheel revolves continuously in one direction like in the ordinary rotative engines.

2.5.16. The most modern system of rotative rod engines is that of Regnier, an example of which is illustrated in Fig. 108, in which a smaller auxiliary engine is coupled to a crank at right angles to the main crank, so as to aid in carrying the latter over the dead points. Very light flywheels are used with these engines, thereby producing, even at the highest admissible number of strokes per minute, a much retarded rotative speed at the dead points, giving almost the pause of the pumprod motion characteristic of the non-rotative engines, and so beneficial to the action of the pump-valves.

2.5.17. Both the Kley and Regnier systems are, particularly for the larger sizes, usually built as compound condensing engines. As such, they represent the highest perfection in steam machinery used for operating pumps by means of rods.

2.5.18. *Geared Engines.* This class of engines is the most widely used on this coast for operating Cornish pumps, particularly for sinking purposes. Figs. 109 and 110 illustrate usual arrangements. The bob is operated by a pitman from a crank, as in Fig. 110, or a crankpin in the side of the gear, which is driven from the engine, as in Fig. 109. In larger plants, the crankpin is usually carried between two gears. The crankpin can be set at different radii, by which means the pump-stroke can be reduced, which gives to the leverage of the engine a greater proportional value, making it capable of pumping from greater depth. This feature, combined with the variability of work per stroke, which is more limited in other systems, makes this class of engines excel all other rod-pumping engines in the range of pumping-depth for which the same engine may be used, and, therefore, particularly fits them for sinking purposes. The decreased capacity due to reduction of pump-stroke can be partially made up by a greater number of strokes.

2.5.19. The capacity of the engine may be still further varied by changing the proportions of gearing. Other advantages of this class of engines are that they are cheap, and when no longer required can often be readily disposed of, on account of the facility with which they can be altered to adapt them to other conditions or other work than pumping. It is also possible to arrange them with reels or drums to carry a cable, and to fit them with means to throw the drum into gear and disconnect the pumps, so that the engines can be made to serve as a pump-hoist for sinking work and lowering parts of machinery in the shaft.

2.5.20. An objectionable feature of the geared engines is, that as the engine work per pump-stroke is uniform they will have their greatest speed when the resistance work of the pumps is least; that is, at the dead points. This causes greater strains in the pumprods than with

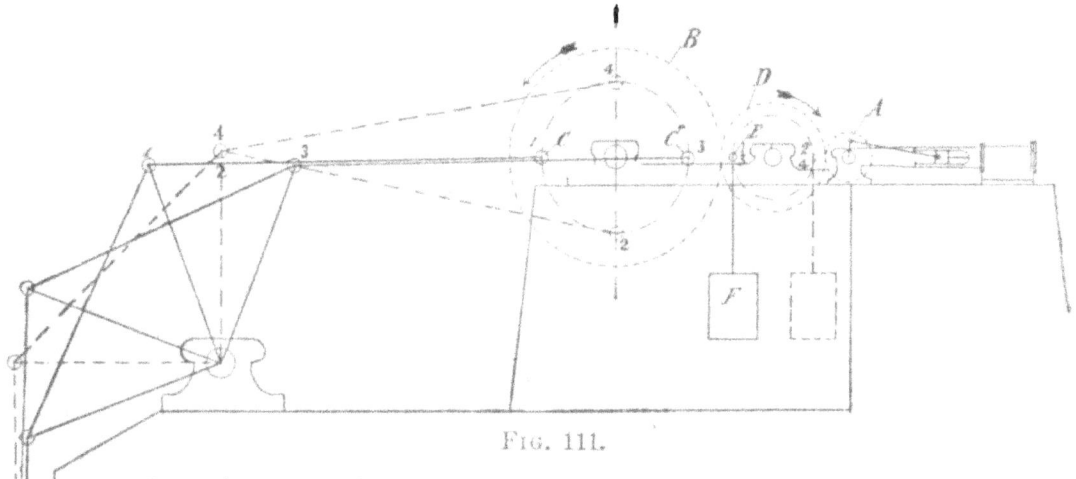

FIG. 111.

the other rotative engines, while at speeds admissible with the latter the action of the pump-valves will be so tardy that shocks will result. The number of strokes that can be allowed with existing types of geared engines is, therefore, appreciably less than with any of the other rotative systems operating pumps under the same conditions, and the geared engines and pumps operated by them have, therefore, to be made larger than if the dead points of the pump-stroke could be turned slowly. Operating a double line of pumprods connected to crankpins at right angles would reduce the acceleration at the dead points, but would not entirely eliminate it; besides, such an arrangement would rarely commend itself, on account of complication in the pump-shaft. To cause the engine to do other work besides pumping, such as air-compression, which would be a maximum at the end of the pump-stroke, is also generally impracticable.

2.5.21. A more perfect method is to cause the engine to perform and store up, near the dead points of the pump-stroke, other work besides pumping, such as raising a weight, and to permit such stored-up work to assist in overcoming the resistance of the pumps near the middle of their stroke. An arrangement for carrying out this idea was patented by Charles Bridges in 1883. The principle of the device, though probably never applied to mining pumps, is shown so applied in Fig. 111. Between the driving-pinion A on the motor-shaft, and the driven gear B, carrying the pin C for operating the pumps by the pitman and bob, there is interposed a third gear D, having half as many teeth as the large gear B, and, therefore, making two revolutions to one of B; a pin E is fixed in the side of gear D, and supports a weight F at its maximum leverage, when the pump crankpin C is at either of its dead points. The direction of rotation must then be such that weight F will be lifted and cause resistance to the engine when that of the pump is lacking. When the pump is at mid-stroke, the pump gear B will have made one quarter of a revolution, and the intermediate gear D will have made one half of a revolution, so that the weight F now descending on the opposite side of the center of D, aids in overcoming the resistance of the pump.

2.5.22. With a pumping-plant of any size, the disturbing effect of the moving mass of the weight F would be very great, and in such cases it is suggested that a piston under a constant pressure of air or steam would be a better contrivance.

2.5.23. The simplest plan would seem to be to automatically vary the steam admission during each stroke, down to zero, or nearly zero, near the dead point. In this manner the engine could be adjusted so as to run slowest when the pumps are near the ends of their stroke, thereby permitting the valve to come to rest quietly, even at an increased number of strokes.

2.5.24. The geared pump engine deserves more consideration toward its improvement than has been accorded to it where rod-pumping is the method to be used. The many existing examples on this coast are mostly of very crude and imperfect design. By arrangements which operate like those described in the three preceding paragraphs, the engines can be made smaller, corresponding to the admissible increase of pump-speed, and the saving in cost could be applied to obtain means for securing greater economy in the use of steam, such as compounding, steam-jacketing, or condensing. More perfect and better constructed gears than those on most of the existing engines of this class are also desirable.

2.5.25. *Remarks.* An important matter in connection with pumping-engines at the surface is their location with reference to the shaft. In general, it is most advantageous to place the engine at that side of the shaft which is farthest from the hoisting-compartments, as in Fig. 112, because then the space around the shaft will be least obstructed. In inclines, however, the pump engines must be placed in a vertical plane parallel to the incline; that is, either on the same or the opposite side of the shaft on which the hoisting engine is located. The nature of the ground and kind of foundation obtainable around the mouth of the shaft may influence the general design of the engine. It may also be influenced by the cost of fuel, the suitability and quantity of water obtainable for condensation purposes. The depth from which water is finally to be raised, and the limit of speed depending thereon, supposing the quantity of water to be approximately known, are the chief elements in fixing upon the size of the engine. Where economy is an object, compounding and steam-jacketing should be resorted to. At high altitudes condensation will be of less advantage than near to sea-level. In case of water which is not suitable for feeding boilers, surface condensation may be of advantage.

2.5.26. As has been stated, a mine requiring pumping should also be equipped with bailing arrangements of the capacity of the pumps, so that, when these or the pump engine requires repairs, the water can be controlled. Where the pumping-plant is of large capacity, geared hoists are generally too slow for handling the water and at the same time taking care of the other hoisting operations. For such cases, large direct-acting hoists should be used which can bring the tanks to the surface rapidly.

2.5.27. The boilers supplying steam to a pump engine should be independent of those from which the hoisting engines are fed, because the intermittent work of the latter causes changes in the steam pressure, which would seriously affect the speed of the pump engine.

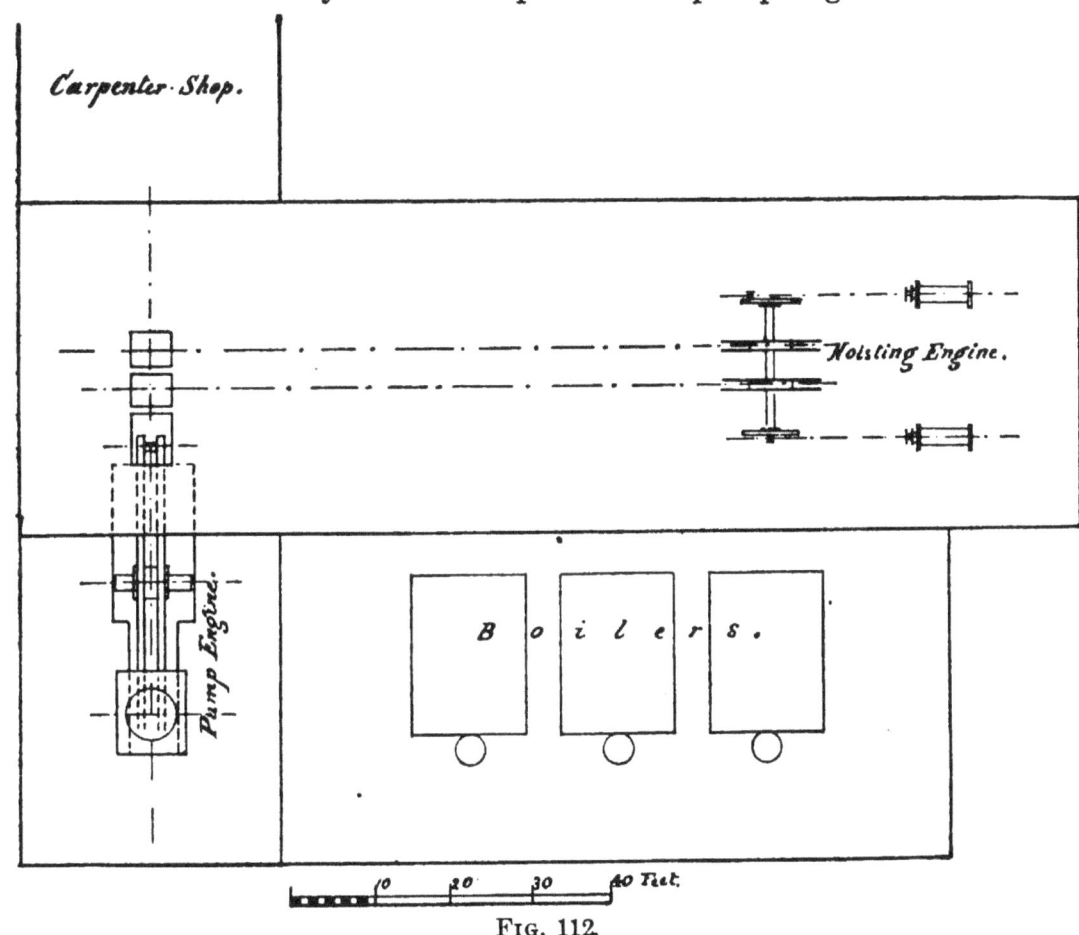

Fig. 112.

2.5.28. The mechanical efficiency of rod-pumping engines depends much on how nearly the pumps at the different levels are proportioned to their work. Under the most favorable conditions their efficiency should reach that of water-works engines of corresponding types. With these, efficiencies of 1 H.P. per hour on 1½ lbs. of good coal have been reached under test. In ordinary operation an engine will not show such results.

2.5.29. *Hydraulic Motors for Pumprods.* These may be either waterwheels or reciprocating-piston engines. Of the former it is only necessary to mention one type, the tangential waterwheel, of which the Pelton, Dodd, and Knight wheels are representatives. Reciprocating hydraulic engines have but a limited application on this coast for operating pumps by means of rods. The ones best known are those designed and constructed by Mr. Knight, of Sutter Creek, California. Neither tangential wheels nor reciprocating engines are suitable for utilizing very low heads of water. These would hardly ever find application for direct pumping, the instances where they can be utilized generally requiring comparatively long transmission by compressed air or electricity.

2.5.30. Cases sometimes occur when it is not easy to decide whether steam or water is the most economical or reliable means of operating

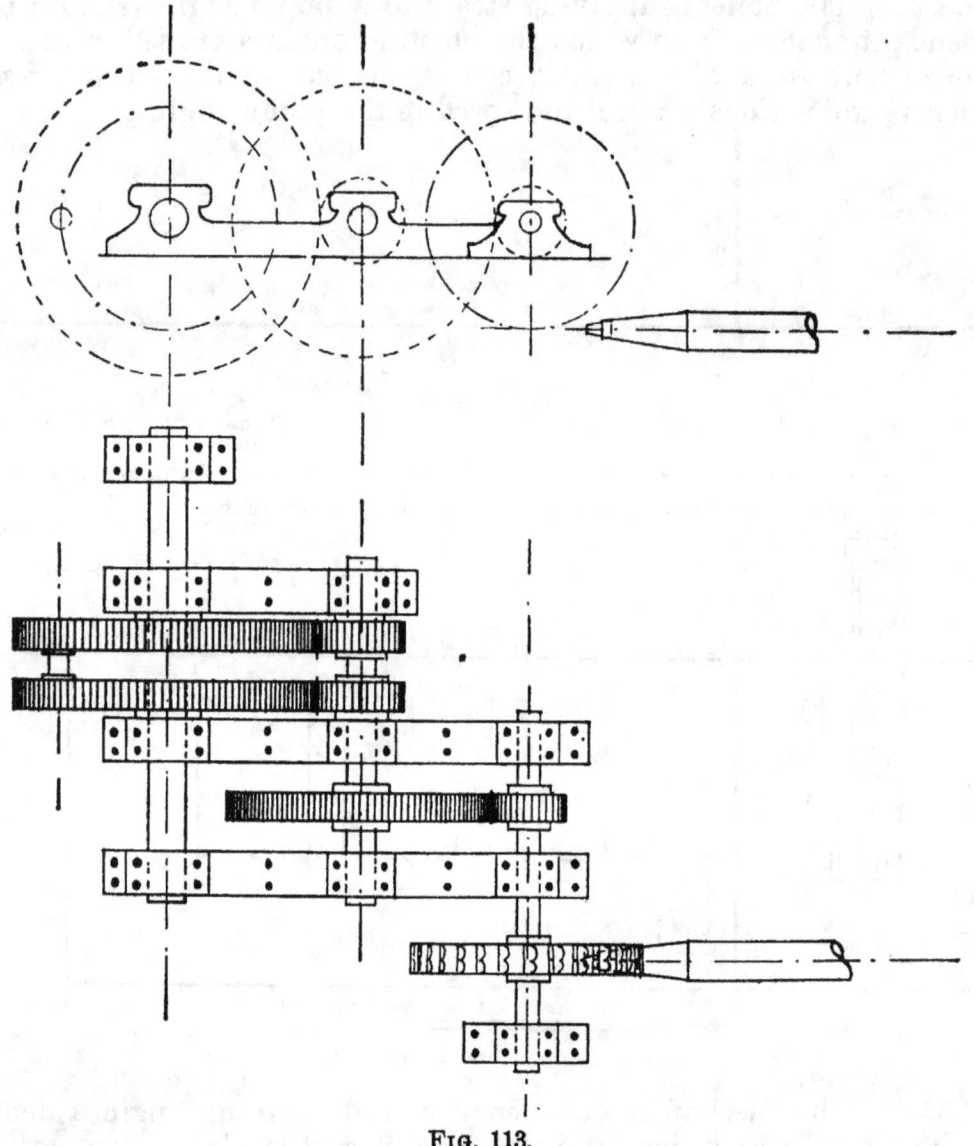

Fig. 113.

the pumps and other machinery at a mine. The cost of installation, the operating expenses, and the permanence of fuel or water supply, or that of its cost, must be considered and compared. It is also necessary to find out if the water supply will fall off appreciably during the latter part of the season.

2.5.31. In using water-power it is generally necessary to provide for stoppage of supply, due to breaks in ditches or pipes, by having a relay of steam-power, at least for the hoisting machinery, which should be of such a capacity that the water in the mine can be controlled entirely by bailing. The steam-relay is also needed where the water supply becomes short in the fall or freezes up in winter.

2.5.32. *Waterwheels.* Those employed for working Cornish pumps are arranged to drive these by means of gearing similar to that used with geared pump engines, the pump-bob receiving motion from a crankpin in the side of the driven gear, as in the geared steam pumping-plant shown in Fig. 109. Gearing is necessary because the pressure usually employed causes the wheel to make too great a number of revolutions, even with the largest practicable wheel diameter, to admit of directly driving the pump-bob. Sometimes even compound gearing is required to obtain the necessary reduction in speed, as in Fig. 113.

2.5.33. In order to utilize the power in the water to the best advantage, waterwheels should run at a fixed number of revolutions. The capacity of the pumps can, therefore, only be changed, aside from changing the pumps themselves, either by varying the radius at which the crankpin acts, or by changing the gearing. Any of these will require corresponding changes in the waterwheel nozzles, to adapt the power to the altered resistance. Increase in depth, the quantity of water remaining constant, must be met by an increase in size or number of nozzles, or by means for varying their discharge.

2.5.34. It is well to have a number of nozzles to the waterwheel, with a gate to each nozzle. With such an arrangement, the power can be adjusted and varied by regulating one of the nozzles, the choking-off amounting then to reduction of efficiency only of the nozzle affected, that of the others remaining undiminished. Fig. 114 illustrates an arrangement of this kind.

2.5.35. Fig. 115 illustrates a nozzle, affording a variable cross-section of jet, designed and patented by Mr. A. Chavanne, of Grass Valley. It consists essentially of an ordinary nozzle, having an opening suitable for the maximum discharge required, the reduction of cross-section being accomplished by the solid mandrel D, which is made with successive abruptly increasing diameters. By pushing the mandrel forward in the nozzle by means of the lever mechanism shown, the area of the opening is reduced by an amount equal to the area of that part of the mandrel which is at that moment within the nozzle-opening, the resulting jet being thereby caused to assume an annular section, of which the inner diameter can be increased so as to reduce the total area. If the mandrel were simply tapered instead of being made to increase by steps, the issuing water would cling to the surface of the mandrel, so as to be partly deflected, thereby causing a disturbed jet. Although the efficiency of the annular jets decreases with the increase of the inner diameter, on account of the greater proportion of wetted perimeter, this reduction is not of so vital importance as to counterbalance the other advantages of the Chavanne nozzle, chief of which are its simplicity and ease of manipulation by either hand or governor.

2.5.36. Waterwheels applied to operating pumps from cranks have the same defect as the geared steam engines; that is, they speed-up near the dead points of the pump-stroke, which results in tardy closing of the valves, and prevents higher speed, on account of the shocks that otherwise occur. What has been suggested in relation to remedying this fault in geared steam pump engines (2.5.21), also has application to waterwheels, with the exception of variation of power supply during each pump-stroke to suit variation of resistance, which would be impracticable with waterwheels.

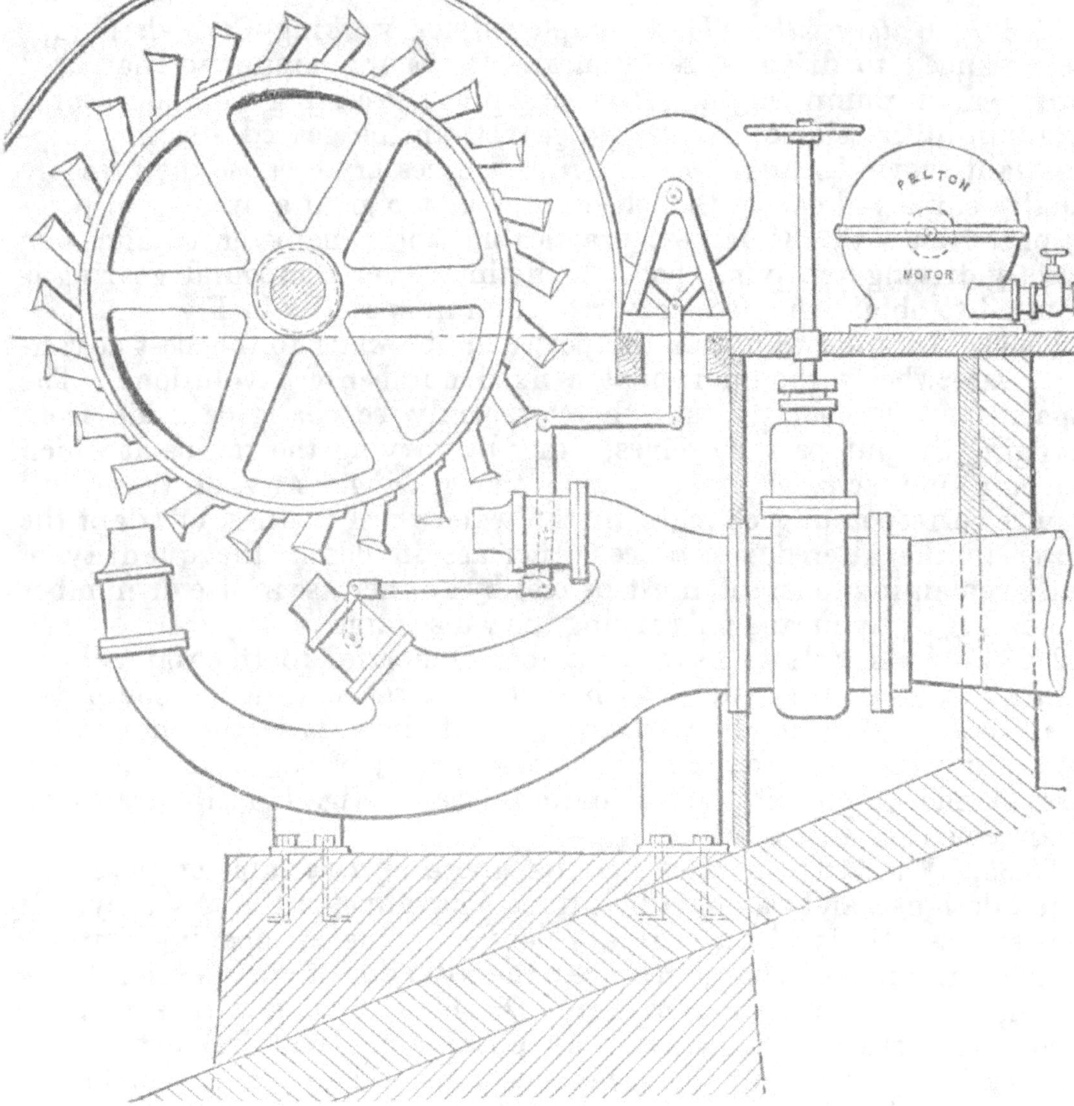

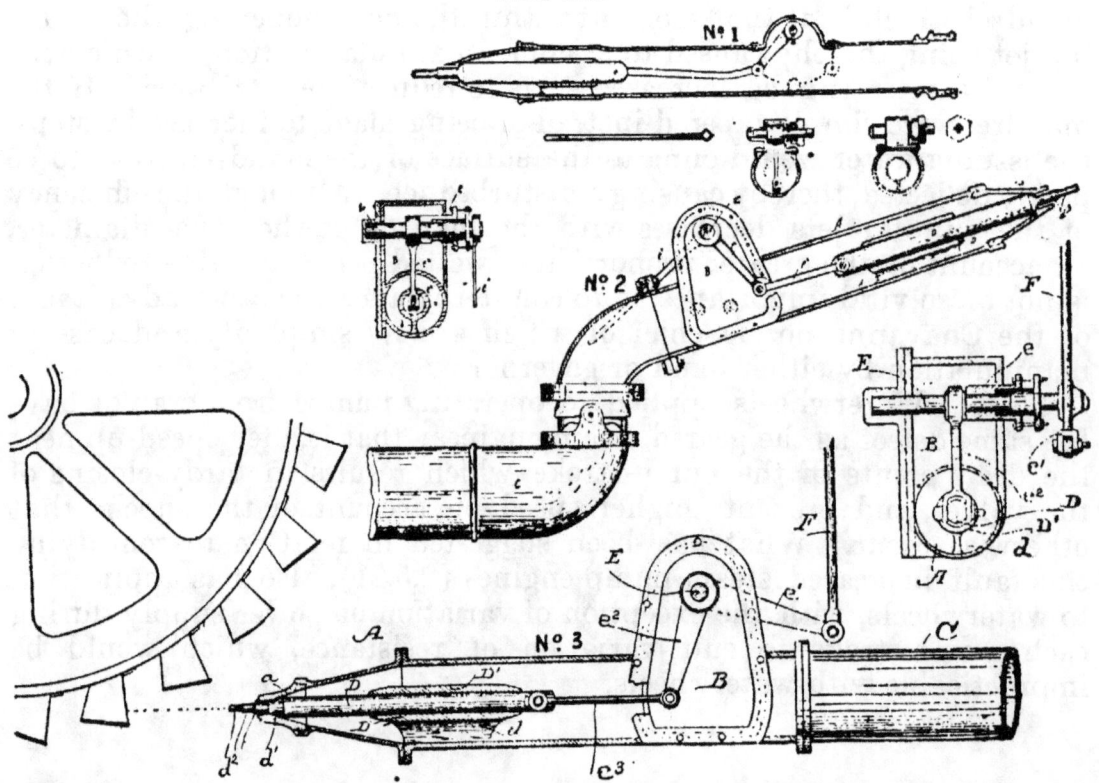

Fig. 114.

Fig. 115.

2.5.37. *Hydraulic Pumping-Engines.* These are suitable for operating under both high and moderate pressures. For the Cornish system, they are arranged either to work the rod direct, or they are

Fig. 116.

connected to it through beams or bobs, like the non-rotative steam engines. Like these, also, they are made single- or double-acting. A type of the latter is shown in Fig. 116. The engine illustrated is working at the Plumas-Eureka Mine, in Plumas County, and is one of those designed and built by Messrs. Knight & Co., of Sutter Creek, Amador County. The engine is operating under a 720' head.

2.5.38. The admission of water to, and discharge from main cylinder is performed by flat D-valves moved by an auxiliary cylinder, seen above the main one. This auxiliary cylinder is controlled by a small slide-valve actuated by the upper tappet-rod. The exhaust passes from the main valves through two balanced piston-valves in the short cylinders seen in front just above the main frame. The piston-valves in the exhaust are controlled by the lower tappet-rod, closing slowly after the main piston has reached its maximum velocity, and retarding and cushioning the flow of the water and the heavy pumprods.

2.5.39. Another of these Knight engines is at the Wildman Mine, at Sutter Creek. This takes hold of the rod directly without the intervention of a bob.

2.5.40. Like the non-rotative steam engines, hydraulic engines can operate at any number of strokes below their maximum. The pauses at the end of the stroke can be made of any duration independently of each other, thereby securing almost perfect action of the pump-valves. In addition, it is possible to reduce the length of stroke, if required. The discharge-pipe should empty under water, or be turned upward, so that the pressure of the atmosphere will keep the working-cylinder full of water during the pauses. Air-outlets and drain-cocks to wash out sediment should be provided. It is also well to have relief-valves at different parts of the engine where shocks are liable to be severe. Where plungers are used in the power-cylinder, they are best made of brass or bronze, for the same reasons as in plunger pumps. With piston engines the cylinders should be of brass or lined with it.

2.5.41. It is important that the power-water delivered to the engine be as clean as possible. Ample provisions for settling and screening should therefore be made at the supply-reservoir at the head of the pressure-pipe.

2.5.42. The pressure-pipe should be large, so that the water in it will have a low velocity, because the flow is intermittent and the column of water alternately started and stopped like in a pump-column. The number of times that this can be permitted to occur per minute is limited, and therefore the engine cannot make a great number of strokes. The admissible number of strokes is less than with non-rotative steam engines, particularly for high heads, on account of the great mass of water in the pressure-pipe, which is usually of great length compared to the pressure-head.

2.5.43. For the sake of reducing shocks, air-chambers are often placed on the pipe near the engine. It is still better to use a number of air-chambers along the pipe-line, as has been done by Mr. Knight.

2.5.44. Hydraulic-pressure engines are not generally used for sinking purposes, because the pressure per square inch is constant, and the total pressure on the piston or plunger can therefore only be varied by changing its diameter.

2.5.45. The field for hydraulic-pressure engines as prime-movers at the surface has been much reduced by the much cheaper and more durable high-pressure tangential waterwheels. These are not only cheaper in themselves, but can have much lighter and also smaller power-mains for the same head, because the flow of water in them is continuous and not intermittent, as in hydraulic engines.

2.5.46. An advantage which the hydraulic engines have over the waterwheel is that they can easily be set up below ground, in which case the discharged power-water is forced up with the mine-water through the column-pipe. The engine need not be larger for this purpose, as the additional water to be raised is balanced by the increase in driving pressure.

2.5.47. By placing the engine thus below in the shaft in the line of the pumprod at the middle of its length, the effect of elasticity of the rod on the stroke of the pumps most distant from the application of power will be much reduced, and the rod system can in such a way be used for greater depths than where the engine is at the surface. Hydraulic engines working pumps direct have also been built by the Risdon Iron Works. Their description will be referred to in the section on direct-driven pumps.

CHAPTER VI.

Operation and Care of Pumps.

2.6.01. *Starting and Adjustment.* Before putting pumps in operation they must first be primed or filled with water, and to make this possible, an escape for the air must be provided by means of cocks located at high points. The manner of priming was described in 1.1.12 and 1.1.13.

2.6.02. In order to adapt the total capacity of a Cornish pumping-plant to the varying water production of a mine, the speed of the pumps must be made adjustable. In geared motors the stroke is also generally made capable of variation, as described in 2.5.18. This was shown in 2.5.33 to be particularly necessary with waterwheels, in which it is not economical to change their speed.

2.6.03. The individual pumps forming a series operated by a rod, will often have to handle quantities of water differing greatly in amount, which amounts again may vary considerably, independently of each other and at different times. Since the relative volume-displacement of the different pumps is fixed by virtue of their attachment to one rod, some of the station-tanks would overflow, while others would be entirely drained and the pumps would draw in air. To provide against overflow the pumps must be speeded up, while those pumps which would then drain their tanks must return to these a portion of the water pumped out, so that the suction-pipe will always be kept covered. To accomplish this the arrangement described in 2.4.13, and illustrated by Fig. 94, is used, in which a float a in the tank, on the sinking of the water-level, operates a lever b, whereby the cock c is opened so as to let water flow from the column-pipe into the tank.

2.6.04. When sinking is suspended and the inflow of water has diminished to a very small amount, it is often better to provide a deep sump, in which the water is allowed to accumulate and from which it is pumped out from time to time.

2.6.05. *Speed of Pumps.* As sinking proceeds and the mine becomes deeper and the masses to be moved greater, the allowable speed becomes less, so that the capacity of the plant is reduced as the depth increases. (See 1.2.38, 2.2.41, and 2.4.24.)

2.6.06. *Lubrication.* The plungers should be kept well greased, the kind of lubricant used depending somewhat upon the temperature of the water in the mine. (See 2.4.04 and 2.4.05.)

2.6.07. *Air.* The pump and suction-pipe must be tight, so that no air will leak in from the outside. There is always some air in the water, a part of which will be liberated in the pump on the suction-stroke, and probably only be slightly reabsorbed by the water on the working-stroke. Where the pumps are set at some distance below the station-tank, so that the water will flow into them, less air will be liberated. Some air in the water, if in small bubbles, is not objectionable, as it reduces shocks by acting as a cushion.

2.6.08. *Putting Pumps Out of Operation.* If water discontinues to come in at the lower levels of a mine, the pumps at such points must be put out of operation. This can be done by disconnecting the plungers or lift pumps; but then the work of the engine will generally be out of balance, because there is less resistance to be overcome on one stroke. If a lift pump be disconnected, the work on the up-stroke is decreased; if a plunger, then the work on the down-stroke is decreased. Lift pumps are easily disconnected. With plungers it is generally better to leave them connected, and to put them out of operation by propping-up the discharge-valve, so that it cannot close. A special arrangement is required for this, consisting of a lever on a shaft passing out through a stuffing-box in the side of the clack-chamber, and having a handle on the outside. In this way the pump-work will also be out of balance if the column-pipe is full, but for the stroke opposite to that for which it is out of balance when the pump is disconnected. By keeping the column-pipe filled to half its height, the work of both strokes will be equally reduced.

2.6.09. *Repairs; Stoppages.* Whenever it becomes necessary to stop for some time in order to make repairs, the bailing-tanks must be put in operation. For short stoppages and moderate inflow of water, it may be necessary to only speed up and pump down the tanks previous to stopping, so that the water from the upper levels will not come down and the sump water-level rise too much.

2.6.10. If the discharge-valve of a plunger pump requires repairs or changing, and there is no stop-valve above it, the water must be run out of the column-pipe; then the pump-work will be out of balance on starting up until the column is filled again, unless a supply for refilling is available on the surface. For this purpose a connection can be made to the column-pipe from the station-tank into which it discharges. A water supply at the surface can be used to make up the amount by overflowing successively from the different higher tanks until the one drained is filled. If the suction-valve of a plunger pump is to be inspected, the water must be let out of the space above it, and the inlet to the suction-pipe in the station-tank be closed by a water-tight cover.

2.6.11. A leaky valve can be detected by the diminished delivery of the pump. If the suction-valve of a plunger pump leaks badly, there is water-ram at higher speeds, because the pump does not fill and the plunger strikes the water with considerable velocity.

2.6.12. For warm water, and where the speed is great, and also in high altitudes, the station-tanks should be placed higher in relation to the plungers than where the water is cold and the speed low, or where the height above sea-level is not great.

2.6.13. The velocity of water in column-pipes at mid-stroke should not exceed 5' per second. (See 1.2.38.) If large column-pipes are used, the pumps may run somewhat faster on that account.

2.6.14. Cornish pumps are frequently run at speeds which, under ordinary circumstances, are not allowable, but which may be justified in controlling, temporarily, a sudden large influx of water. At the Ontario Mine, Park City, Utah, the rotative engine operated two 16" plunger-sets of 10' stroke, having a combined total lift of 455', at thirteen strokes per minute. The engine is situated 7,500' above sea-level, and the pump-chambers would not fill quickly enough, so that the resulting water-hammer frequently broke the column-pipes, which consisted of 15" diameter hydraulic-tubing. The velocity of flow in the column-pipe was, at mid-stroke, nearly 7' per second. Later, the same engine operated by means of a 16" wooden rod 1,060' long, two sets of 20" plunger pumps of 10' stroke, each set having a lift of 200', the total lift being 400' to the level of the drain-tunnel. Under these conditions the maximum speed of the pumps was eight strokes per minute, while the smoothest running was obtained at about six strokes. At the Combination shaft on the Comstock, before the hydraulic pumps were put in, there was a Davie non-rotative engine, operating a double line of 14" plunger pumps by means of a pumprod over 3,000' long; the maximum speed was about six strokes per minute, at which frequent breakdowns occurred.

SECTION III.

DIRECT-DRIVEN RECIPROCATING PUMPS.

CHAPTER I.

General Features.

3.1.01. The desirability, particularly in deep mines, of means other than the cumbersome pumprods for transmitting power to underground pumps, is continually leading to improvement in methods of transmission by steam, air, water, or electricity. The perfection of these has already done much to narrow the field of the old rod-pumping system. The greater simplicity and the cheapness of most of the direct-driven pumps have been sufficient incentives for their introduction in many places, the lower economy with which they operate being in many cases compensated for by the smaller capital invested in the plant. At present, however, considerable attention is being paid to improvement in economy, particularly of steam and compressed-air transmission, and it is to be expected that the direct-driven pumps will still further encroach upon the domain of the Cornish pump.

3.1.02. By direct-driven reciprocating pumps are meant those in which the pump-piston or plunger is rigidly coupled to, and moves coincidently with, the piston or plunger of a motor cylinder connected with the frame of the pump; the motive power may be steam, compressed air, or water. The pumps are always double-acting, and often duplex, like the Worthington, in which the two engines mutually control each other's valve-gear. They may be non-rotative or rotative, with or without a flywheel. Except in the case of sinking-pumps, they are nearly always horizontal.

3.1.03. Being double-acting, and having a comparatively short stroke, with very much smaller masses in motion than in the rod system, the direct-driven pumps are, unless the speed is limited by that of a hydraulic-pressure engine, capable of making a much greater number of strokes. They are, therefore, in the best position to utilize the advantages to be derived from the application of the air-chamber, with which they are commonly fitted on the discharge, or pressure, side, and often also on the suction-pipe. If the pump-stroke is long the admissible piston-speed will again be greater, because the number of reversals of motion is less.

3.1.04. On account of the double action in connection with the effect of the air-chambers, which tends to equalize the flow in the discharge-pipe, the water in the latter is kept continuously in motion in the same direction. The column-pipe can, therefore, be of much less diameter than with the single-acting Cornish pumps. As there is no stopping or back-flow of the water in the pipe, there will be fewer shocks, and higher speed of the pump will be admissible. Duplex pumps, which complete their strokes in regular rotation, maintain an almost uniform flow in the column-pipe. The higher speed and consequently smaller size are features which commend direct-driven pumps for use in mines. In the suction-pipe, similarly, if an air-chamber is used, the column will be more uniformly kept in motion, and will, therefore, not lag back so

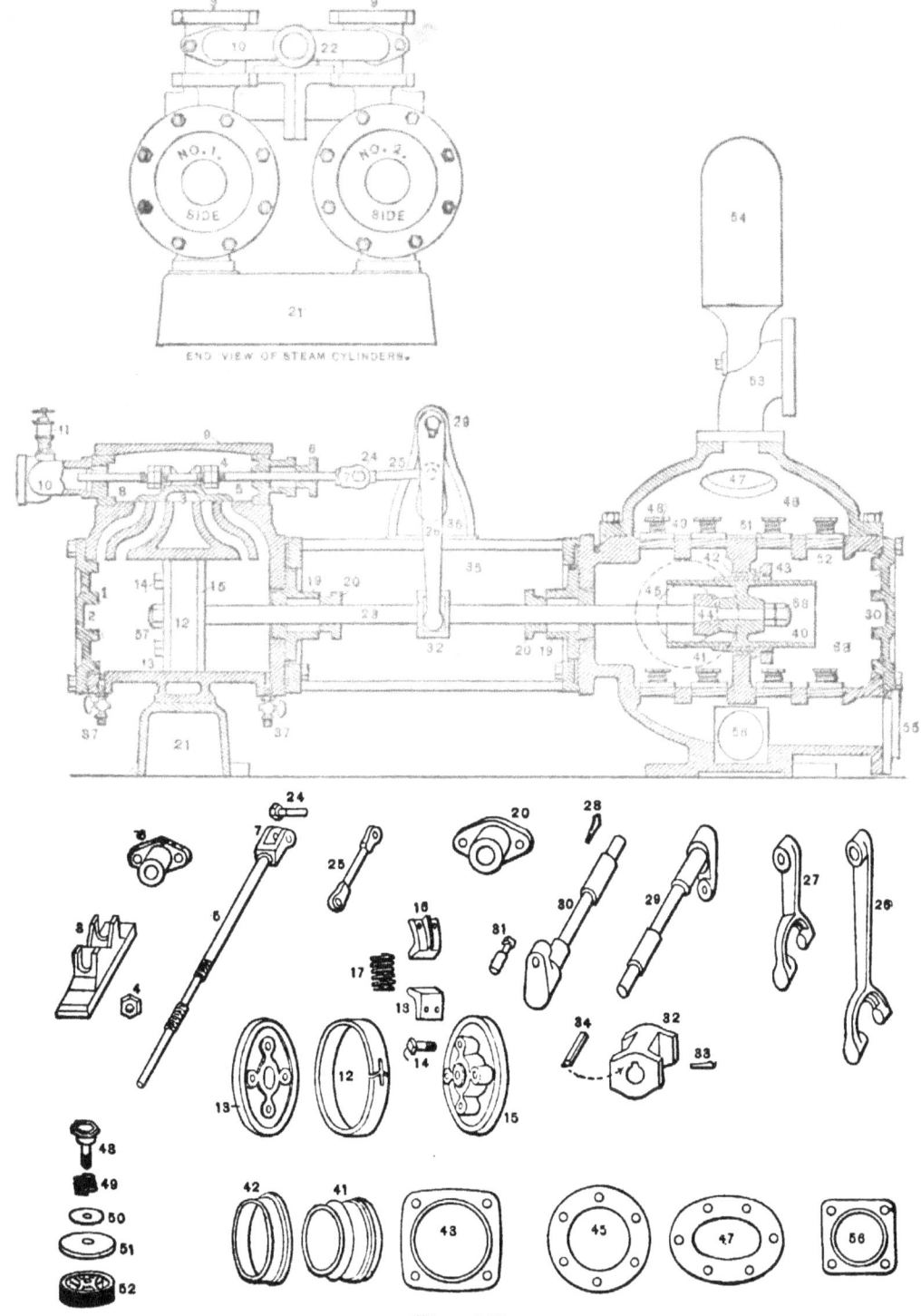

Fig. 117.

readily for higher suction-lifts or greater speeds, and thereby fail to fill the pump-barrel. Higher suction-lift or longer or smaller suction-pipe may therefore be used.

3.1.05. Some pumps have such large waste-spaces that when the suction-lift is great they will not prime themselves. In such cases a foot-valve should be placed at the lower end of the suction-pipe, and a pipe connection made for filling the suction and also the pump from the column-pipe or other water supply. Expelling the air by admitting steam, which then condenses, will also prime the pump; this can be conveniently done where the exhaust is condensed in the suction-pipe.

3.1.06. *Valves.* Direct-driven pumps are nearly always fitted with straight-lift valves. These are made of rubber for low pressures, or of rubber-composition for pressures up to 200'. Above this pressure, and up to 400' or 500', rubber-composition with bronze cages, as in Fig. 36, are used. (See 1.3.09.) Beyond 500' only metal valves will answer. For the larger sizes the valves are placed in sets of several; except in the case of mechanically operated valves, which are almost necessarily single. They are usually spring-loaded, and their area, or the aggregate area of those of one nest, is great, so that they have a small lift and close quickly on completion of the pump-stroke. Fig. 117 shows a pump

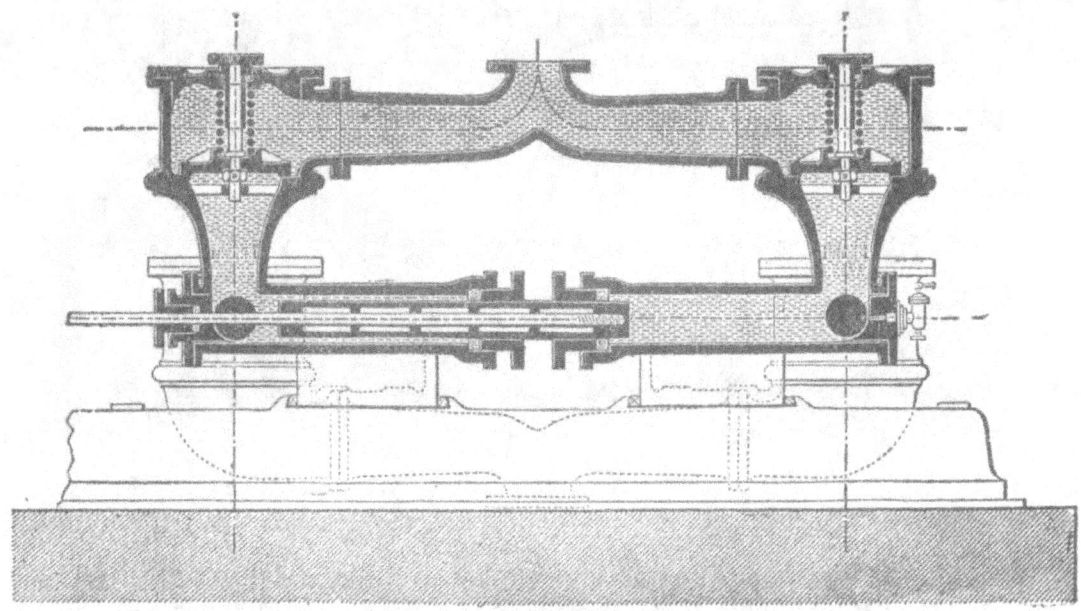

FIG. 118.

in section with multiple suction- and discharge-valves. Notwithstanding the large valve-area in the commonly used forms of pumps, the suction and discharge currents induced by the piston or plunger generally act locally with the greatest force on some only of the valves of a nest, causing these in particular to lift higher and to close more tardily than they should, while others remain almost closed, so that shocks are not avoided to the extent that might be expected by the use of a large number of valves alone. The reason of this is that the current is not diffused into a uniform stream of lower velocity corresponding to the valve-area, before it reaches the valves, but it rather breaks through portions of comparatively still or sluggish water. To overcome this defect Mr. G. Hanarte constructs pumps with valve-chambers designed so that the current is gradually and continuously increased in velocity from that with which it passes the suction-valves to that of the plunger, by the conoidal form of the elbow-chamber above the valves, and again reduced in velocity in a similar manner by the conoidal form of the chamber below the discharge-valves, as shown in Fig. 118. From the discharge-valves the velocity is again gradually increased by the conoidal entrance piece to the discharge-pipe. Such a pump, with piston

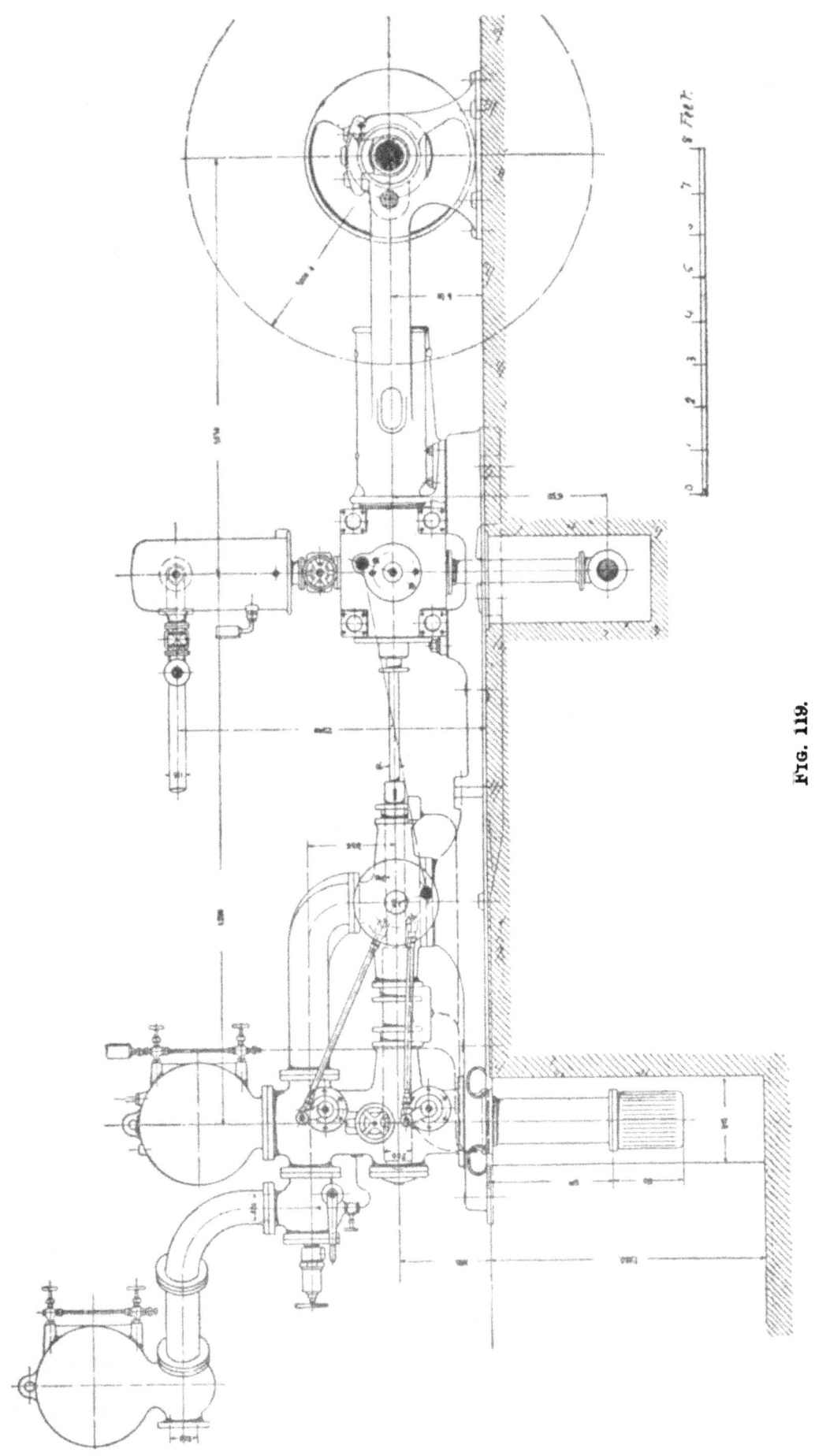

Fig. 119.

3" in diameter and 12" stroke, Mr. Hanarte has run at 400 double strokes per minute without the least shock. This gives 800' piston-speed, which is remarkable, particularly for so short a stroke. In experiments made at 200 revolutions, and at 10' lift, the pump gave 10% greater discharge than the piston-displacement; at 100' lift the discharge was equal to the

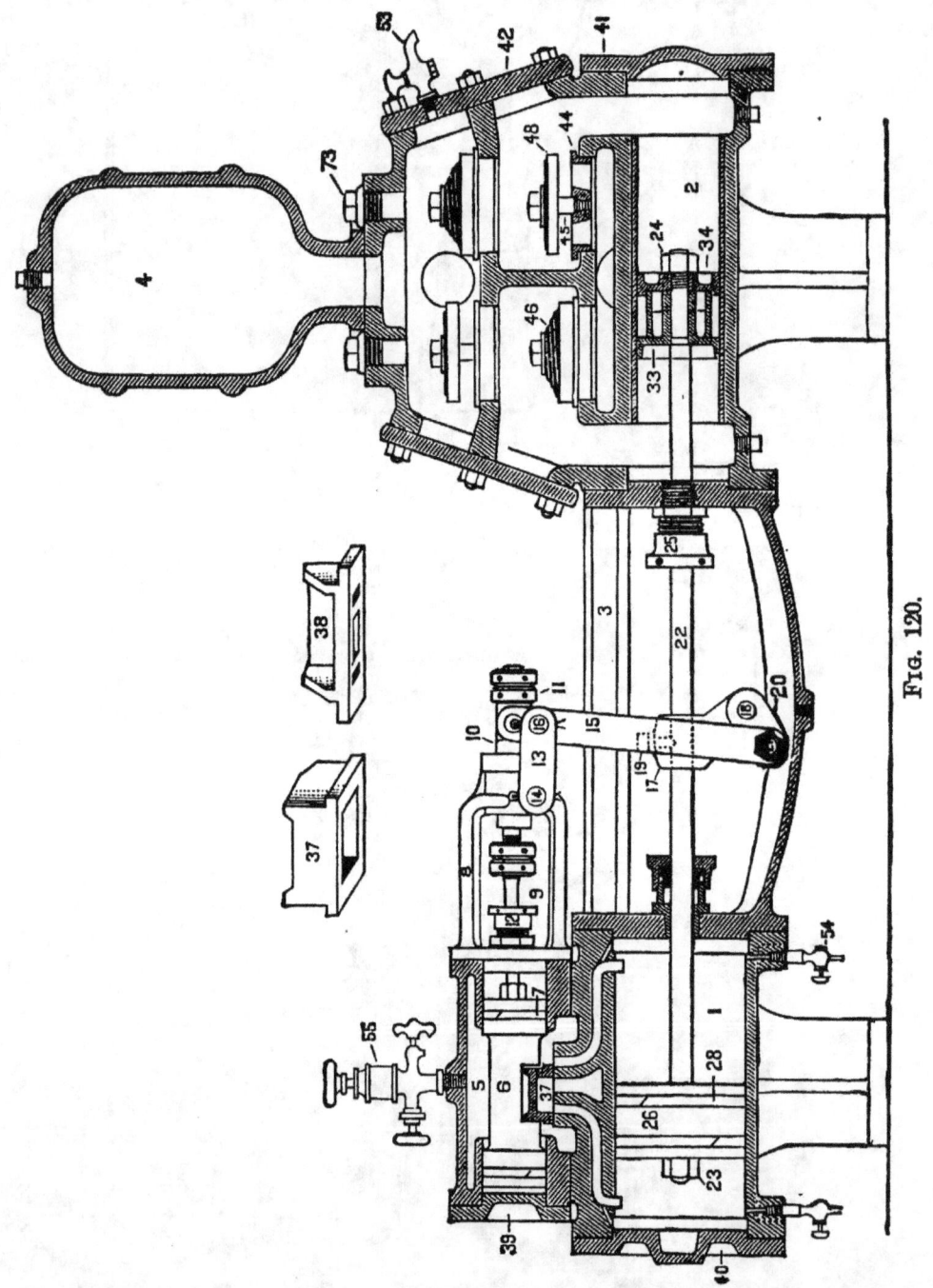

Fig. 120.

displacement, and at 200' lift it amounted to 92% of it. The mechanically actuated valves of Professor Riedler, described in 1.3.18, and illustrated in Fig. 42, have not admitted of such high pump-speeds. The Riedler pumps have, on the other hand, been successfully used to overcome very high lifts, a pump of 6" diameter of plunger and 20" stroke having been run at 80 revolutions per minute without appreciable water-ram while pumping against a head of about 1,300'. In Fig. 119 is shown a complete Riedler pump.

3.1.07. *Piston Pumps.* Figs. 120 and 121 show common forms of these pumps. They are applicable for only moderate lifts, and where durability is desired they should be used only for pumping clean water.

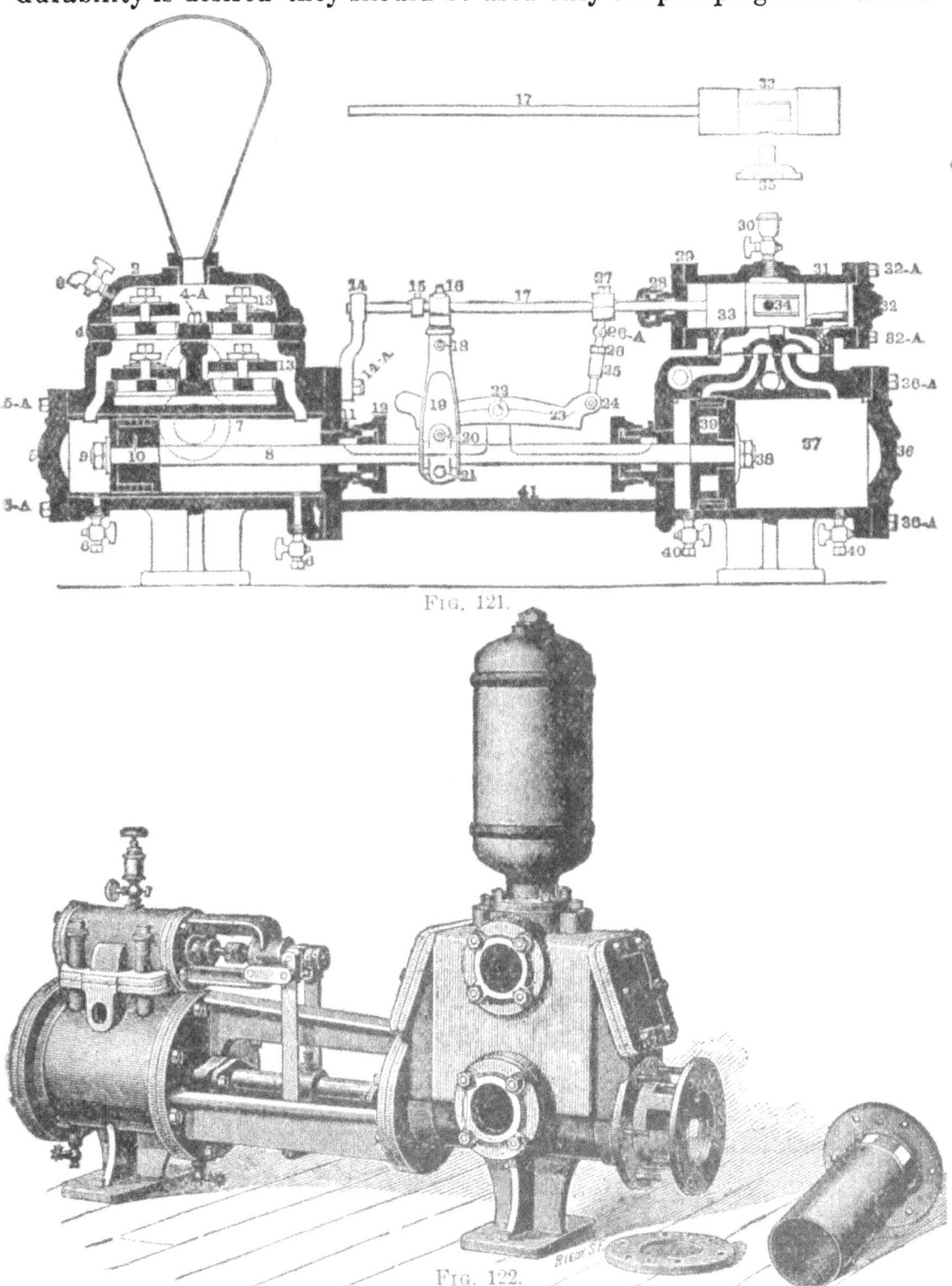

Fig. 121.

Fig. 122.

Common use is, however, made of the non-rotative form for feeder pumps to the station-tanks in mines; and in such duty they have to pump very dirty water, and also receive hard treatment otherwise. They are used for this purpose on account of their small weight, compactness, and low cost, and are run until they give out, when they are sent to the surface for repairs and replaced by others. Some of these pumps, like the one shown in Fig. 122, are made with exchangeable cylinder-liners, so that the whole pump-cylinder will not have to be thrown away when its surface is worn out.

3.1.08. The piston-packing is usually similar to that used for packing plungers: hemp soaked in Albany compound for cold water, or square-braided cotton intermixed with plumbago for hot water. Such packing is driven in tight, and held in place by a follower. Rings of square rubber packing or double cup leathers, as in Fig. 123, are also sometimes used.

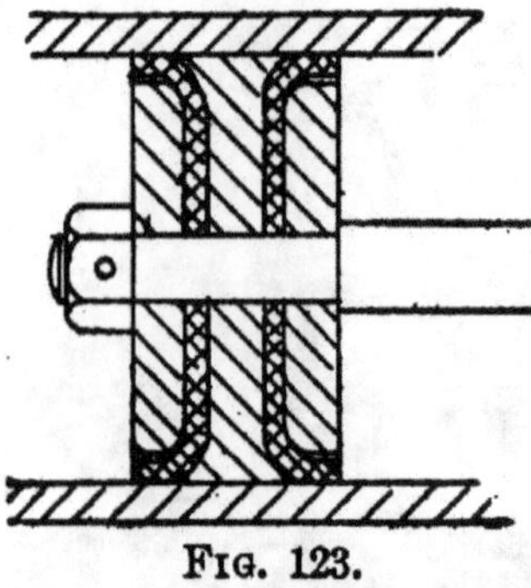

Fig. 123.

3.1.09. Leakage of pistons is not easy to detect, and the packing cannot be tightened without stopping and taking apart the pump. Lubrication of pistons is difficult. In Fig. 124 is shown a means of lubrication through the hollow piston-rod as applied to a Knowles pump. The pump-cylinders are best made with brass linings, but, on account of cheapness, the plain iron cylinders are mostly used.

3.1.10. The piston pumps used underground are usually operated by compressed air in a manner which leaves much to be desired on the score of economy.

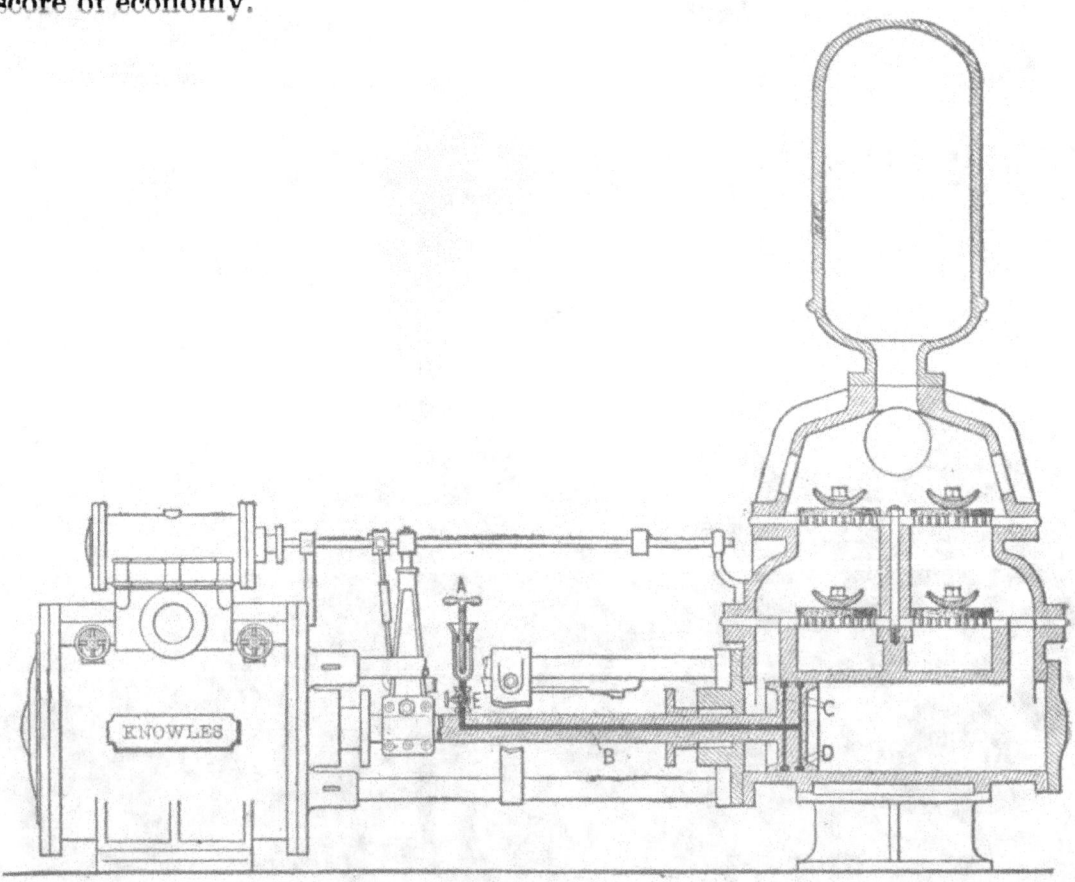

Fig. 124.

3.1.11. *Plunger Pumps.* Plungers have, in direct-driven pumps, the same advantages as in Cornish pumps, *i.e.*, the packing can be tightened while the pump is running, and they can be used for acid water and also for water carrying sand, though when horizontal, not with the same freedom from wear as vertical plungers. In direct-driven pumps they can be used for pumping against very high heads. The packing is the same kind as used with the Cornish plungers. Figs. 125, 126, 127, 128, and 129 show common forms of high-pressure, double-acting plunger pumps. Brass plungers are often used, on account of the reduced friction and better resistance to acid water. In regard to plunger packing and lubrication, the same applies as remarked in 2.4.04 and 2.4.05.

3.1.12. In order to make the pump double-acting, either two plungers are connected oppositely, as in Fig. 125, or a double-ended plunger works in two oppositely located pump-barrels, as in Fig. 126.

3.1.13. A very compact form of pump results from the use of a so-called differential plunger. Such a pump is illustrated in Fig. 128. The area of the smaller part of the plunger, which, in reality, is only a large piston-rod, is half that of the large part. The pump has only one suction- and one discharge-valve, but is, nevertheless, double-acting, as far as resistance to motion is concerned, because only half of the water delivered through the discharge-valve is forced into the column-pipe, while the remaining half is drawn into the space surrounding the smaller part of the plunger, to be forced out again on the return-stroke, while the larger end of the plunger is drawing in water through the suction-valve.

3.1.14. A plunger without packing is shown in Fig. 117; the plunger slides simply with reasonable fit in a long sleeve, the

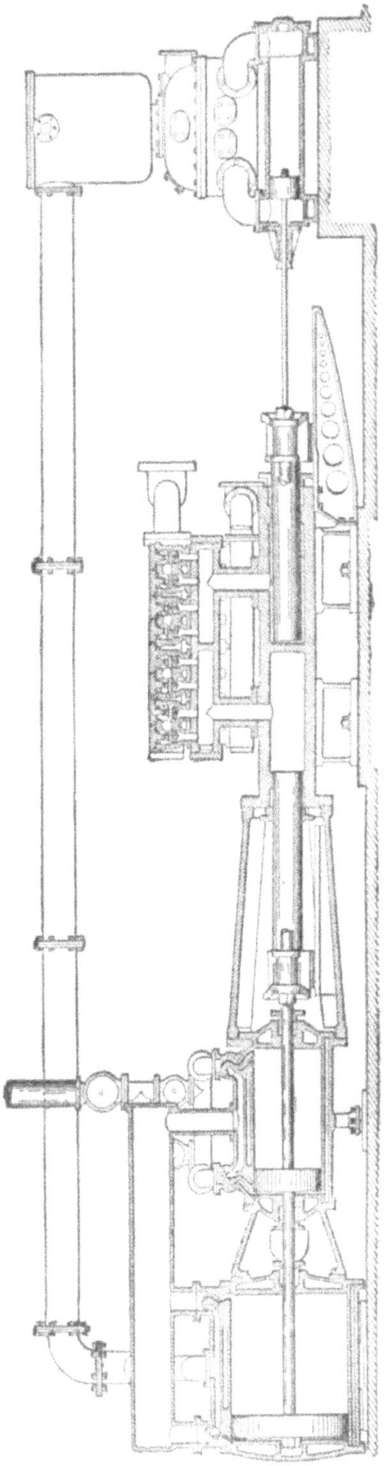

Fig. 125.

Fig. 126.

lubricant serving as a sort of packing. The plunger is made hollow, and of such thickness that it will be of the same weight as an equal volume of water, thereby causing it to exert no pressure on the sleeve,

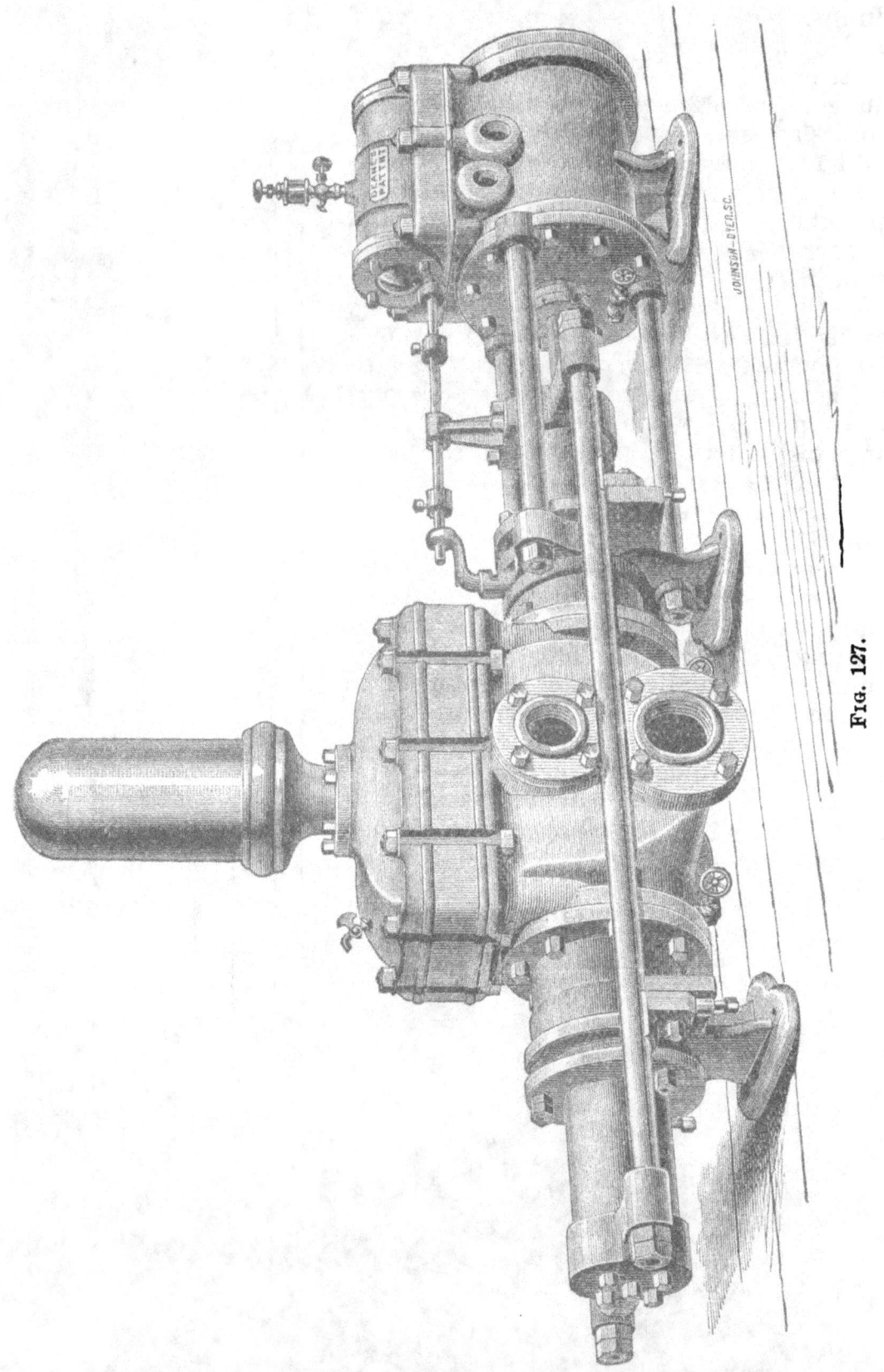

Fig. 127.

thus reducing the wear. The sleeve should be made so as to be readily interchangeable. For high pressures this form cannot be kept sufficiently tight, and it is not suitable where the water contains much grit.

3.1.15. Another form is the bucket-plunger (Fig. 130), which is suitable only for vertical pumps, such as sinking-pumps, and for clean water. The water here passes through the plunger and the discharge-valves, which are located on top of it. The pump shown also utilizes the differential principle, described in 3.1.13.

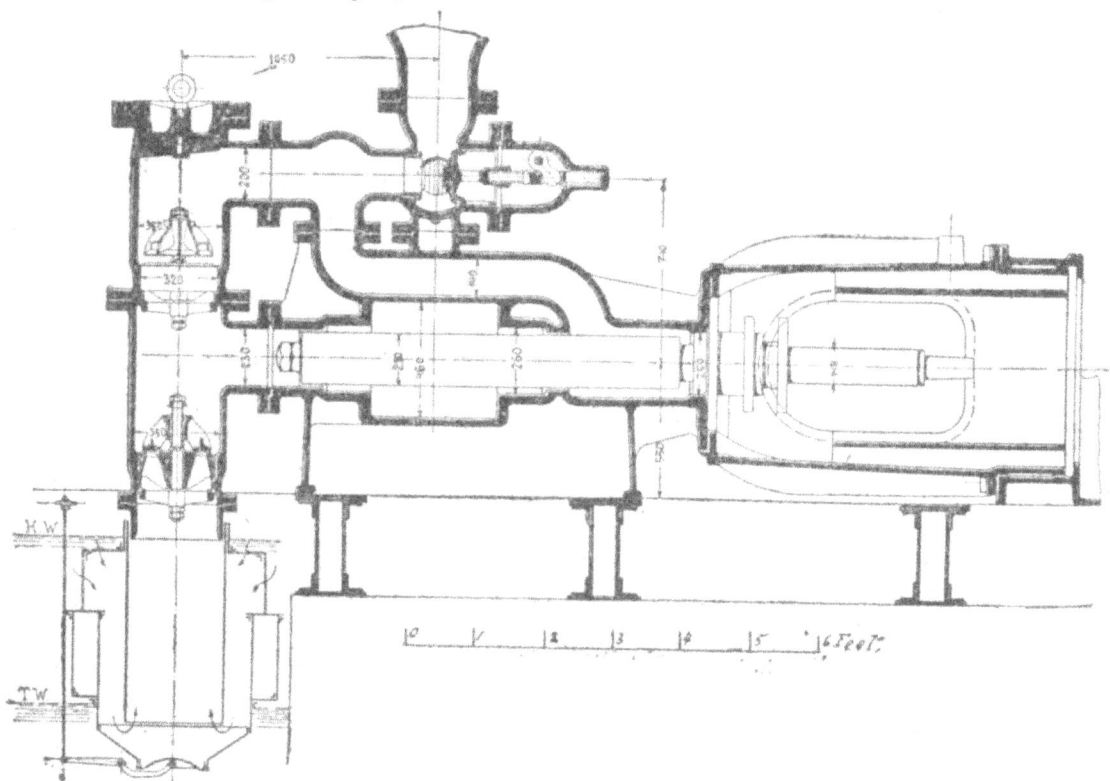

Fig. 128.

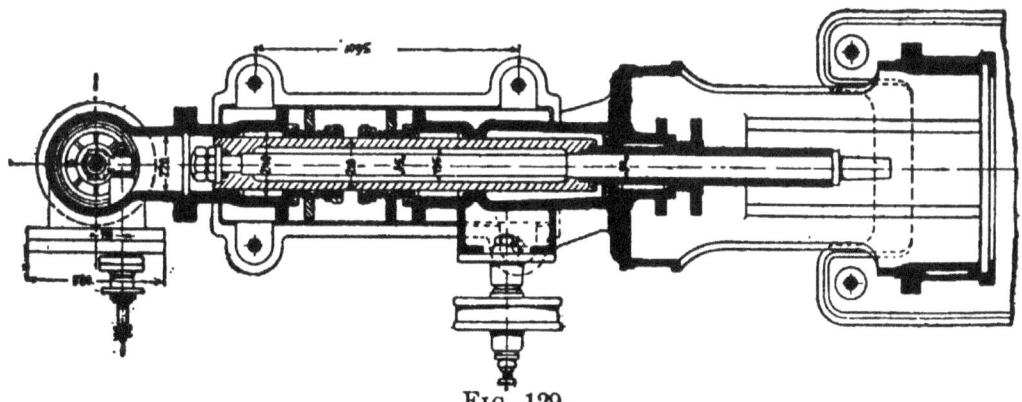

Fig. 129.

3.1.16. *Air-Chambers.* The object of air-chambers, as stated in 1.2.42 and 1.2.44, is, firstly, to change the intermittent motion of the water moving with the pump-piston or plunger, into a flow as uniform as possible in the discharge-pipe; and, secondly, to reduce the shocks or water-ram. Pumps fitted with properly proportioned air-chambers can be run at greater speed and against higher heads than those not so provided, because the mass of water reciprocated by the pump is comparatively small in the former. It follows, in order to keep this mass a minimum, that the distance between the discharge-valve and the air-chamber should be as short as possible; therefore, an air-chamber should, in large pumps, be placed directly over each set of discharge-

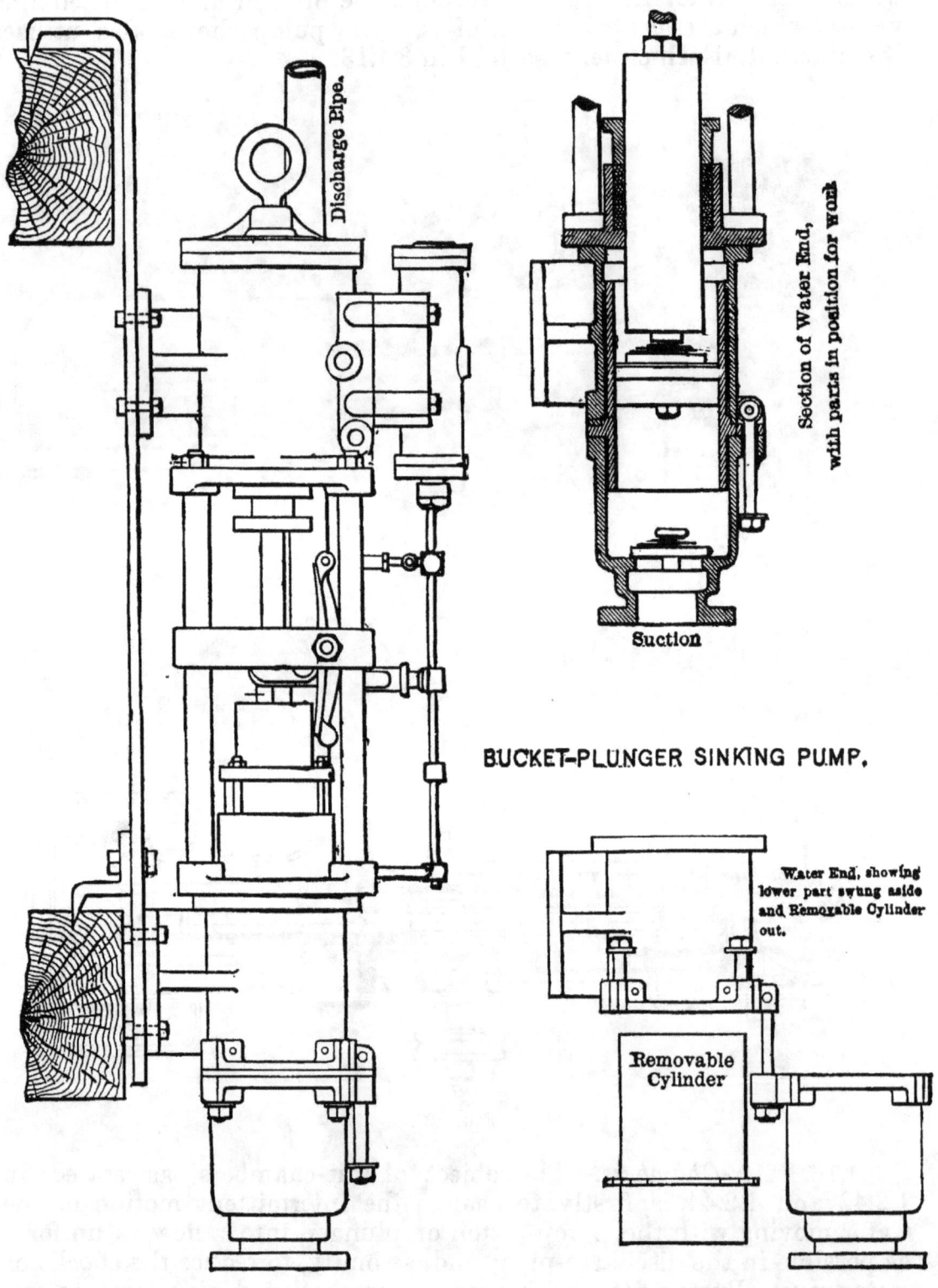

Fig. 130.

valves. Rarified-air-chambers are often used below the suction-valves to equalize the flow in the suction-pipe.

3.1.17. The requisite volume of air-chamber is largest for single-acting pumps, much less for double-acting ones, still less for duplex double-acting, and least for triple pumps.

3.1.18. In the pressure air-chambers the air is generally absorbed by the water; in the suction air-chamber it is liberated. Pressure air-chambers, therefore, generally require replenishing from time to time. This may be done periodically by a small hand air-pump, or automatically by one operated by the pump. The usual plan is to admit a small quantity of air at each suction-stroke into the space between the suction- and delivery-valve by means of a small pipe provided with a cock to regulate the quantity of air to be admitted, and also, so as to prevent outflow on the working-stroke, with a check-valve.

3.1.19. Air-chambers, particularly those of cast-iron, should be tested for tightness under full pressure, and then painted on the inside. For light pressures gauge-glasses will answer to indicate the water-level, but for higher pressures try-cocks must be used. It is also advantageous to have a pressure-gauge on the air-chambers, which will indicate the fluctuations of pressure. Of course, the use of such appliances is warranted only with larger pumps.

3.1.20. Instead of air-chambers, spring-loaded plungers or pistons have been applied in some recent high-lift pumps. (See 1.2.44.)

CHAPTER II.

Non-Rotative Pumps.

3.2.01. Non-rotative pumps, commonly termed "direct-acting pumps," are the type of the direct-driven pumps most generally used in this country. They are cheaper, occupy less space, are more easily erected, and can be run at much slower speeds than single steam or compressed-air pumps fitted with cranks and flywheels. They are designed for operation by steam, by compressed air, and, in some cases, by hydraulic pressure. They are less economical in operation by steam or compressed air than the rotative type, because they cannot utilize the benefit of expansive working to any extent and have to work with considerable clearance in the steam cylinder, being, in this respect, in a position similar to that of the rod-pumping steam engines of the Cornish, Ehrhardt, or Davie types, described in 2.5.05 to 2.5.08. As in these, compounding improves their economy. For larger units the station pumps are generally constructed on the duplex plan, as illustrated in Fig. 126, first introduced by Henry Worthington. Duplex pumps admit of higher piston speeds than single pumps, because with them the column of water is kept in more uniform motion; they are also more easily started.

3.2.02. The station pumps are always horizontal, while the sinking-pumps are generally vertical or inclined in the line of the shaft.

3.2.03. The direct-acting pumps built by the various manufacturers differ chiefly in their mechanism for effecting the distribution of steam. Those illustrated and previously referred to show some of the great variety in existence.

3.2.04. *Sinking-Pumps.* For direct-driven sinking-pumps the non-rotative type is the only suitable one. Figs. 130 and 131 show types of these. Those to be operated by steam usually have a condenser for the exhaust steam located in the suction-pipe, as in Fig. 132, which illustrates in section a large pump of a style much used on this coast. Duplex sinking-pumps, of which the Worthington is a type, are not used so extensively, on account of the amount of space they occupy in the shaft.

3.2.05. The suction-pipes are always of hose, and often have a foot-valve just above the strainer, in order to keep the suction full of water whenever the pump is stopped for lowering or raising or for repairs to the suction-valves. This foot-valve should properly remain open during the operation of the pump, and not close with the suction-valves, so that the suction-resistance may not thereby be unnecessarily increased. There should be a relief-valve in some part of the suction-pipe, whenever a foot-valve is used at the lower end of the hose, so that any leakage past the suction-valve, while the pump is stopped, will not burst the hose, but will be permitted to escape under a moderate pressure. The suction-hose should be wrapped with rope to protect it during blasting.

3.2.06. Sinking-pumps are generally obliged to handle water full of grit, it being impracticable to settle it in a large reservoir, as is done with the station pumps. The larger the cross-section of suction-pipe and the less, therefore, the velocity of the water in it, the less also will be the amount of sand drawn up into the pump. But this expedient is generally not sufficient to protect the pump and to prevent the necessity of its early removal for repairs. This difficulty, in one case of a large pump, led to the design of a settling-chamber attached below the pump, as shown in Fig. 133. The two suction-branches *a a* entering the chamber are bent over so as to discharge circumferentially and cause the water to assume a rotary motion, whereby the sand is driven by centrifugal action against the wall of the chamber and falls to the

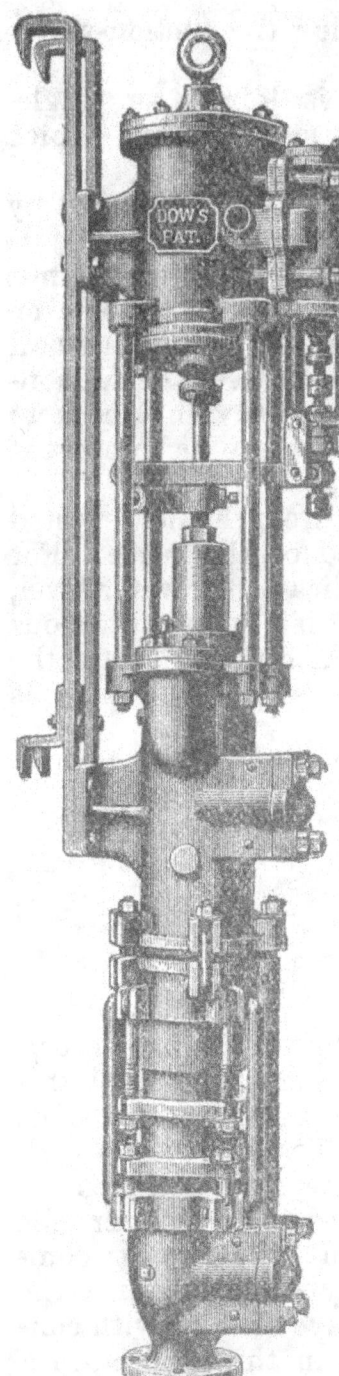

Fig. 131.

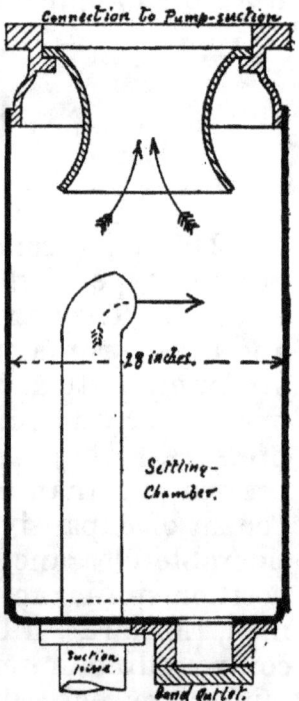

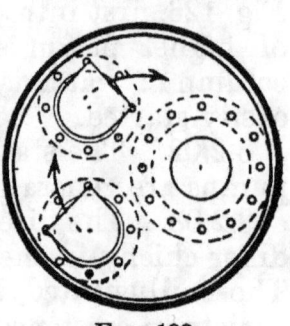

Fig. 133.

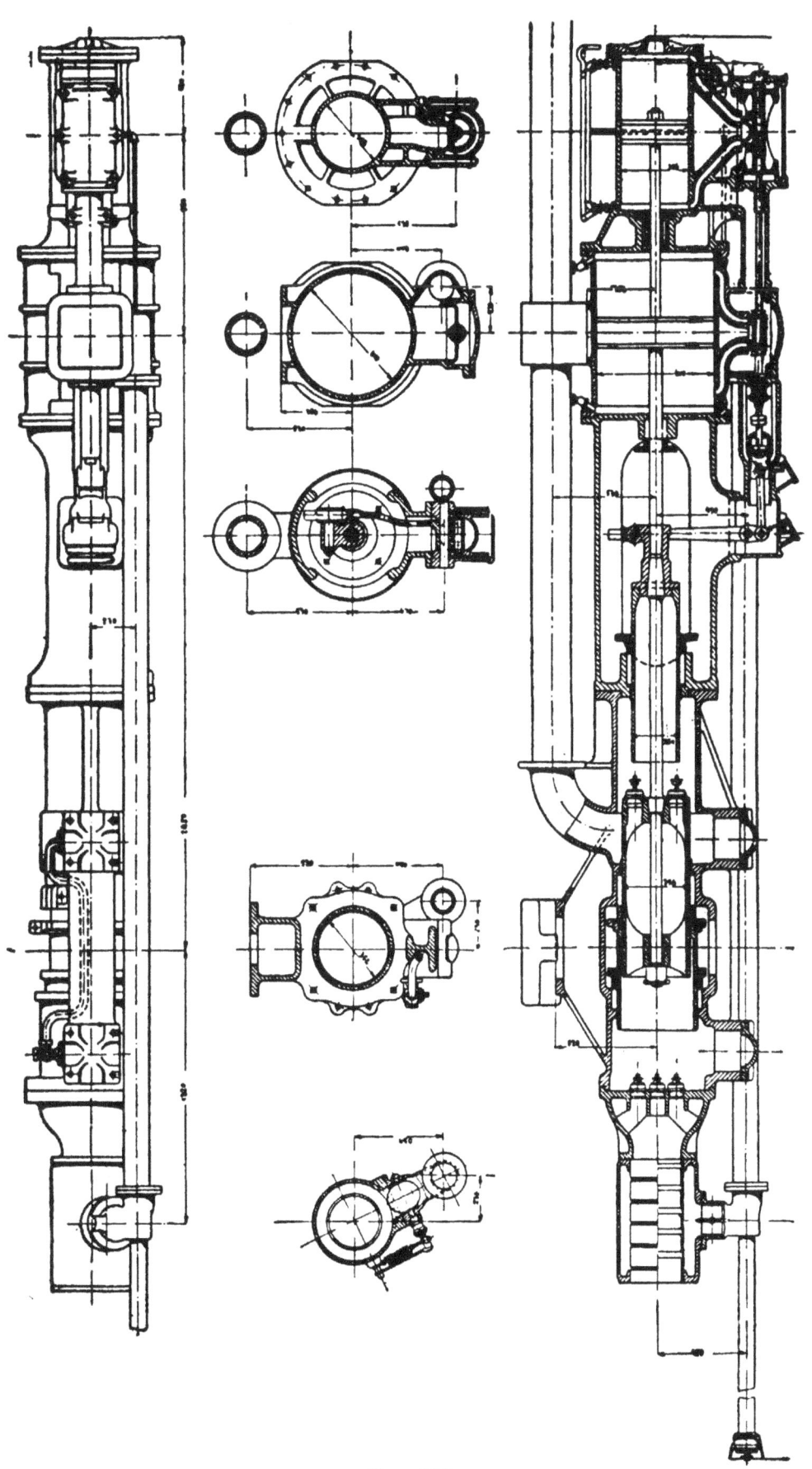

Fig. 132.

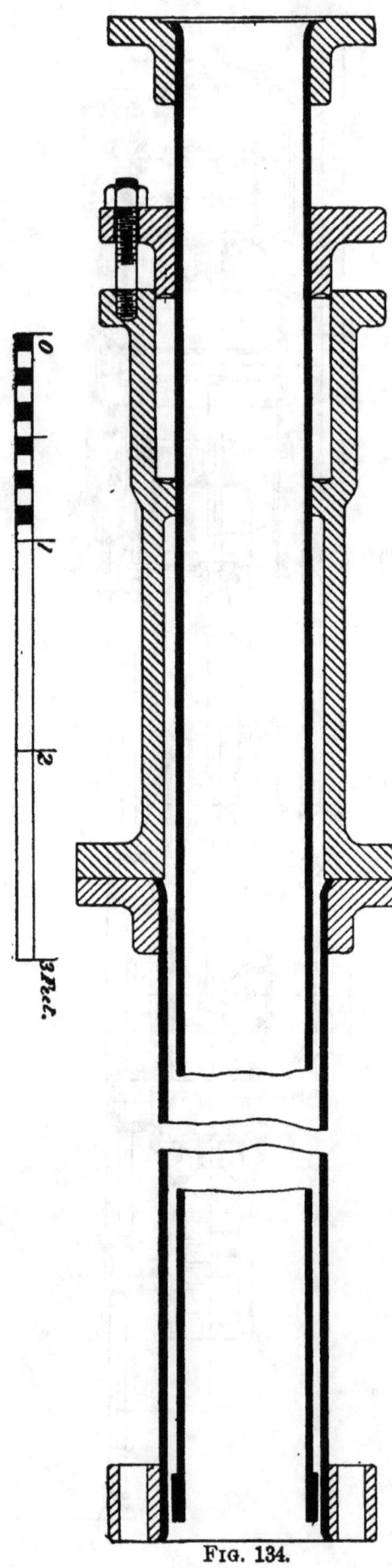

Fig. 134.

bottom, whence it is periodically withdrawn through the outlet b, for which purpose the pump is stopped. This device is said to have operated satisfactorily. It was jointly designed by Mr. W. R. Eckart and Mr. G. Dow.

3.2.07. In case of a sudden, large inrush of water the sinking-pump is sometimes raised and the water permitted to accumulate to a depth of about 10', in order to obtain a large body of comparatively quiet water, in which the sand will be more liable to come to rest and settle. The pump then draws from near the surface where the water is cleanest.

3.2.08. In order to be able to lower the pumps conveniently and with the least delay, the steam-pipe and water-column are generally made with slip-joints, often of a length sufficient to admit of lowering the pump for a distance equal to the length of a full section of steam- or column-pipe. When shorter slip-joints are used, a short section of pipe must be kept ready to be put in and taken out alternately between permanent insertions of full sections. Fig. 134 shows the construction of a long slip-joint or telescope.

3.2.09. Instead of inserting pieces in the column-pipe above the slip-joint, which necessitates the emptying of the entire pipe, it is generally better to lower the entire column when the pump has gone down as far as the slip-joint will permit, and then to add the necessary length to the upper end. Sometimes, also, the column is lowered with the pump, and then no slip-joint or telescope is required. For the steam-pipe the telescope is always required. There should be a gate-valve at the discharge connection of the pump, so that, when internal repairs or adjustment of the pump becomes necessary, the water will not have to be drained out of the column. This applies also to station pumps. (See also 1.3.28.)

3.2.10. The pump is generally raised and lowered by means of a chain-block suspended from a beam thrown across the shaft timbers. Where it is desirable to lower the pump for some distance, say equal to the length of a pipe-section, the chain should be lengthened, as the blocks usually admit only of a lift of about 10'. Large pumps are often handled by hoists from the surface.

3.2.11. The smaller pumps are usually secured by heavy iron claws attached to the pump, which hook into the top of the shaft sets. Large pumps are best provided with regular guides, to keep them in line with the steam- and column-pipes. Incline pumps of large size are usually mounted on flanged wheels guided on a track of wood or on iron rails.

CHAPTER III.

Rotative Pumps.

3.3.01. In these the motor- and pump-cylinders are also connected, so as to move coincidently, like in the non-rotative pumps, but they are further arranged with a crank coupled by a connecting-rod to a cross-head moving with the pistons or plungers. They can be operated by steam or compressed air; for operation by water pressure the rotative engines are not well adapted.

3.3.02. Single pumps require a flywheel, but the duplex pumps can often dispense with one. On account of the crank and flywheel, rotative pumps can complete their stroke close up to the steam-cylinder-heads, and therefore have little clearance as compared with the non-rotative pumps. For this reason, and also because they can utilize the expansion work of steam or of reheated compressed-air, they operate more economically than non-rotative pumps. By varying the point of cut-off of the steam or air, the work per stroke can, as in the rotative rod-pumping engines, be varied within much wider limits than in the non-rotative pumps. This is useful in adapting the pumps to increase of lift.

3.3.03. Single rotative pumps cannot be run below a certain speed. Duplex rotative pumps can be made to run very slowly, and are therefore capable of a wide range of capacity.

3.3.04. Rotative pumps are the only ones which admit of the use of mechanically actuated pump-valves. The well-known Riedler pumps, referred to in 1.3.18, are rotative.

3.3.05. The rotative principle, as stated in 3.2.04, cannot well be applied to sinking-pumps, as the space occupied would be too great, and the rough treatment to which such pumps are subject would soon unfit them for service.

3.3.06. Rotative station pumps require much better and more extensive foundations than the non-rotative pumps. The stations must also be larger. On the other hand, the rotative flywheel pump generally admits of higher speed than the non-rotative, and can therefore be made of smaller size. Examples of speeds and lifts attained in practice with the best modern types, like the Riedler and Hanarte pumps, have been given in the preceding pages.

CHAPTER IV.

Underground Pumps Driven by Steam.

3.4.01. *Steam Supply.* It is not often possible to place steam boilers under ground, as they require large excavations, around which the ground must be well supported; the smoke and waste gases must be led to the surface; the fuel must be brought down, and the ashes raised or transported; generally, also, the mine-water is unfit for boiler use, and suitable water has to be led down from the surface. For these reasons underground steam pumps are nearly always supplied with steam by means of pipes leading from boilers located at the surface. Such pipes have been described in Section I, Chapter II. It was there stated that they should be well protected by non-conducting covering to prevent excessive loss of heat; that it was an advantage to have a reservoir or drain interposed between the pump engine and the steam-pipe, in order to produce a more uniform flow in the steam-pipe and to keep up the initial pressure in the steam-cylinder; and that there should always be a valve in the pipe at the surface, besides one at each pump.

3.4.02. *Types of Steam Pumps.* In the United States underground pumps are mostly of the direct-acting type, although recently rotative Riedler pumps have come into use in a few places. The reason for the preference of the non-rotative pumps has been stated in 3.1.01 to be due to their greater compactness, simplicity, cheapness in the case of smaller sizes, and minimum of attendance required; also, because economy in the use of fuel, particularly in smaller plants, is generally of less importance than other considerations. The necessarily non-rotative sinking-pumps are not economical in the use of steam, even when arranged on the compound principle, because they have to meet great variations of pressure, and, as they are proportioned to work with full pressure at their limiting lift, the steam has to be very much throttled while they are operated under lower lifts. Direct-acting station pumps can be better proportioned to their work than sinking-pumps, and can utilize the advantages of compounding. They can also be fitted with steam-jackets and independent exhaust-valves, which materially adds to the economy in use of steam or air.

3.4.03. The Cross compound rotative engines, driving Riedler pumps at the mines of the Boston and Montana Gold and Silver Mining Company, are compound Corliss condensing engines. Being arranged on the duplex plan, they can run at a high speed; and as the cranks are at right angles to each other, the speed can also be reduced to a very low limit.

3.4.04. Pumping engines should be fitted with a governing device to keep the speed below the permissible maximum, which might otherwise be exceeded in case of breakage of the column-pipe near the pump, whereby the resistance would be thrown off the pump and engine. There should also be a control of the steam admission by a float in the station-tank, so that the speed of the pump will adapt itself to the flow of water coming into the tank. These remarks apply also to pump engines operated by compressed air or water.

3.4.05. With compound engines the steam pressure should not be lower than 100 lbs., in order to secure sufficient benefit from its expansion. With single-cylinder engines the pressure is usually 70 to 80 lbs. With triple-expansion engines the steam pressure should be still higher than with the compound engine, in order to secure sufficient additional economy to warrant their extra cost.

3.4.06. Large compound engines should have their valve-gear so arranged that the points at which cut-off occurs in the high- and low-pressure cylinders can be adjusted in relation to each other. It is also proper to have the amount of compression adjustable. With smaller pumps, such refinement would be too expensive, and the mechanism also liable to get out of order through lack of attention. Large engines are usually under careful supervision, and there steam-saving appliances will pay.

3.4.07. *Condensation of Exhaust-Steam.* Steam pumps, when used at moderate depth, can have their exhaust-steam conducted to the sur-

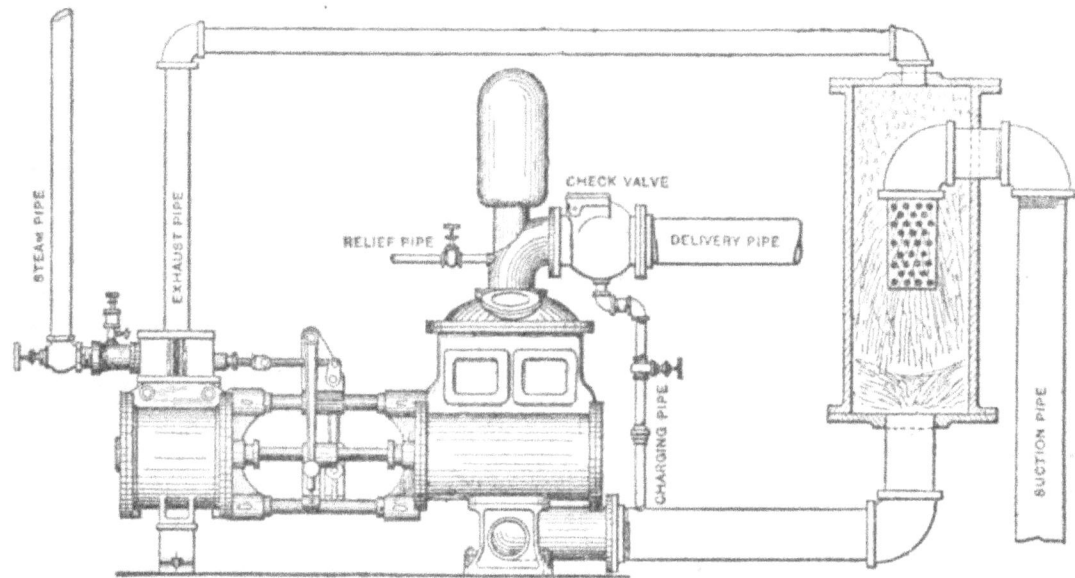

Fig. 135.

face. Condensation of the steam exhausted from deeply located underground pump engines is always necessary in order to get rid of the vapor and heat by conveying these to the surface in the water pumped. Sometimes such condensation is carried on at atmospheric pressure, in which case no economy to the engine results. It may be necessary to do this where the mine-water is very warm, and advisable in very high altitudes where the barometric pressure is so low as to afford little advantage in extra effective steam pressure. Generally, however, the steam is condensed under a very low pressure in the ordinary manner.

3.4.08. Condensation of the steam may be carried on either in the suction-pipe, which is the universal practice with sinking-pumps, or by means of an independent condenser with air-pump, like in the case of most of the large station pumping-plants.

3.4.09. Where condensation is effected in the suction-pipe, the steam should enter the latter in a direction almost parallel to the flow of water, so that it will act like an injector and thus aid in accelerating and lifting the water. Such a condenser is attached to the sinking-pump in Fig. 132.

3.4.10. The amount of vacuum obtainable by this method of condensation depends upon the suction-lift; the greater the latter, the better is the vacuum obtained.

3.4.11. With horizontal pumps, such as are used at the stations, the suction-lift is usually low, and there is danger of the water rising into the steam-cylinder, particularly when the pump is stopped. A small cock at the upper part of the pipe, opened as soon as the pump is stopped, would prevent this by admitting air and destroying the vacuum. Such cocks are, however, liable to be forgotten, and in some places it has been found safer to drill a small hole into the pipe, which will continually admit some air, at the sacrifice of a part of the vacuum. A better plan is to carry up the exhaust-pipe sufficiently high, and then drop it to the condenser, as illustrated in Fig. 135, so that the bend will be above the top of the column of water due to barometric pressure. The lowest point of the exhaust-pipe can readily be drained by a small pipe running down in a corner of the shaft, with its lower end dipping into a vessel of water placed at such a depth below the exhaust-outlet on the cylinder that the outside air can not force the water up into the exhaust-pipe. With vertical sinking-pumps, where the steam-cylinder is located at a considerable height above the condenser, there is usually no danger of the water ever reaching so high, unless the pump is working close down to the water-level.

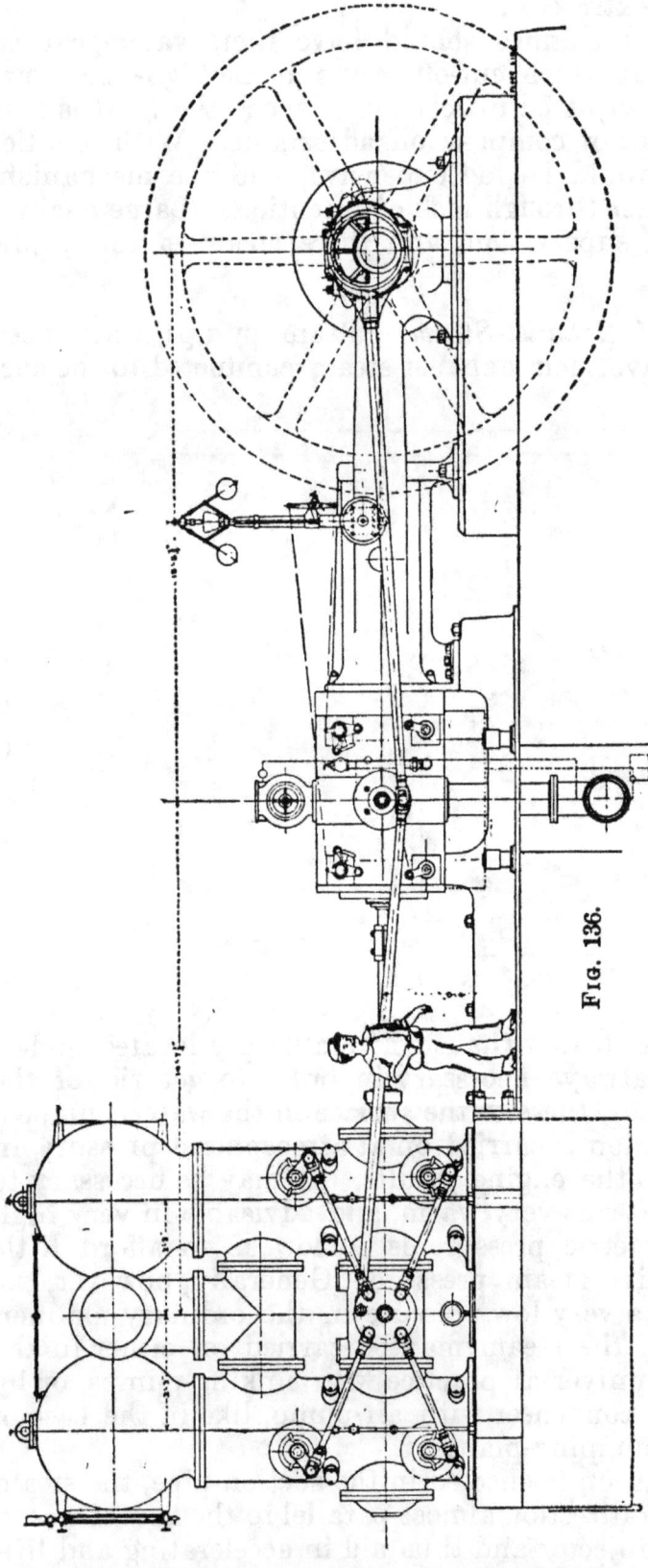

Fig. 136.

3.4.12. Station pumps are generally arranged with air-pumps and condensers independent of the suction-pipe. In this case also the injection-valve for the condensing water must be closed as soon as the engine is stopped, so that water may not rise into the steam-cylinder. This danger can be avoided by the submerged drain extending down the shaft, as described in the preceding paragraph. The air-pump is either driven from the pump engine, or operated by an independent engine; the former plan is the most common with rotative engines. Figs. 136 and 137 show a Riedler pump equipped in this manner. Direct-acting pumps are usually arranged with an independent air-pump, as in Fig. 138, and the latter often serves for a number of pumps.

3.4.13. Where there is a chance for leading the exhaust to the surface the exhaust-pipe leading to the condenser should have a branch-exhaust into the atmosphere, closed by a tight stop-valve when the condenser is running. A valve to close communication with the condenser, while the engine is exhausting into the atmosphere, should also be inserted, so that repairs can be made without stopping the pump.

3.4.14. *Mechanical Efficiency of Direct-Driven Steam Pumping-Plants.* While large, triple-expansion pumping-engines of the best design and most efficient type have,* when in best adjustment, reached, under test, a pumping effect of 1 H.P. per hour on less than $1\frac{1}{2}$ lbs. of good coal, even the best class of underground steam-operated pumps will probably never be able to approach such a result. If an efficiency of 1 H.P. on $2\frac{1}{2}$ lbs. can be attained, it may be called a very excellent result. The ordinary small, direct-acting, single-cylinder pumps, even in their best condition, probably consume about five or six times that amount of fuel, but when leaking and badly adjusted, as is so often the case, ten times may not be an excessive estimate.

CHAPTER V.

Underground Pumps Driven by Compresséd Air.

3.5.01. *General Remarks.* The transmission of power by compressed air is one of the most convenient, and, if properly carried out, a very economical method of operating pumps and other machinery under ground. It has the advantage over direct steam that it requires no condensers; that its exhaust cools instead of heats the air in the shaft or stations, and can be turned to use in assisting ventilation; and finally, that it is generally essential for operating machine-drills and other machinery at points removed from the shaft, where the use of steam would not be admissible. There is, in addition, very little danger from the rupture of a compressed-air-pipe such as there is with steam-pipes.

3.5.02. *Efficiency of the Old System.* In the majority of the smaller plants equipped with ordinary compressors and pumps driven by single-cylinder engines, which receive the air without previous reheating, the mechanical efficiency is very low. In the first place, there is a loss in the compression of air by part of the energy expended upon it being converted into heat, which is afterwards dissipated, thereby reducing the volume of the air before it reaches the engines which it is to operate.

---

*The case referred to is that of the Milwaukee pumping-engine built by the E. P. Allis Company. Of the coal used, 1 lb. evaporated over 9 lbs. of water.

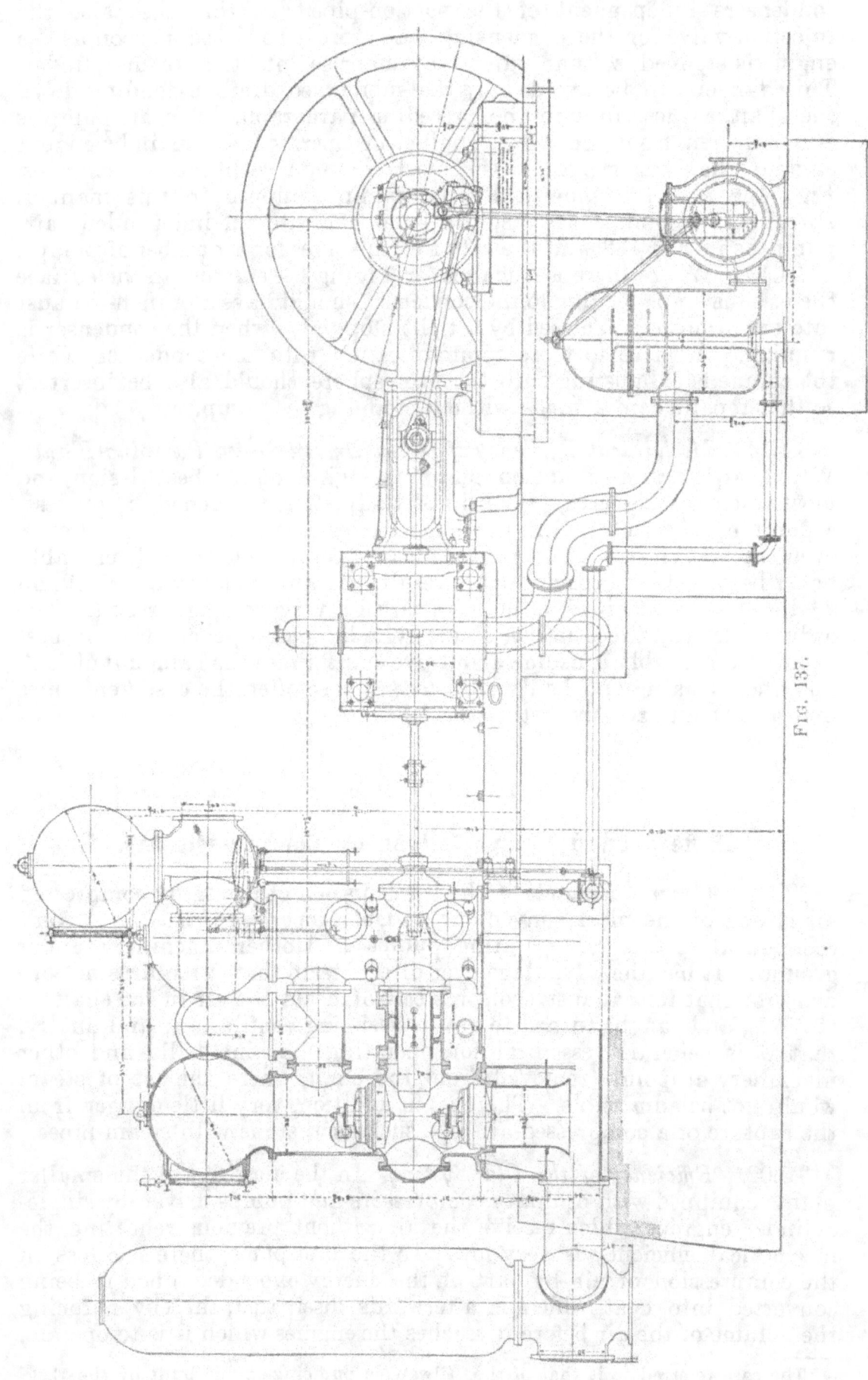

Fig. 137.

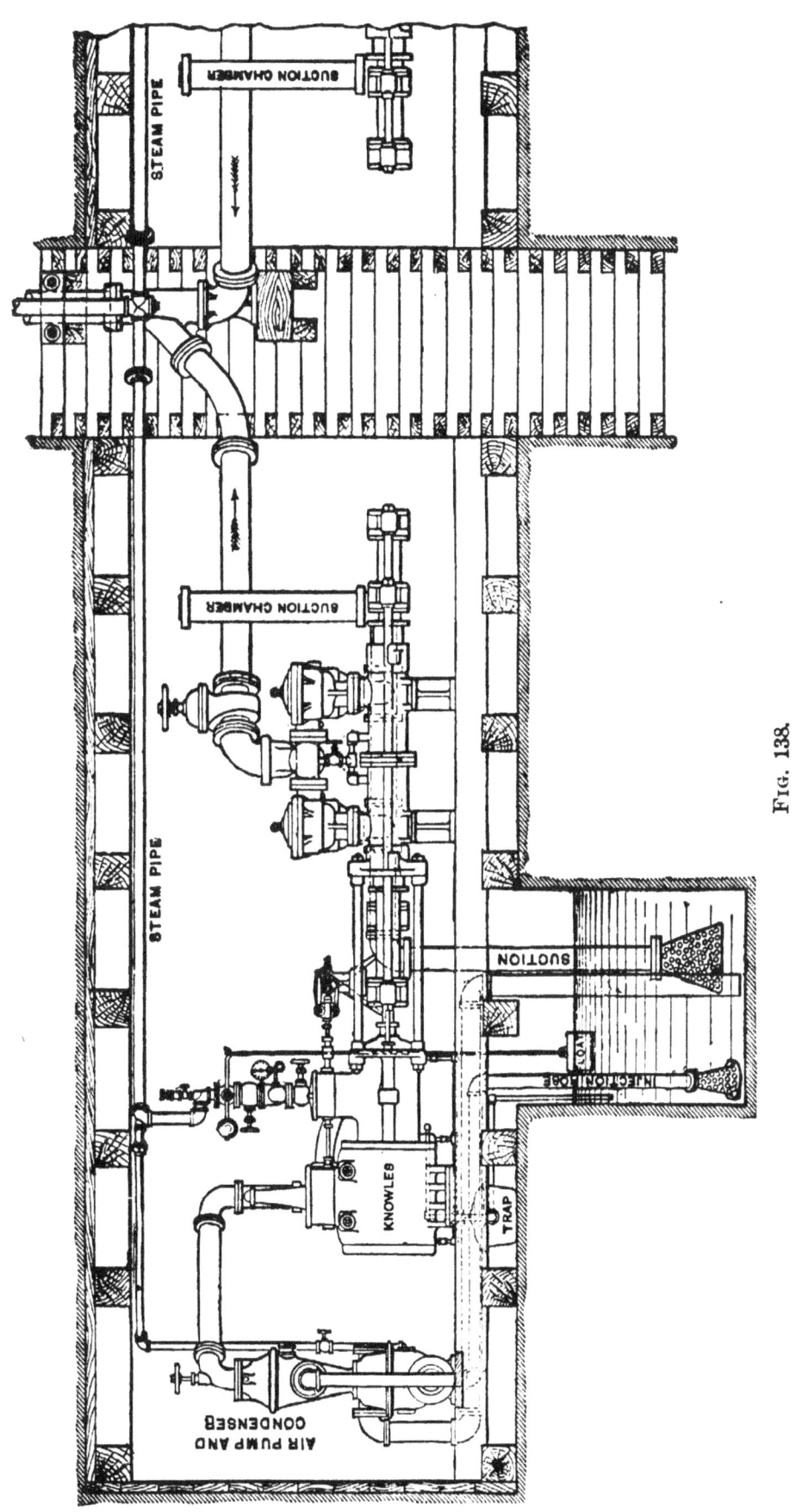

Fig. 138.

In the ordinary types of the latter also, if the air is admitted at the ordinary temperature, the work of expansion cannot be utilized, because the air on expanding is lowered in temperature to such an extent as to freeze the entrained moisture, which thereby blocks the engine in a very short time. The diagram, Fig. 139, illustrates the relation of the work expended in compression to that performed in the engine. It is laid out for a pressure of six atmospheres, or nearly 90 lbs. absolute or 75 lbs. gauge pressure, and a volumetric effect of the compressor of about 90%. The line AB is the compression line, such as is usually obtained in an ordinary single-cylinder compressor without spray-injection. The area ABDE, shaded in horizontal lines, then represents the work performed on the air in the compressor-cylinder. The distance BC represents the reduction in volume of air due to cooling to the temperature of the atmosphere before it reaches the motor engine. The distance DF represents the loss of pressure due to resistance of the pipe-line. The distance from the line IK to the atmospheric line 1 is the back pressure on the motor piston. Finally, the area FJGKI, shaded in vertical lines, represents the work capable of being expended on the piston of the motor in case a moderate expansion and compression into the clearance spaces is admissible, as in the case of rotative pump engines. The proportions of the areas ABDE and FJGKI give the relation of the indicated work expended upon and

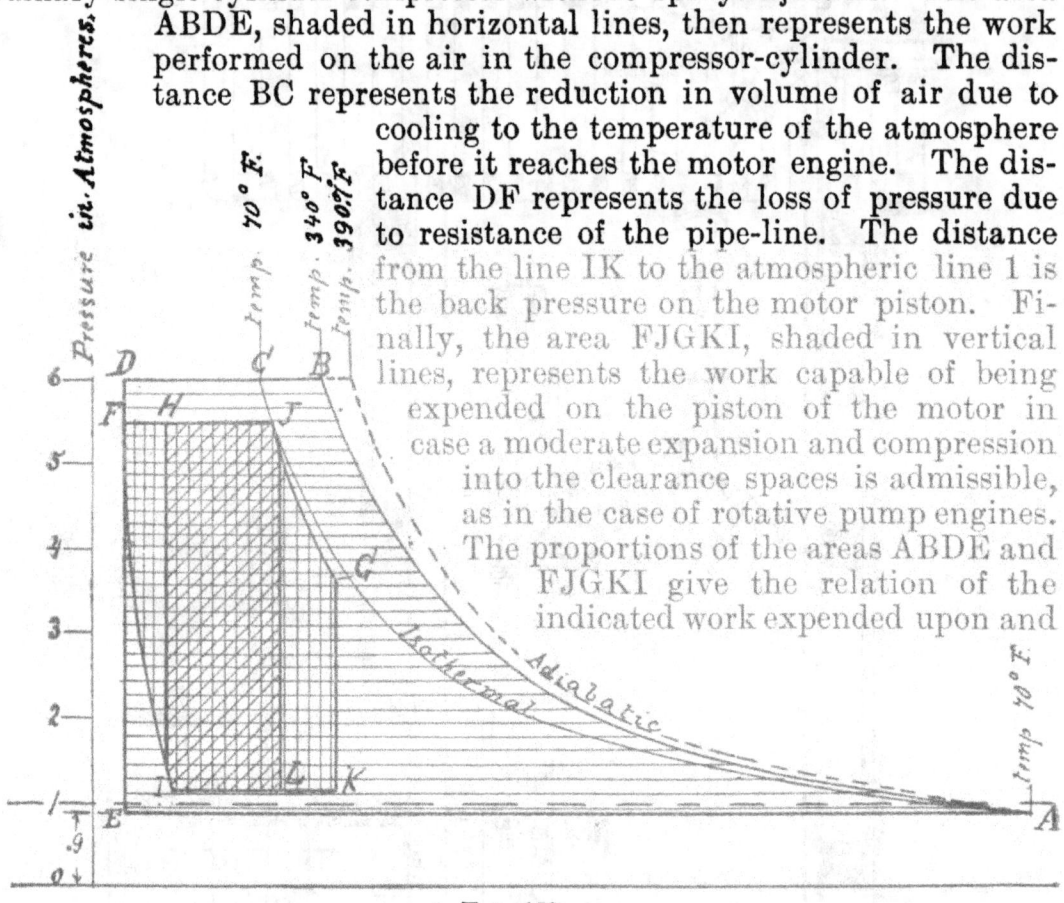

Fig. 139.

given out by the air. With direct-acting pumps, when they run slowly, no expansion or compression will be admissible, and the air in the clearance spaces will be wasted. The area HJLI, filled out with inclined lines, represents about the proportion of work recovered in this case, if, as is generally the case, the clearance space, as represented by FH, is large. No allowance has been made in the diagram for friction of the compressor and that of the engine or other motor driving it, and the friction of the pumping-engine and pump. These several losses will again increase the work of compressing the air and reduce that capable of being done by the pumping-engine; from all of which it is apparent how low is the efficiency of this method of using compressed air.

3.5.03. *Modern Efficient Compressed-Air Transmission.* Improvement in efficiency of compressed-air transmission must be effected, first, by reducing the work of compressing air; and, second, by utilizing the work stored in the compressed air to the best advantage.

3.5.04. The work of producing the compressed air can be improved, firstly, by increasing the volumetric effect or fill of the compressor. This can be effected by large, light valves, preferably operated by mechanism. Secondly, by cooling the air more effectually during compression, either continuously by surface cooling or spray-injection of water, or in stages between the partial compressions, in a series of two or more cylinders, so that the temperature is reduced as much as possible during compression or between stages of compression, with attendant reduction of the work necessary to bring the air to the required condition.

3.5.05. The work obtainable from the compressed air delivered to the driven engine may be increased by enabling the air to work expansively. If the air were perfectly dry, expansion could be utilized directly; but as this is never the case, the air must be reheated, the heat thus expended having the additional effect of increasing, in proportion to the heat added, the volume, and thereby the amount of work obtainable from the air. In compound pumping-engines, the air should be reheated in two stages, and to a more moderate degree. This would be an advantage in a mine, where the heat can generally be imparted more conveniently to the air by means of the limited tempera-

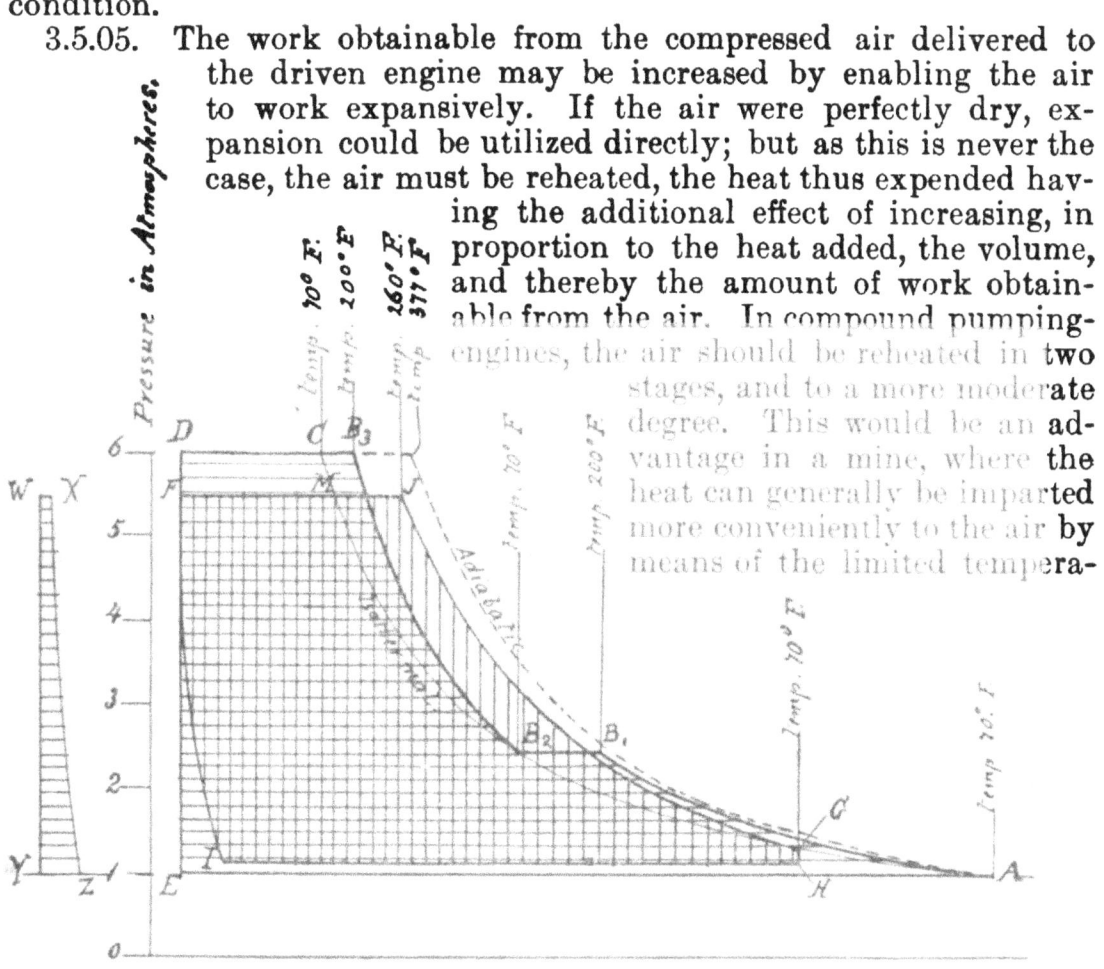

Fig. 140.

ture of steam conveyed to a heater located near the pump underground. The air-pipes can be smaller than in the common system, because less weight of air is required in the pump engine to do a given amount of work.

3.5.06. The reheating of the air, by whatever means effected, is best performed close to the pump engine. If the reheating be carried on at the surface, the air-pipes have to be larger, in order to pass the increased volume of the heated air. Such pipes also require good non-conducting covering.

3.5.07. In order to give the best effect, the reheating should be only just sufficient that the air after expansion in the engine will have the temperature of the surrounding air.

3.5.08. It is to be noted that the expansive work of the air, like that of steam, can only be well utilized in rotative engines, and only imperfectly in direct-acting ones at high speed. Compound, direct-acting pump engines are better situated in this respect than the single-cylinder type, as they admit of a wider range of expansion for the same variation of pressure per stroke.

3.5.09. The diagram, Fig. 140, illustrates the result of increasing the volumetric effect and compounding the compressor, and of reheating the air to about 260° Fahr. before it enters the pump engine, thus increasing its volume from FM to FJ. The pressures are the same as for the diagram, Fig. 139. The horizontally shaded area $AB_1B_2B_3DE$ represents the work spent in compressing the air, while the area FJGIH is that capable of being done by the air when reheated to 260°. The area WXYZ represents the additional work that could be utilized in the compressor for an expenditure of fuel equal to that used in reheating the compressed air. This area must be added to the compressor-diagram $AB_1B_2B_3DE$ before comparing it with the diagram of work capable of being performed by the reheated air. A compound compressor can be run at higher speed, and consequently be of smaller size, than a single-cylinder machine, if the action of the admission-valves be such as to insure a good fill. In such a compressor the chief cooling of the air is not performed during compression, but in the receiver

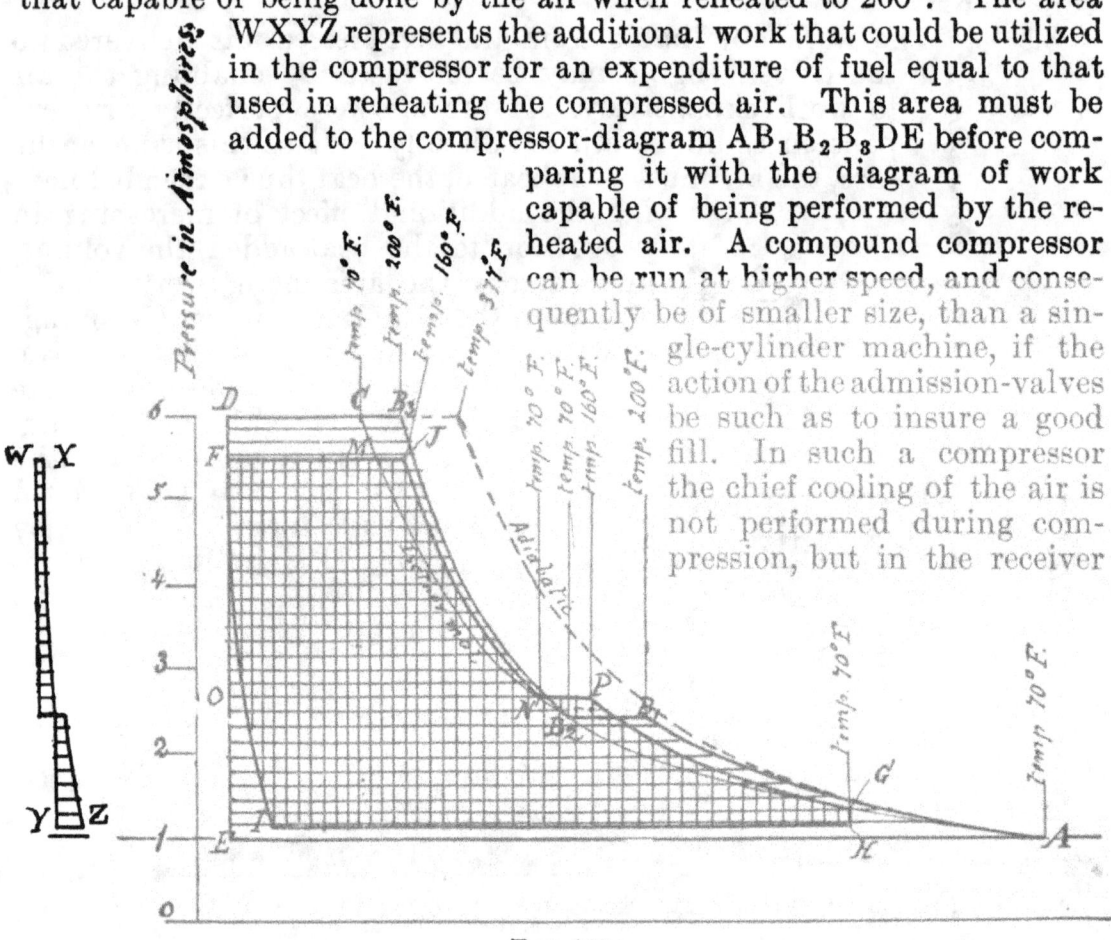

FIG. 141.

between the two cylinders, where there is more time and surface afforded for efficiently lowering the temperature.

3.5.10. The diagram, Fig. 141, shows the relation under the same conditions of compression as for Fig. 140, but for a compound pumping-engine, with a more moderate amount of reheating of the air up to about 160° Fahr. in two successive stages. In the first reheating the volume of the air is increased from FM to FJ; in the second, from ON to OP. The area WXYZ represents the value in compressor work of the amount of reheat.

3.5.11. *Rise in Temperature with Ratio of Extreme Pressures.* The temperature to which air will be heated by compression, if we neglect any cooling effect during this operation, is proportional to the initial absolute temperature* of the air. It increases also with the ratio of

---

*The absolute temperature is that indicated by the thermometer plus 461.2° Fahr.

the final absolute pressure to the initial absolute pressure. In ordinary compressors the pressure of the air drawn in is, on account of valve resistance, initially less than that of the outer air, so that in such machines the ratio of pressures, and therefore the final temperature, is greater than if the air filled the cylinder with atmospheric pressure. For the same reason air will be heated more by compression to the same gauge pressure above the atmosphere in higher altitudes than at sea-level. The formula expressing these relations is:

$$\left(\frac{p_1}{p_2}\right)^{0.2908} = \frac{T_1}{T_2}$$

In which p_1 is the initial absolute pressure, p_2 the final absolute pressure, T_1 the initial absolute temperature, and T_2 the final absolute

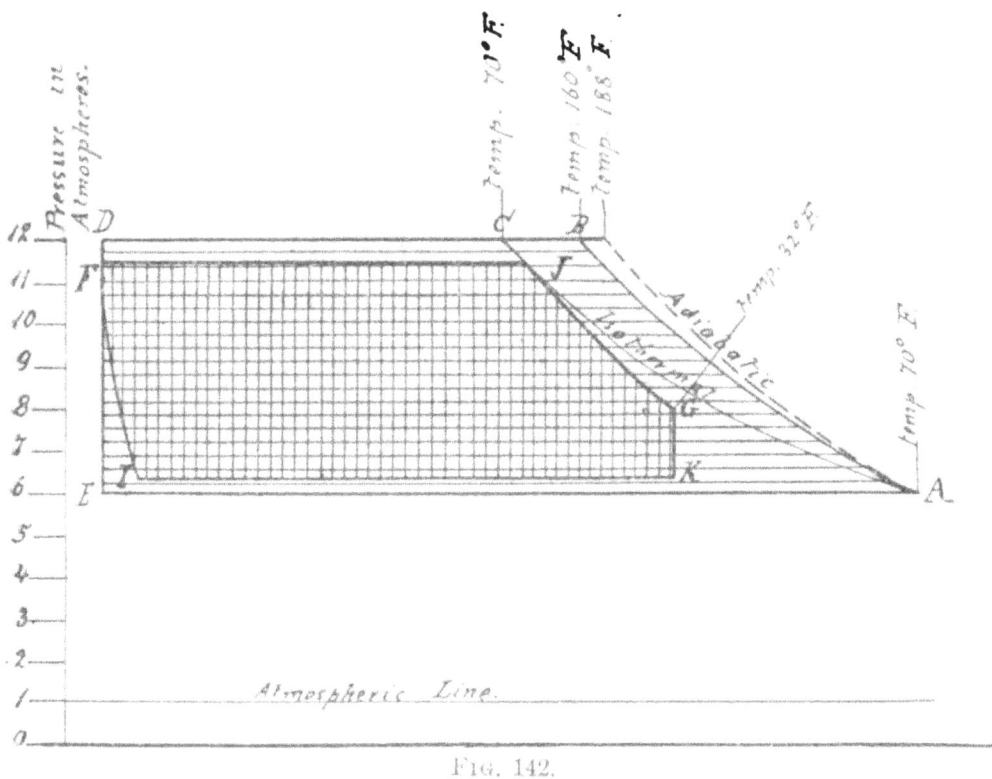

Fig. 142.

temperature.* The heating due to compression will be modified in practice by moisture contained in the air and by cooling, which always takes place to some extent, even if no provision for it is made. If the compression is rapidly performed the effect of cooling during compression will be less. The lower initial air-pressure in high altitudes requires correspondingly larger compressors.

3.5.12. The heating by compression having been shown in the foregoing to be greater with the ratio of final to initial pressure, it is natural to consider a reduction of this ratio. If, however, we should employ atmospheric pressure as the initial one for such reduced ratio, the compressor and air engines would require very large cylinders.

3.5.13. *The Cummings System of Compressed-Air Transmission.* Considerations like the foregoing have led Mr. Charles Cummings to devise a system of air-power transmission, in which the initial absolute press-

---

* Absolute pressure is gauge pressure plus atmospheric pressure.

ure of the air entering the compressor and the equal final pressure of the air leaving the engine are high, say 80 to 100 lbs.; while the final pressure in the compressor, or the initial pressure in the engine, are about twice this amount. These conditions necessitate a return pipe for the lower pressure, the compressor, air engine, and pipe-lines forming a closed system, in which the same air is used over and over again. The diagram, Fig. 142, shows by areas the relation of the work necessary to compress the air, to the work capable of being utilized in the air engine. The compressor work is indicated in horizontal, and that of the engine in vertical shading.

3.5.14. The weight of air required to pass the system per unit of time, for the same power, is greater in this system than when using the ordinary reheating system. As a result, the return pipe particularly

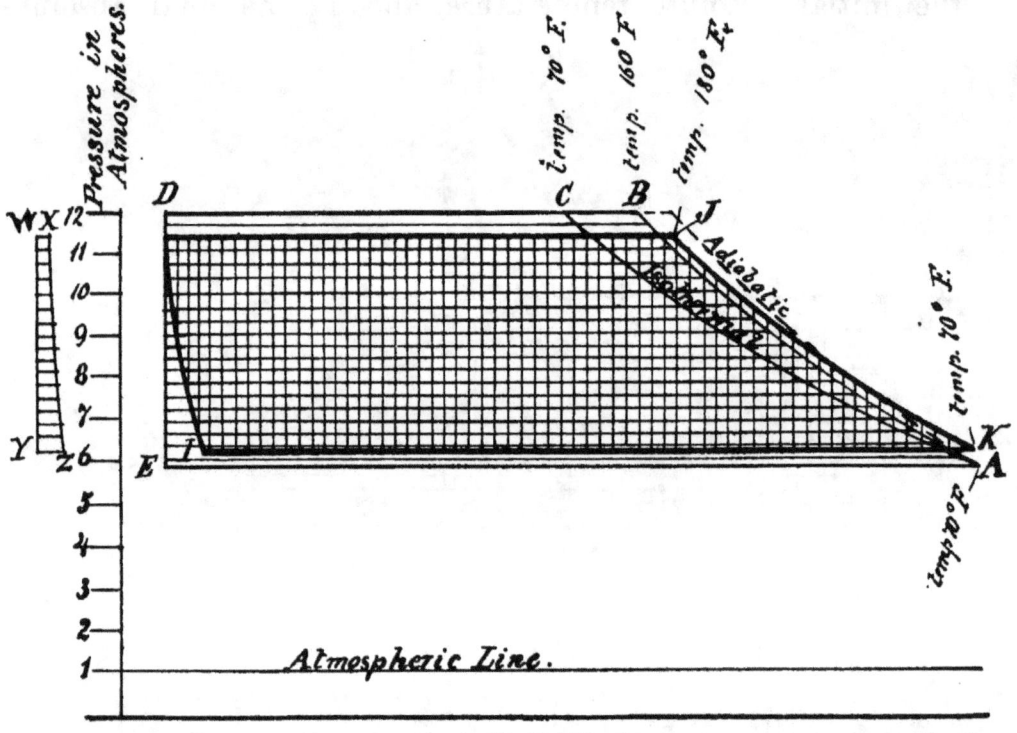

Fig. 143.

must be of larger size than the single pipe of the reheating system, if the velocity of air in the pipes be assumed to be the same in both cases. The power pipe-line will not vary much from that required for the reheating-plant, only in that it is subjected to higher pressure.

3.5.15. Notwithstanding the extra cost for transmission pipe, the Cummings system has very much to recommend it for operating pumps underground, especially where reheating would introduce complication. With it the advantages of compound compression and cooling during compression are much less marked than in the systems using initial atmospheric pressure, which permits dispensing with considerable complication. As the air cannot heat by compression beyond a moderate amount, the compressor can be allowed to run faster, and therefore be made of correspondingly smaller size. Reheating also permits increase of efficiency, though, on account of the originally high efficiency, it is of less proportional value than in the ordinary system. Fig. 143 shows the effect of reheating to the mechanically economic limit. The average pressure in the compressor and also in the engine varies less from the extremes than in the atmospheric-pressure system. An incidental advantage is, that the pump engines, particularly if direct-acting, can be operated under water until they give out. This feature adds a valuable guarantee to the safety of the mine. The compressor and air engine are cheaper, but the transmission pipes are more expensive than in other systems. To properly estimate the value of this system it must be compared with the reheating system as to mechanical efficiency, first cost, and simplicity of construction and manipulation.

3.5.16. *Compressed-Air Pump Engines.* The pump engines driven by compressed air are similar to those driven by steam. Lubrication of slide-valves is somewhat more difficult if the air is reheated. Puppet-valves would probably be better for reheated-air engines. Rotative fly-wheel engines are the ones adapted for utilizing the expansive work of the air. Direct-acting pump engines are, however, much operated by air, particularly the small pumps used in winzes and parts of the mine removed from the shaft, which are nearly always operated by this means. Air-driven station pump engines can be regulated in the same manner as steam engines by control of the air supply from a float in the station-tank.

3.5.17. *Reheaters.* These may be designed for heating by direct fire or by steam. The latter would be the most convenient underground. They should be arranged with the pipes so that the steam travels a considerable distance in them while the air travels in a course opposite to that of the steam. This gives the best possible effect of heat expended. The water of condensation should be removed automatically by a trap at the lower end. In the compressed-air pumping-plant at the Magalia Mine, Butte County, California, the air is reheated close to the pump by steam conducted down the shaft. Reheating at the surface is the method adopted at the North Star Mine, Grass Valley, and the air-pipe is covered by non-conducting material to retain as much of the heat as possible.

3.5.18. On first thought it may appear to many that the steam might be used to better advantage in driving a pump engine than to reheat air, but this is not the case, because the steam parts with its latent heat of vaporization in heating, nearly all of which heat can be converted into work in the air engine, which is not possible in the steam engine. In reheating compressed air the steam is used about five times as efficiently as in performing work directly. Where an air engine is operated at no great distance from the compressor, and where the latter delivers the air at a considerable temperature, it may be possible to keep the heat in the air by non-conducting covering applied to the pipe. In this way good efficiency could be realized without reheating.

3.5.19. *Receivers.* Long air-pipes serve in a measure as receivers and storage for the compressed air. Separate receivers are, however, generally also located near the compressor at the surface, and sometimes near the engine underground, to serve as regulators. (See 1.2.60.) They serve incidentally also for trapping part of the moisture contained in the air. They should be fitted with a waterglass to indicate the level of this water, a pressure-gauge, safety-valve, and means for draining off water.

3.5.20. *Compressors.* Compressors must adapt their output to the requirements of the underground machinery. For this reason they have to operate at all speeds, and frequently must stop altogether. Their regulation in this respect is made dependent upon the air-pressure in the receiver, which, by suitable mechanism, causes a shutting off of the power supply and slowing down of the compressor when the pressure rises above a given limit, and inversely causes an increase of power supply and speed when the pressure falls. Where the draught upon the compressor is so irregular that it has to stop frequently, a duplex compressor presents the advantage of being self-starting.

3.5.21. The irregular duty and speed of the compressor are much more favorable, as regards mechanical efficiency, for operation by steam than by water-power. This may affect the choice of power where the water has to be bought. It is, however, to be remarked, that for pumping the work is generally more regular than for hoisting or rock-drilling, and where a separate compressor operates the pumps, it may be possible to utilize water-power with some degree of efficiency.

3.5.22. Water-injection cooling is much less employed now than formerly. If water is used in the compressor-cylinder, it should be perfectly clean, otherwise the cylinder will soon wear out. It is generally difficult to obtain clean water in mining regions. If so much water is injected that the temperature of the air is kept down to $20°$ or $30°$ above that which it had on entering, there is quite an amount of work required to force in the larger volume of water, for which no useful returns are had. As the volume of water injected per stroke is liable to be greater than that of the clearance space, the speed of the compressor must be kept lower than where no injection is used, to avoid risk of breaking the compressor-heads. Another objection to the use of injection-water is that it interferes with lubrication by floating the oil away from the rubbing surfaces.

3.5.23. The volumetric effect of a compressor is reduced not only by the resistance of the suction-valves, but also by the heat of the metal of these valves and passages, which impart some of this heat to the air drawn in on the suction-stroke, causing it to expand and thereby fill the cylinder with a volume of air of less weight and higher temperature. The effect is a double one: first, a lowered output of the compressor, then a higher final temperature due to compression, both these combining to lower the efficiency.

3.5.24. A high-class, modern, steam-operated, compressed-air pumping-plant will compare favorably in commercial efficiency with underground steam pumping-engines. The high-duty steam pumping-engines, whether rotative or direct-acting, are more expensive than those which give good efficiency when operated by compressed air. The former also require more and better attendance, which increases the operating expense if there are pumps, as is usual, at several points in the shaft. It is also possible to use a larger engine at the surface, which can be more easily made of high efficiency than underground engines, particularly if the underground machinery is cut up into several units.

CHAPTER VI.

Pumps Operated by Attached Hydraulic-Pressure Engines.

3.6.01. Hydraulic operation of underground pump engines may be accomplished in several ways. Firstly, the pumping-engines underground may be directly operated by a natural head of water, the engine

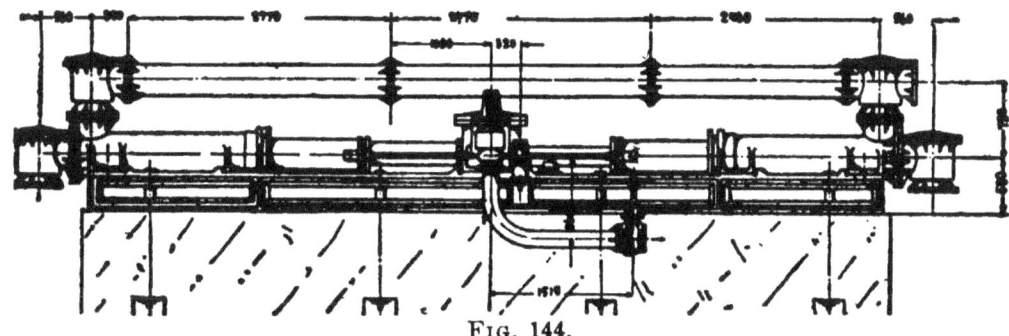

Fig. 144.

delivering the spent power-water, together with the mine-water, at the point of discharge. Secondly, the pressure necessary to drive the underground hydraulic pumping-engine may be artificially generated, either entirely or as supplemental to an insufficient head, by a steam engine driving pumps at the surface. In this case, also, the power-water is

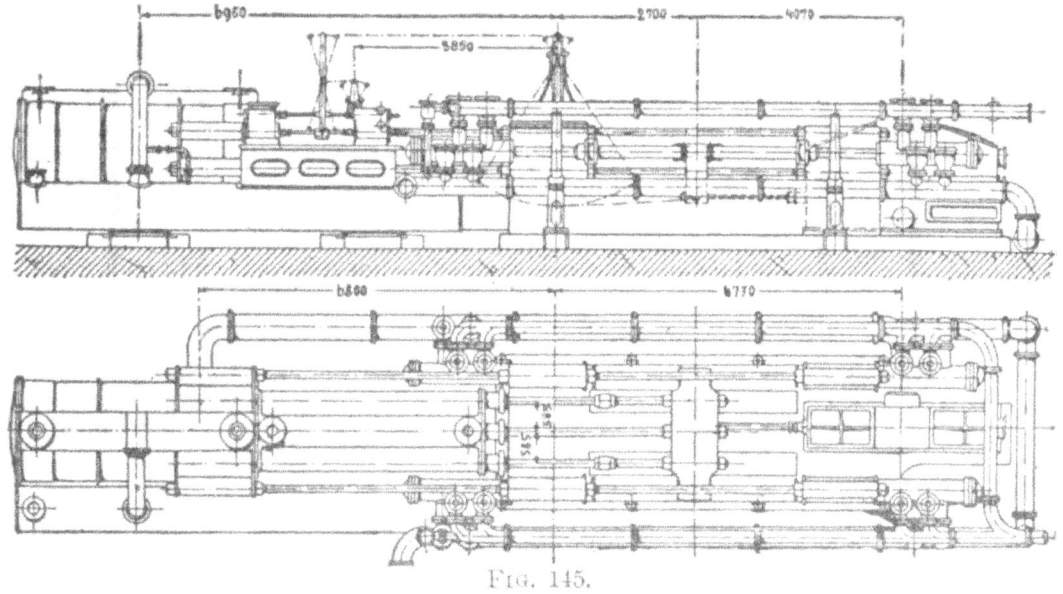

Fig. 145.

delivered, together with the mine-water, at the point of discharge, but if this point is at the surface, the power-water may be used over and over again. A third method of operation, different from the two foregoing ones, is by means of the so-called hydraulic pumprods. In this plan one or two columns of water are reciprocated by a valveless pump driven by a prime-mover, and impart corresponding motion to a piston or plunger connected with the mine pump.

3.6.02. *Hydraulic Engines Controlled by Valves.* The underground hydraulic engines employed to operate pumps in the manner first named are similar to those used at the surface for working pumps through rods, except that the pump-plungers are directly connected to the plungers or piston-rods of the engine. Fig. 144 illustrates the Davie hydraulic pumping-engine, which is of this class. A modification of this type was used in the Combination shaft, Virginia City, Nev. The pumps at the lowest level raised the water over 1,400' to the level of the Sutro Tunnel. There were two independent pumping-engines at the station, each capable of making, at a maximum, about ten

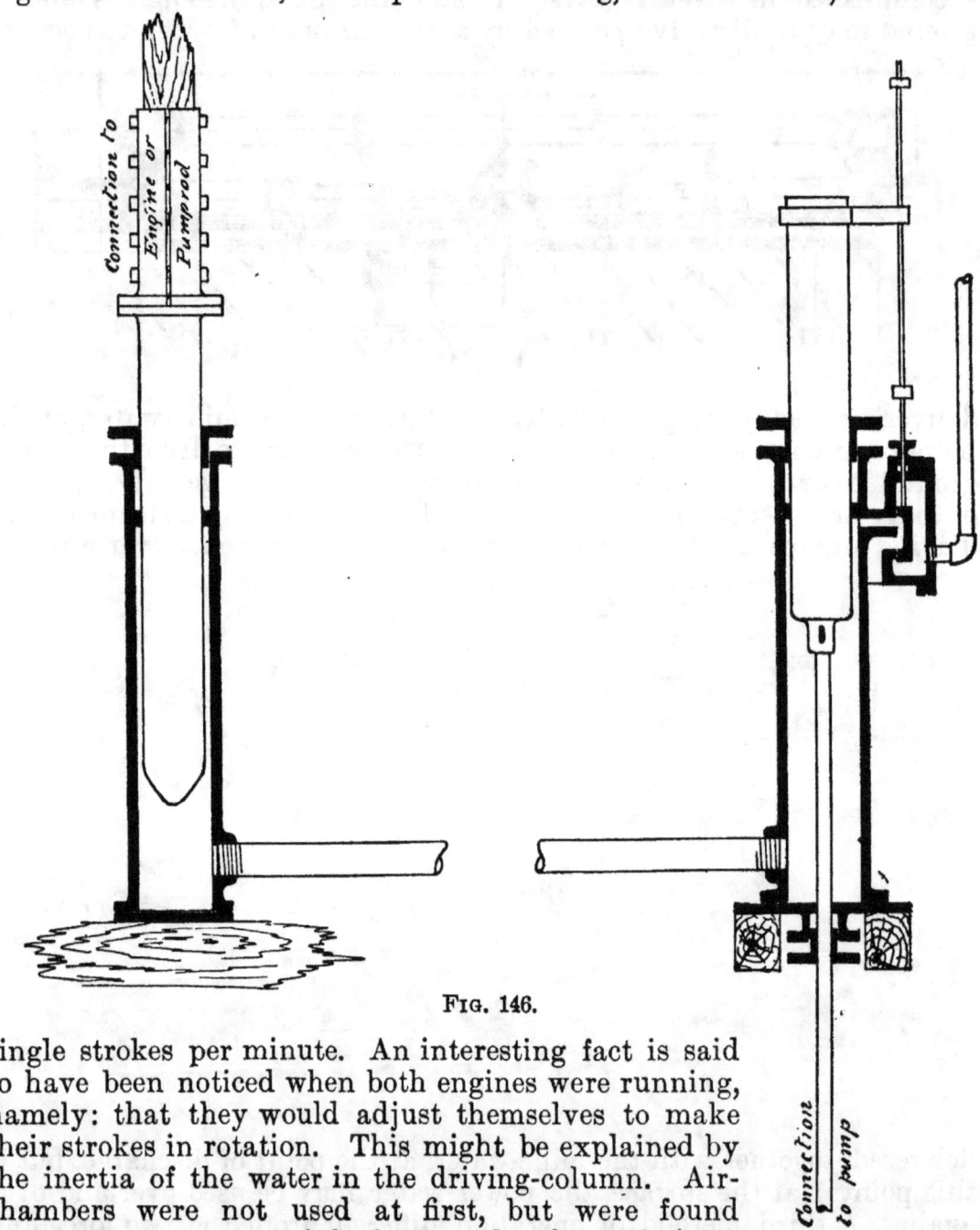

Fig. 146.

single strokes per minute. An interesting fact is said to have been noticed when both engines were running, namely: that they would adjust themselves to make their strokes in rotation. This might be explained by the inertia of the water in the driving-column. Air-chambers were not used at first, but were found necessary, on account of the frequent breakage of pipes.

3.6.03. The hydraulic pumps of the Combination shaft at first were operated by artificial excess of pressure of about 1,000 lbs. generated by a steam engine driving pumps at the surface and forcing the water into an accumulator-chamber charged with compressed air. The surface pumping-engine was of the Davie type, and is illustrated in Fig. 145. By this arrangement the pressure-water was used over and over again, but the expense of operation is said to have been over $6,000 per month, while the excessive pressure caused frequent breakages. The natural pressure of the city water was afterwards used to drive the pumps, and the total cost of operation, including that of power-water, was reduced to about $1,000 per month. The plant was perhaps the largest hydraulic mine pumping-plant ever built. Its first cost amounted to about one quarter of a million of dollars. The Knight hydraulic pumping-engine described in 2.5.37, 2.5.38, and 2.5.39, and illustrated in Fig. 116, could also be coupled directly to underground pumps.

3.6.04. *Valveless Engines.* An example of the hydraulic rod system, though not operating a direct-acting pump-engine, exists at the New Almaden Quicksilver Mine, Santa Clara County, California. Fig. 146 illustrates the principle of the arrangement. It is a horizontal transmission, and therefore requires only a single reciprocating column, the weight of rod and other parts being sufficient to accomplish the return-stroke of the water-column.

3.6.05. Where there is no weight of pumprods or other parts to be raised, and also where the transmission is vertical, two balanced reciprocating columns of water are used.

3.6.06. No admission- and discharge-valves are required with engines operated by reciprocating columns of water, nor with the plungers used to reciprocate the water. It is, however, necessary to keep the volume of water in the reciprocating columns constant, so that the stroke of the underground engine and pump will not exceed the proper limits at either end. Leakage will reduce the amount of water in the columns, and this must be made up by forcing in a small quantity, either with each stroke or continuously. As it is not possible to adjust the quantity to be added so as to be exactly equal to the leakage, it is necessary to force in an amount slightly greater than the leakage, and to get rid of the excess by causing the underground engine to open a valve at the end of the stroke whenever this passes the prescribed limit. By adjusting these valves so that the volume swept through by the working piston or plunger of the underground pump will be somewhat less than that of the driving-pump at the surface, the latter can draw in the necessary replenishing water during the suction-stroke through a small suction- or check-valve. In the single-column hydraulic transmission, illustrated in Fig. 146, and referred to before, the stroke of the underground engine is regulated by admitting a small quantity of water under greater pressure than that existing in the pump, by means of some such arrangement as the slide-valve at the side of the working-barrel. The same arrangement permits the escape of a small quantity of water near the upper end of the stroke. The slide-valve is shown to be operated by means of a tappet-rod from an arm at the top of the plunger.

3.6.07. *General Remarks.* Hydraulic-pressure operation of pumps requires, in most cases, very expensive plants for any considerable depth, as the parts have necessarily to be made heavy so as to resist not only the static pressure, but also the strains due to water-ram. In the matter of water-ram, the artificial-pressure systems are in a better position than those employing a natural fall, because the mass of water moved and arrested is less. Where two hydraulic engines are supplied with pressure from the same column, and are operating so that their strokes occur in rotation, the column of water is kept continuously in motion, and the danger from water-ram is much reduced; so that in general, the engines can run faster under such conditions.

136   MINE DRAINAGE, PUMPS, ETC.

3.6.08. Hydraulic pressure is not suitable for sinking purposes, except perhaps for moderate depths, and in such cases generally artificial pressure will be better adapted, because the pressure can be suited to the increasing depth.

3.6.09. Clean water only should be used, but drain-cocks should be fitted at points where sediment is likely to accumulate in the engine. Plungers are much preferable to pistons, for the same reasons as with pumps.

CHAPTER VII.

Pump-Stations.

3.7.01. The stations for direct-driven pumps require a greater amount of excavation than for Cornish pumps, because they have to accommo-

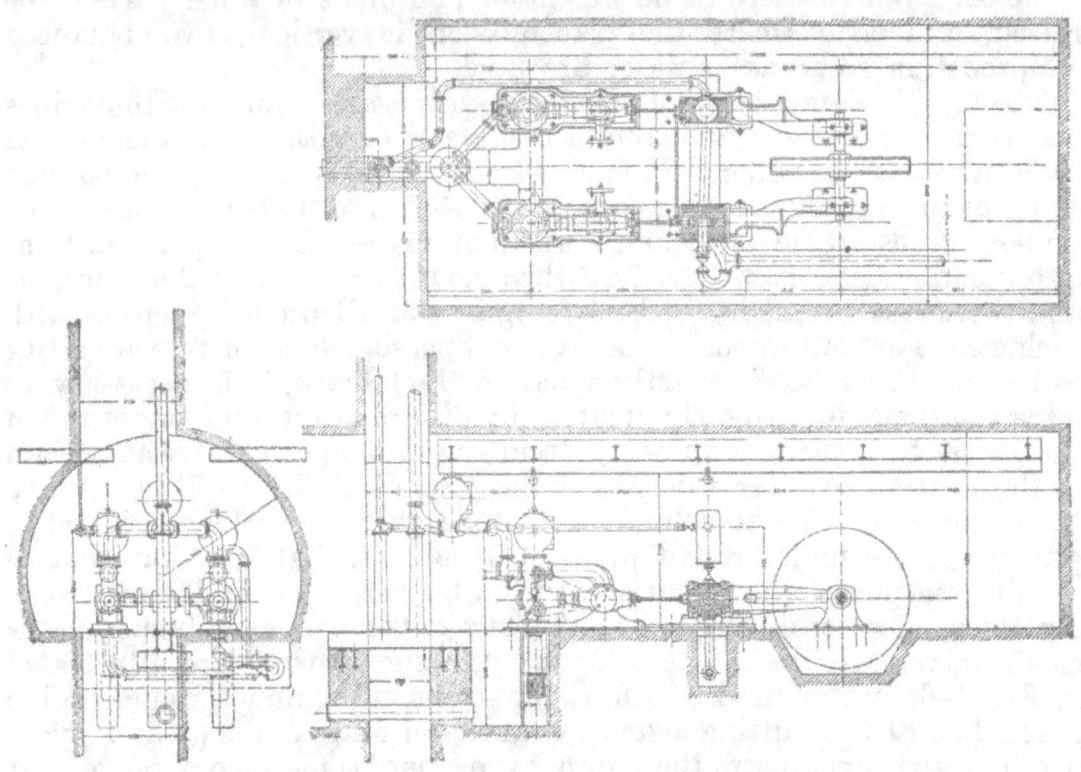

Fig. 147.

date the pumps as well as the tanks, and must also leave room to get around the pumps conveniently. Where several large pumps are in one station, and the latter is in ground requiring support, it is generally best to arrange the pumps in line and not side by side, so that the station assumes more the shape of a tunnel, in which the ground can be more easily and cheaply supported.

3.7.02. If the rock is very hard, so that the excavation cannot be extended at a later period without blasting, it is a good plan to excavate it larger than required at first, so that additional pumps can be quickly added in case of requirement.

3.7.03. The tanks are often placed below the pumps, as in Figs. 138 and 147. They should, if possible, be large, to afford a chance for settling of sand. Fig. 147 illustrates a pump-station with rotative pumps at the mines of the Boston and Montana Gold and Silver Mining Company, of Butte, Montana. In the pump-station of the Crown Point incline shaft, at Virginia City, Nevada, shown in Fig. 148, the tank was placed at a level above the pump.

3.7.04. The facility with which direct-driven pumps can be regu-

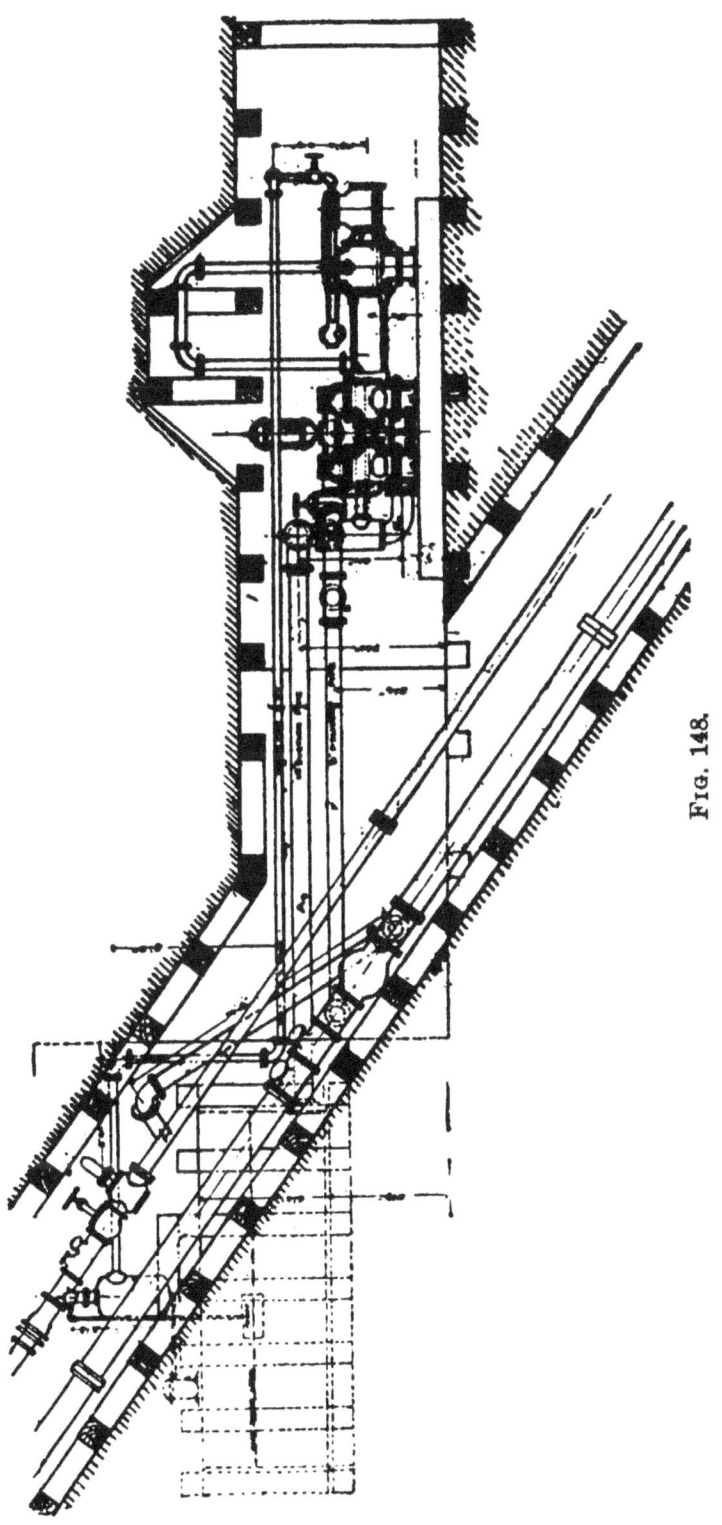

Fig. 148.

lated, so as to adapt their work without great reduction in efficiency to the inflow of water, constitutes, where pumps have to be used at different levels, one of the most striking advantages over the rod-pumping systems, in which, as described in 2.4.12 and 2.6.03, the relative regulation of the pumps is effected by returning a part of the water pumped by those pumps for which the speed is greater than required, but whose speed is necessarily the same as that of the pump having the greatest quantity of water to handle.

3.7.05. As direct-driven pumps can be constructed for very high lifts, the capacity of pumps at any station within the limits of admissible lift need only be sufficient to handle the water received into the tank at that station, and not, as in the Cornish system, arranged so that the upper pumps handle the water coming in at their stations, as well as that supplied by the lower pumps. Only the column-pipe and power-pipe have to be increased above the points where the pumps connect, to adapt the pipes to the greater quantity of flow.

3.7.06. Subdivision of the pumping capacity of a station is generally advisable, as it affords a chance of making repairs on one of the pumps, while the other can be forced a little to do the entire pumping duty during this time. Where two pumps are connected to cranks at right angles, they should be arranged so that either one of them can be cut off from communication with power- and delivery-pipes. With steam-driven pumps, in which low speed is attended with greater steam loss from the initial condensation, the subdivision into several units affords the chance of a more advantageous rate of operation by shutting down one part of the plant when the inflow of water becomes reduced.

3.7.07. Direct-driven station pumps obstruct the shaft only to the extent of their piping, thus leaving the largest part of the shaft free for the operation of a cage, which is of advantage in affording rapid communication between the different stations. The free space in the shaft is also needed for hoisting the sinking-pump in case of requirement. Crooked, small, or inclined shafts present disadvantages to the rod-pumping system, while direct-driven pumps are little affected by such conditions.

SECTION IV.

CHAPTER I.

Underground Geared and Belted Crank-Driven Pumps.

4.1.01. These are chiefly used where electric motors, or waterwheels under very high heads, constitute the driving power, though they can also be operated economically by steam or air engines. Electric motors

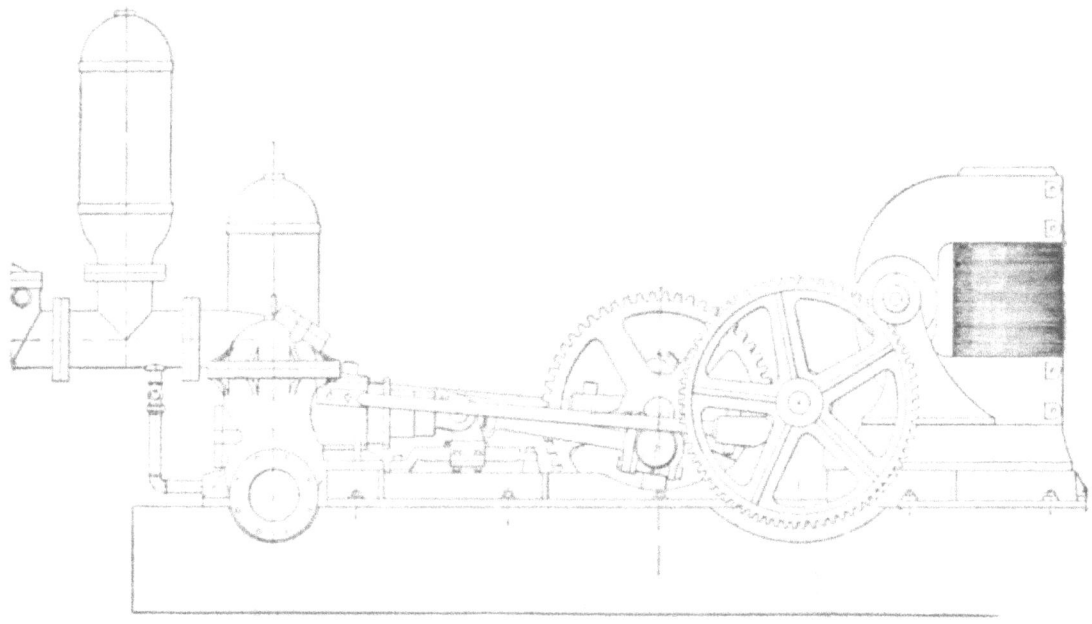

Fig. 149.

and waterwheels used underground generally require to make a great number of revolutions per minute. Although the short-stroke pumps employed permit a greater number of revolutions than those of long stroke, pumps driven by these two kinds of motors must generally be compound-geared, as the electrically-driven pump shown in Fig. 149, in order to get the proper reduction to the speed admissible for the pump.

4.1.02. It is important that the resistance during one revolution of the pump-crank be as uniform as possible, where waterwheels or electric motors furnish the driving power; otherwise, the power will be applied with loss of efficiency. For this reason such pumps are usually of the duplex or triple form. The duplex form may be used where the pumps are double-acting, like piston pumps, or differential plungers, as in Fig. 128. The most uniform resistance is afforded by the triple, single-acting, plunger pump, a type of which is illustrated by Fig. 150. This arrangement gives six maxima and six minima of pressure during one revolution, the maxima being only about $5\frac{1}{4}\%$ above the average pressure, and the minima, which are of shorter duration, $10\frac{1}{2}\%$ below it. A flywheel mounted on the rapidly revolving motor- or wheel-shaft would further aid in equalizing the resistance. Belt-driven pumps should also have a very uniform resistance, so as to prevent the whipping of the belt, which hastens its destruction.

4.1.03. The high-speed gears should be carefully cut in a gear-cutter. The pinions should, if possible, be made of more durable material than the gear, because the teeth are subjected to greater wear. Rawhide or

Fig. 150.

bronze pinions are generally used on the armature- or waterwheel-shaft. Sometimes also the gear driven by a high-speed metal pinion is made with wooden cogs, as in Fig. 151. For the slow gearing, double-angle teeth (as in Fig. 152) are, if well made, the best, because they can be of finer pitch for the same strength. The finer division, and also the continuous contact on the pitch line, cause such gears to run with very little noise. If, however, angle-tooth gears are badly made, so that, for example, the angles of the teeth do not lie in the same plane of rotation, then such gears are worse than simple spur gears, because they will be thrown from side to side as they revolve and cause noise in working. For operation by steam or compressed-air engines, which make a less number of revolutions than either waterwheels or electric

Fig. 151.

motors, the pumps are either single-geared or only driven by belts on a pulley fixed on the pump crank-shaft.

4.1.04. Waterwheels used for driving pumps underground require a head much greater than at the surface, because they have to drive pumps which must deliver at the surface not only the water of the mine, but also that discharged from the waterwheel. Such a method of operation would therefore only find application for a moderate pumping-depth, and in this case the pumprod system would generally be preferable. There is, therefore, not much chance for the application of waterwheels to driving pumps underground.

4.1.05. As ordinarily constructed and coupled up with their driving-engines or -motors, the crank-driven pumps admit of no great variation in capacity. The crank-length is fixed, so that no adjustment of length of pump-stroke can be made. Electric motors cannot be varied in speed in very wide limits. With waterwheels it can be done, but not without very considerably reducing the efficiency. Steam and compressed-air engines, however, if of the duplex or compound form, with cranks at right angles, can be run at almost any speed, and are, therefore, better adapted for operating crank-pumps at variable speed.

4.1.06. Electrically-driven sinking-pumps are necessarily of the geared-crank type. The writer knows of no case of their use in practice. They would scarcely seem to be adapted to the hard usage to which they are subject in this kind of work.

Fig. 152.

4.1.07. An efficient pumping-plant can be made with geared pumps driven by high-duty engines or motors. Some three-crank plunger pumps have been run at piston speeds of 460' per minute without water-ram. This is due to the very uniform motion of the water in the suction- and discharge-pipes. Large valves and proper air-chambers naturally contribute to attain the best results.

4.1.08. Electricity is not well adapted as an economical means of transmitting power to reciprocating pumps which have a variable duty, unless they can be run intermittently, permitting the water to accumulate in large reservoirs or tanks during the time of stoppage. The method of applying electricity to pumping machinery will have to be much improved before it can bear out the claims of efficiency made by electrical companies.*

4.1.09. The XIIth Report of the State Mineralogist, published in 1894, contains an interesting chapter, entitled "Electric Power-Transmission Plants, and the Use of Electricity in Mining Operations," by Thomas Haight Leggett, in which a few electric pumping installations are mentioned. The reader is referred to that ably written article for further details on this subject.

---

*Two electrically-driven pumps are in use at the Golden Banner Mine, near Oroville, Butte County, California, where other electrical appliances have been installed.

SECTION V.

BAILING-TANKS.

5.1.01. The simplest method of raising water from deep mines is by means of bailing-tanks, which may take the water either from station

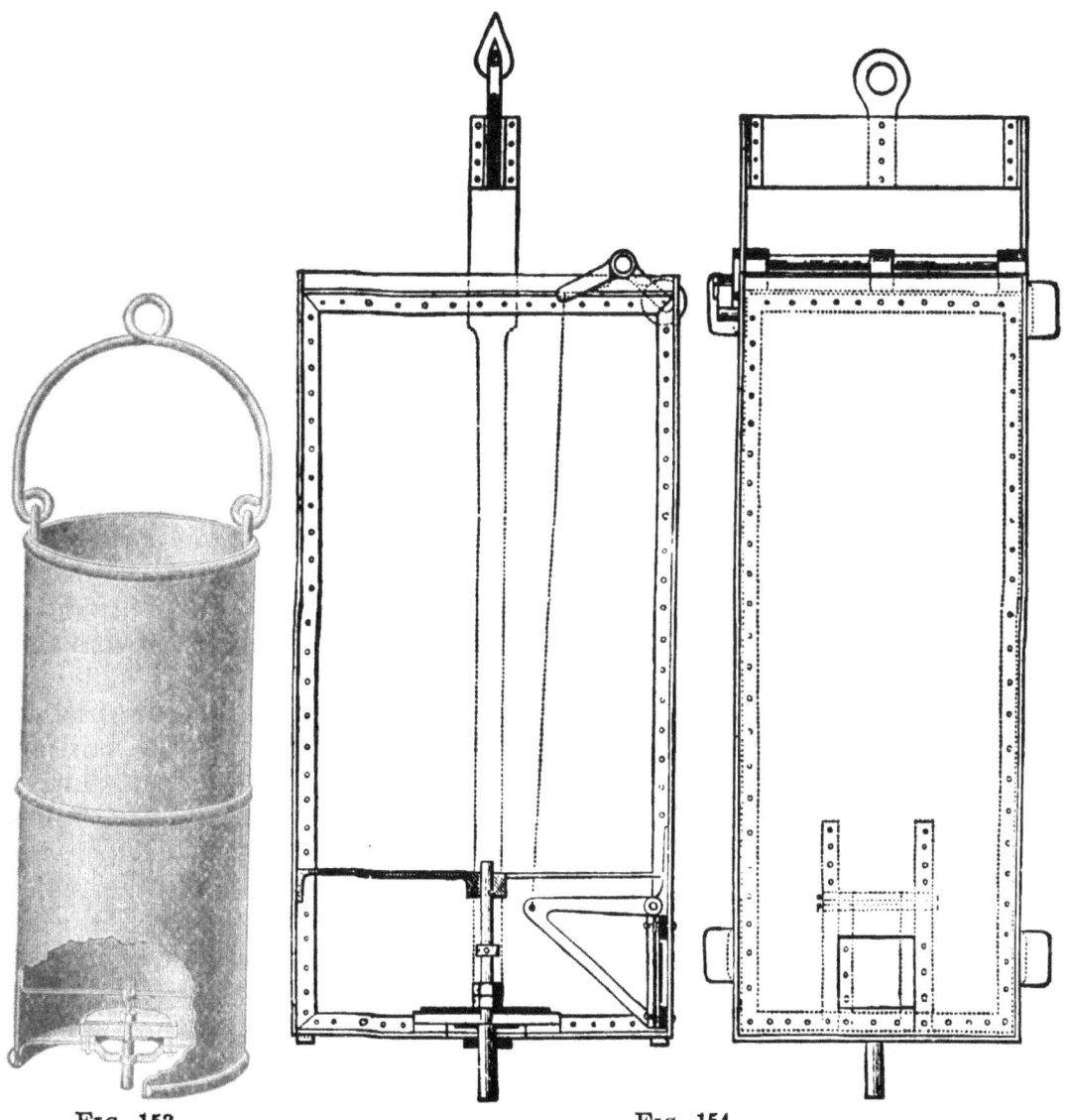

Fig. 153.   Fig. 154.

reservoirs or from the sump. In the latter case they are either made self-filling or are filled by means of pumps or other contrivances. Rapid filling of bailing-tanks is very important. Figs. 153 and 154 illustrate types of tanks fitted with a valve in the bottom, which opens of itself when the tank sinks into the sump-water. Such a method requires a considerable depth of sump in order to fill them, and tanks are therefore not adapted for sinking purposes unless an artificial method of filling them is used. Means for thus filling the tank may be movable steam or compressed-air pumps, injectors, or pulsometers. These should be of ample capacity so as to fill the tank as rapidly and with as little delay as possible. The bailing-tank in this method need not approach quite to the bottom, so as to cause no interference with the other operations of sinking. Since, however, in case of a considerable rate of inflow, the water may rise in the intervals of the trips of the bailing-tanks to such a depth as to interfere with the other work, an artificial sump is sometimes used. This consists of a tank suspended in the shaft and sufficiently larger than the bailing-tank, to admit of the latter dipping entirely into it for the purpose of filling itself. Pumps or other water-raising appliances can then be

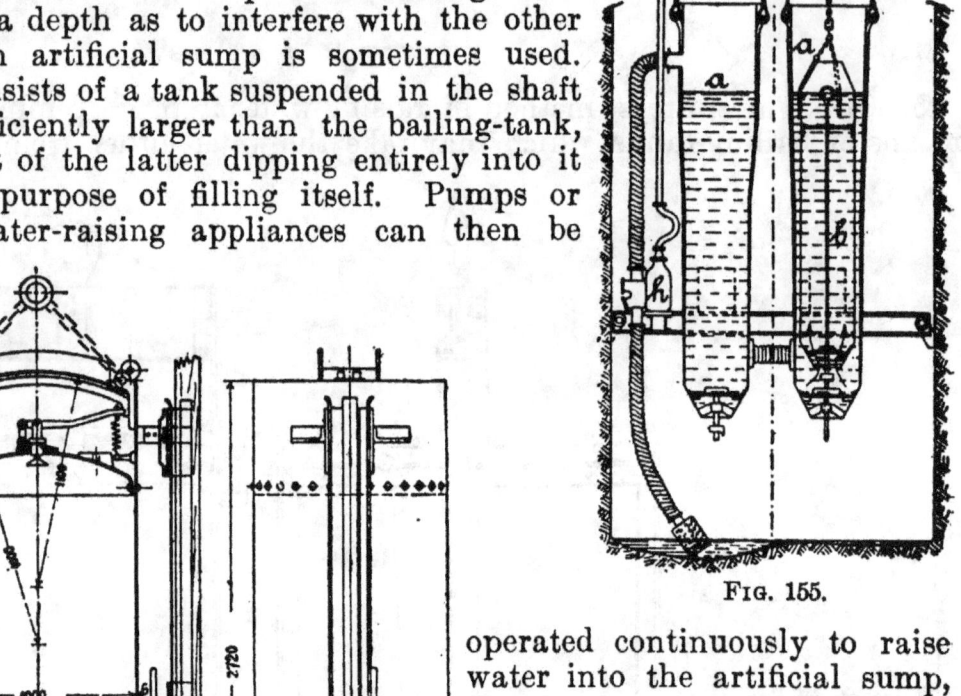

Fig. 155.

operated continuously to raise water into the artificial sump, thereby keeping the bottom of the shaft as free as possible for the men to work. The artificial sump is periodically lowered as the shaft goes down. This can be most quickly accomplished by attaching it to the bailing-tank. Fig. 155 shows a double arrangement of bailing-tanks and artificial sumps for a very large circular shaft like those used in Europe. In the case illustrated the artificial sumps are suspended by cables from the surface, which serve also as guides for the bailing-tank.

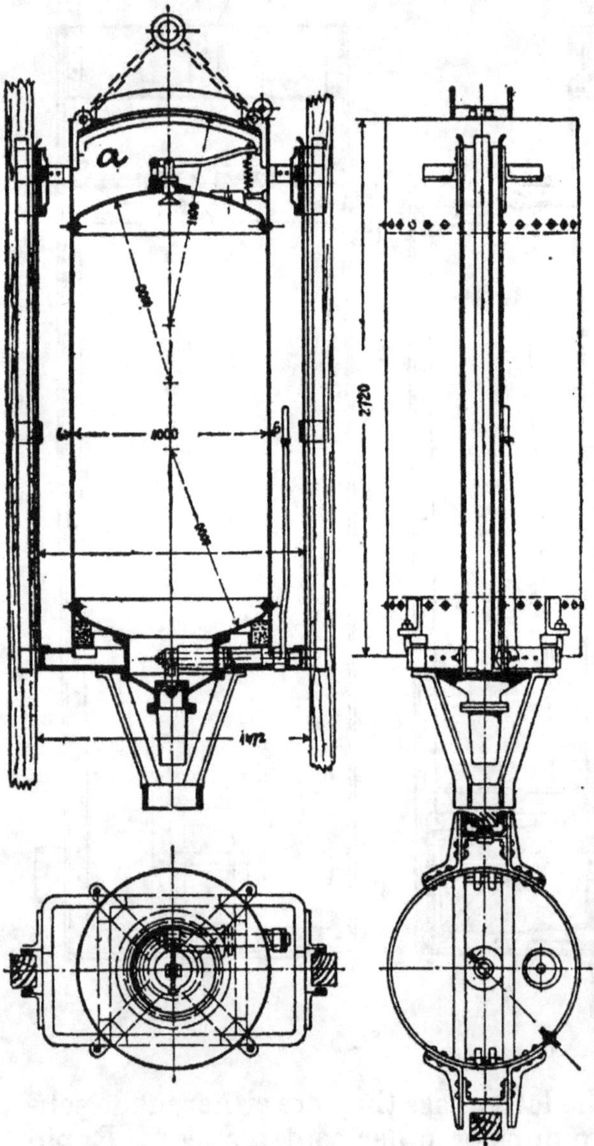

5.1.02. Bailing-tanks are usually made of iron. Square wooden tanks, held together by bolts, are sometimes constructed at the mine for an emergency. They are not very durable. Bailing-tanks on the Comstock were sometimes very large, some holding about 150 cu. ft. Sometimes bailing-tanks are suspended below a cage. In such cases they should be fitted with safety-catches, so that the cage-catches will not be called upon to catch the combined load of cage and tank.

Fig. 156.

5.1.03. *Vacuum-Tank.* An interesting and presumably efficient method of filling a bailing-tank rapidly was suggested by Mr. Bacher for the sinking of the Max shaft at Kladno, Bohemia, referred to before in this paper. The bailing-tank is illustrated in Fig. 156, and is called a vacuum-tank. It consists of a closed vessel, which is filled with steam at the surface in order to expel the air, and to cause, by its subsequent condensation, a sufficient vacuum to draw in water, and thus fill itself at a distance above the sump. The bottom of the tank has an opening closed by a valve, and communicating with a suction-hose of suitable length. When the tank arrives at the surface the discharge-valve at the bottom is opened by means of the lever at the side, and at the same

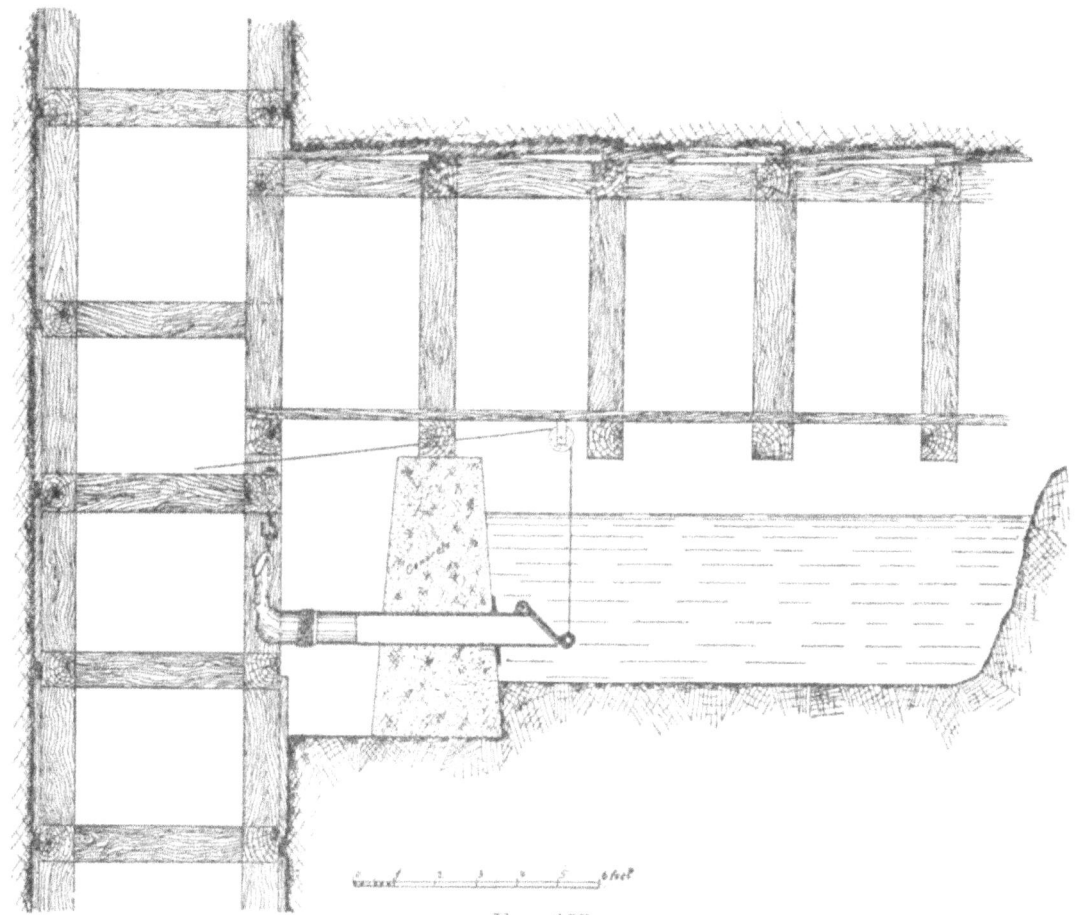

FIG. 157.

time the steam supply is coupled on at the top of the tank, so that the pressure will aid in expelling the water more rapidly. As soon as steam issues at the discharge both openings are closed, and a valve, communicating with the small reservoir of water, a, at the top of the tank, is opened to admit a spray for condensing the steam. Arrived at the bottom the lower valve at the upper end of the suction-hose is opened to fill the tank, and closed again before the tank starts to the surface.

5.1.04. *Bailing-Tank Stations.* Where the water is taken from some of the higher levels, reservoirs or tanks are built to collect the inflow at that point. The reservoirs are fitted with a discharge-pipe carrying a valve on the inside, which is operated by a rope and communicates with a discharge-hose to lead into the bailing-tank on the outside. Fig. 157 illustrates a reservoir formed by a masonry bulkhead placed across an excavation, with the discharge-pipe built in.

146   MINE DRAINAGE, PUMPS, ETC.

5.1.05. *Tank Discharges.* The discharge from bailing-tanks at the surface or at a drain tunnel must be effected in such a manner that none of the water will fall down the shaft again. The ordinary water-buckets used in small workings are drawn aside and the water simply poured out. The bucket in Fig. 153 has a downwardly projecting stem on the valve, which strikes the floor and lifts the valve when the

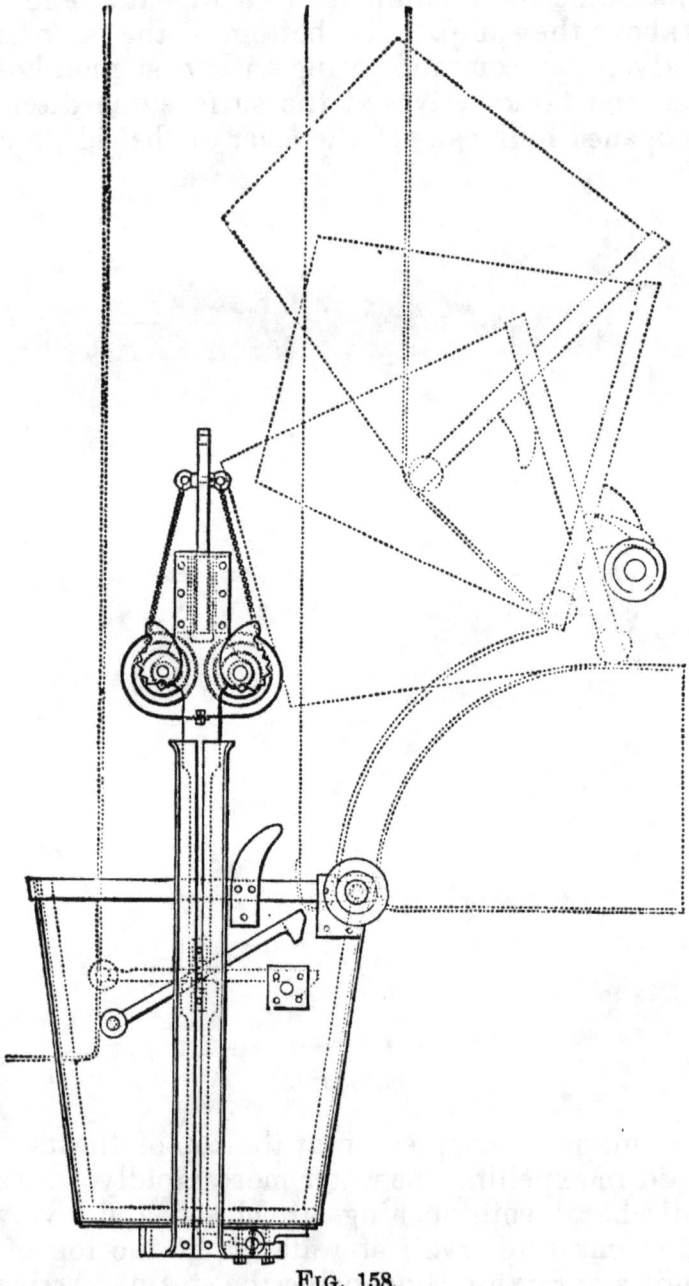

Fig. 158.

bucket is lowered into the discharge-sluice, thus permitting the water to escape. Bailing-tanks guided in vertical shafts usually have side-valves like in Fig. 154, and these sometimes have attached a hose which leads the water into the discharge-sluice; or a short sluice mounted on rollers is pushed under the tank to conduct away its discharge. In inclines the discharge of the tank can always be brought over the sluice or tank without any other arrangements.

5.1.06. The most rapid manner of effecting the discharge of bailing-tanks is to construct them like skips, so that they will dump automatically as they are hoisted above the collar of the shaft. Fig. 158 shows a self-dumping skip for a vertical shaft, which is used for hoisting both rock and water. Fig. 159 illustrates a simple skip used for inclines. The manner of effecting the dumping appears readily from the illustration.

5.1.07. As a permanent method of controlling the water of a mine, bailing is generally economical only for smaller quantity of inflow. Such conditions prevail at some of the mines on the Mother Lode, in Amador County, California, as at the Kennedy Mine, before referred to, where the water amounts to about 75,000 gals. per day during the dry season, and about double that amount during the wet season. For large capacity, bailing is justifiable only as an emergency measure to supplement pumps during their repair, or to aid them temporarily during a large influx, or, as in the case of the Susque-

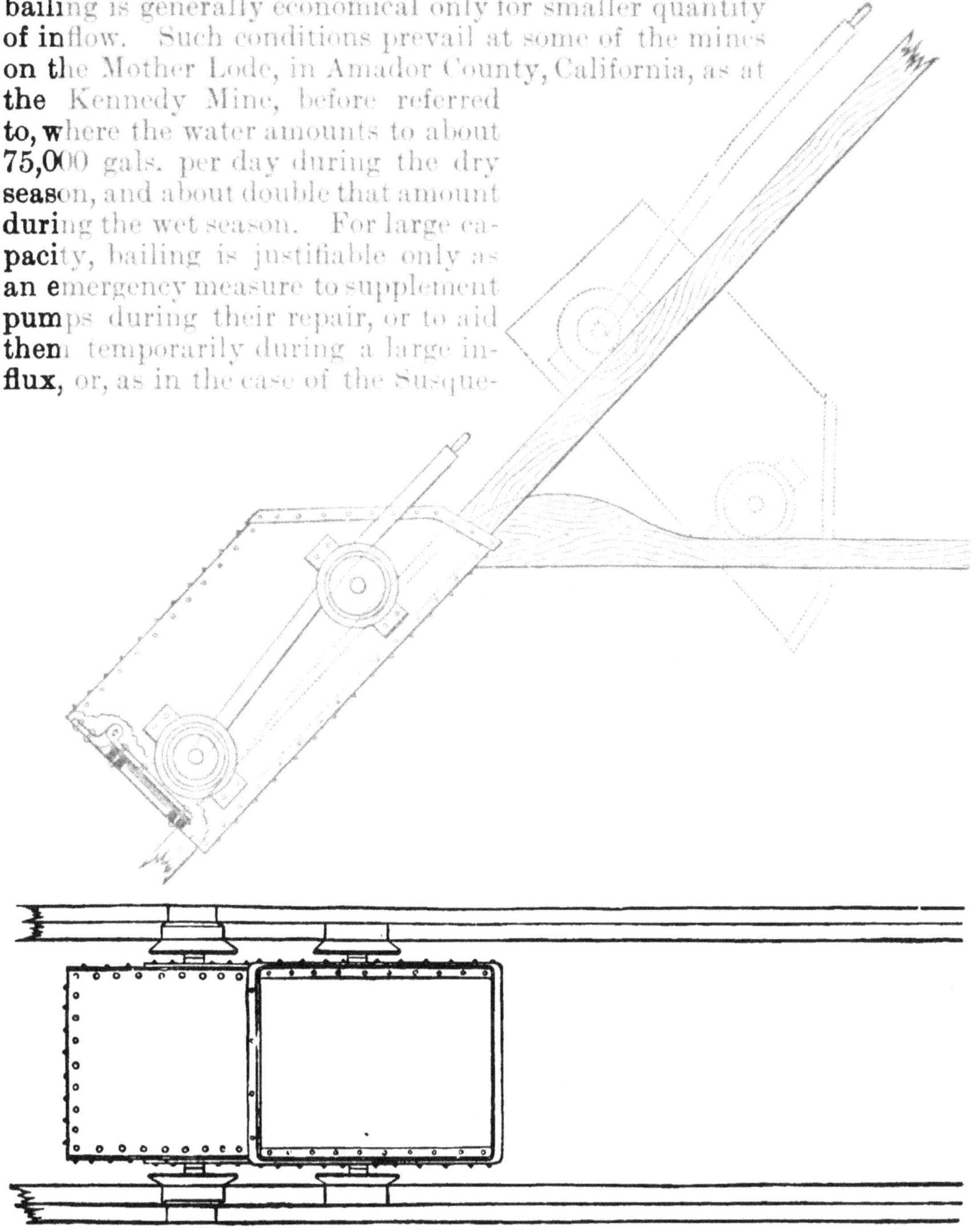

FIG. 159.

hanna Coal Company's mines, to rapidly drain a flooded mine with the large hoisting-plant on hand; bailing-tanks being in this case most rapidly got ready, and the method being also the most economical on account of the small first cost of plant and the short period that it would be required to be used. The details of construction and method of operating are shown by Figs. 160 and 161, and the following more detailed

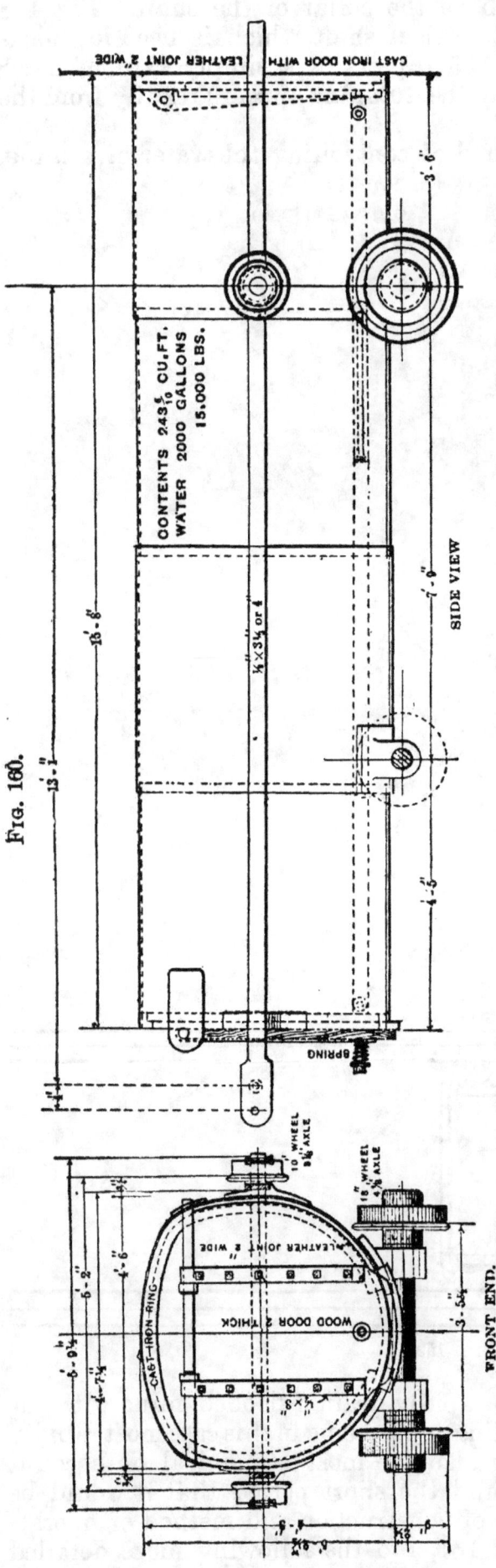

Fig. 160.

description. At the end of each tank is a large iron door of almost the full size of the end of the tank, opening inward, so that when immersed the tanks fill almost instantly. To provide for holding the water while it is hoisted up flat pitches, a wooden door is attached to the front of each tank, opening outward. Each front door is attached to the door at the back by an iron rod, provided with a sliding link, so that the back door can open independently of the front, but the latter is held closed as long as the rear door is closed. This connecting rod, as shown in Fig. 160, passes through the front door and through a spiral spring in front of it, so that the amount of pressure necessary to keep the water from leaking out may be readily applied. The tanks are mounted on self-oiling, closed wheels, so arranged as to exclude water from the bearings while the tanks are immersed, and to retain the lubricant. Each tank is provided also with side-wheels, vertically over the rear axle, which have a gauge sufficiently wide to clear all other portions of the tank; and on the surface an elevated track is provided, upon which these dumping-wheels run and thus raise the rear end of each tank as much as may be necessary to dump the water into a trough between the tracks, the tilting forward of the tanks opening the back door and releasing the front one. The tanks while emptying rest on their forward wheels and on the dumping-wheels. By having the tracks at the surface slightly up-grade, the tanks will run back when empty, as soon as the rope is slackened. To allow this dumping, the hoisting-rope is attached to the tanks by a yoke reaching back on the sides and pivoting on the axle of the dumping-wheels, the tanks back of the first one being attached by eye-bars reaching from axle to axle of the dumping-wheels on the tanks. A stop is provided,

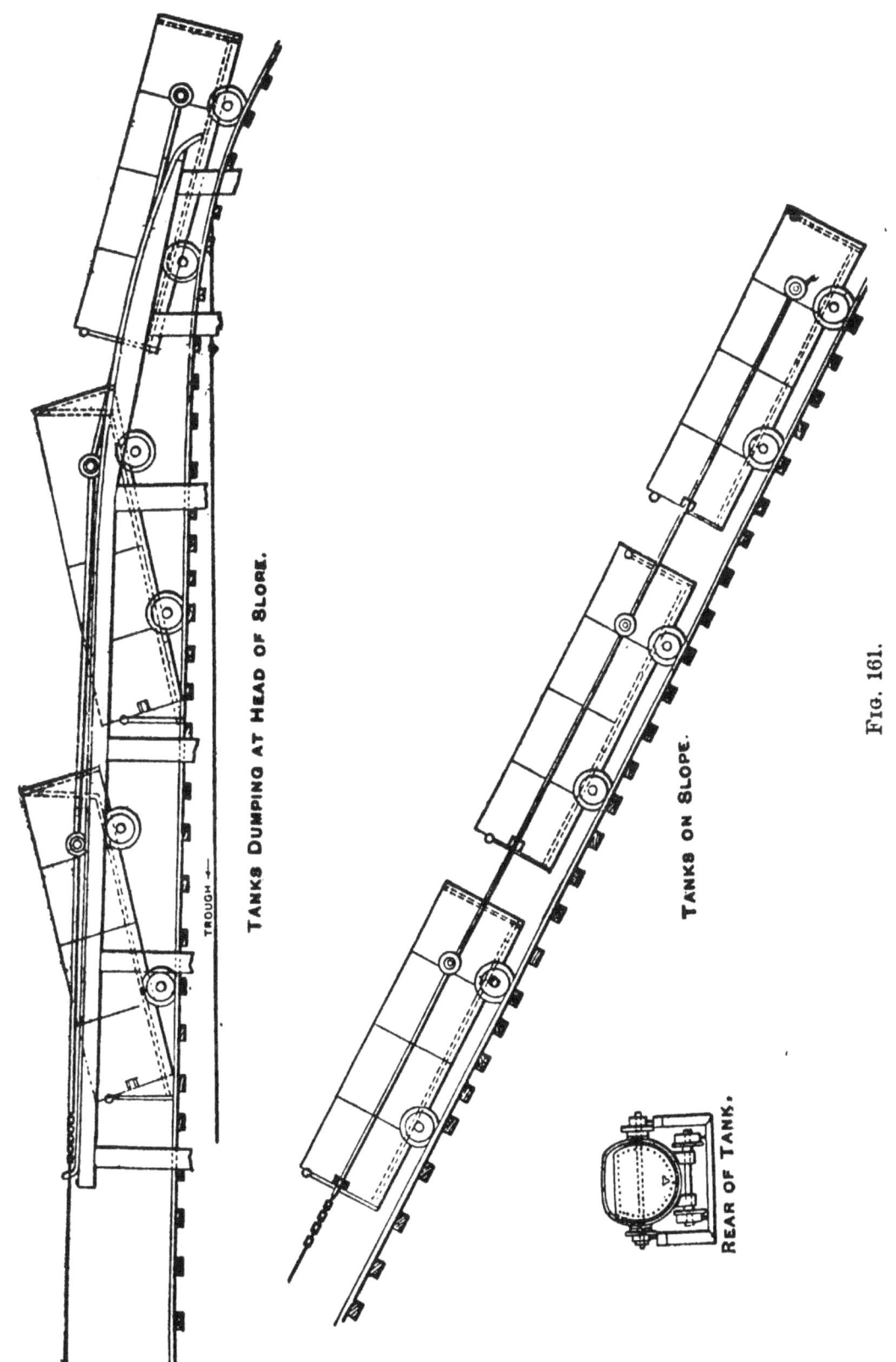

Fig. 161.

to prevent the yoke on the forward tank from dropping and catching in the track when the rope is slackened. This plan of "tandem tanks" was designed and used to hoist about 25,000,000 gallons of water which had been admitted to extinguish a mine fire in one of the Susquehanna Coal Company's mines. The slope was small in section, and 3,200' long, with single track, and with pitches varying from 4° to 20°. The hoisting-plant consisted of a pair of 26"x 60" direct-acting engines with cast coned-drum, 9' to 12' in diameter, carrying $1\frac{3}{8}$" steel rope. These engines had been previously hoisting five cars, weighing about four tons each when loaded.

5.1.08. Bailing appliances should be in readiness for immediate use at every mine operated through shafts or inclines, to relieve or aid pumps. This fact, and the necessity of being able to control large bodies of water, have an important bearing on the necessary hoisting capacity of a mine. This should generally be adequate to handle not only the rock and ore output, but also all the water that may be expected to be encountered. With deep mines the hoists should raise the load to the surface as rapidly as possible, and the use of direct-acting hoists is therefore advisable. In inclines which follow the vein, and are therefore generally crooked, rapid hoisting is not admissible, and the hoisting engine should then be capable of handling very heavy loads.

5.1.09. Bailing is deficient in economy for several reasons: First, the weight of the tank and cable must be raised, together with the water, at each operation; second, hoisting-engines, if operated by steam, on account of their frequent starting and stopping, being thereby alternately heated and cooled off, are not economical steam-users. If operated by air, they can work more economically, as long as the compressed air can be produced cheaply. Water-power hoists also are inefficient, on account of starting and stopping, and operating at all speeds varying from that required for best efficiency. The effect on economy of the weight of the tank and cable will not exist in those not frequent cases where two tanks are run so as to balance each other.

5.1.10. Gasoline hoisting-engines have of late come into use to some extent to operate bailing-tanks of small capacity, especially in arid regions or places where solid fuel is scarce.

SECTION VI.

CHAPTER I.

Pumps and Other Appliances for Raising Water from Moderate Depths in Mines.

6.1.01. *General Remarks.* Appliances for raising water to moderate elevations are useful in many ways for mining. They may be used in order to avoid too frequent moving of the heavy sinking-pump by providing a lighter low-lift apparatus, which pumps into an artificial sump provided for the sinking-pump, or they may pump into bailing-tanks, or into reservoirs from which the tanks are filled, as described in 5.1.01, so that no great depth of water is required to be maintained in the sump for filling the bailing-tank. The generally low efficiency of apparatus available for such purposes affects the total efficiency of the system but slightly, because the amount of low-lift work is small compared with the rest of the lift. Low-lift appliances also find application for draining open workings or levels in a mine, or in drift mines in which the channel rises and falls, or where the operating-tunnel has been driven too high. Where the water has merely to be raised over and dropped down the other side of an elevation less than the barometric height to which water will rise, siphons may sometimes be used. The siphon action should also be utilized to aid the pumps in all cases where water has to be raised over and dropped down the other side of an eminence. This is much more

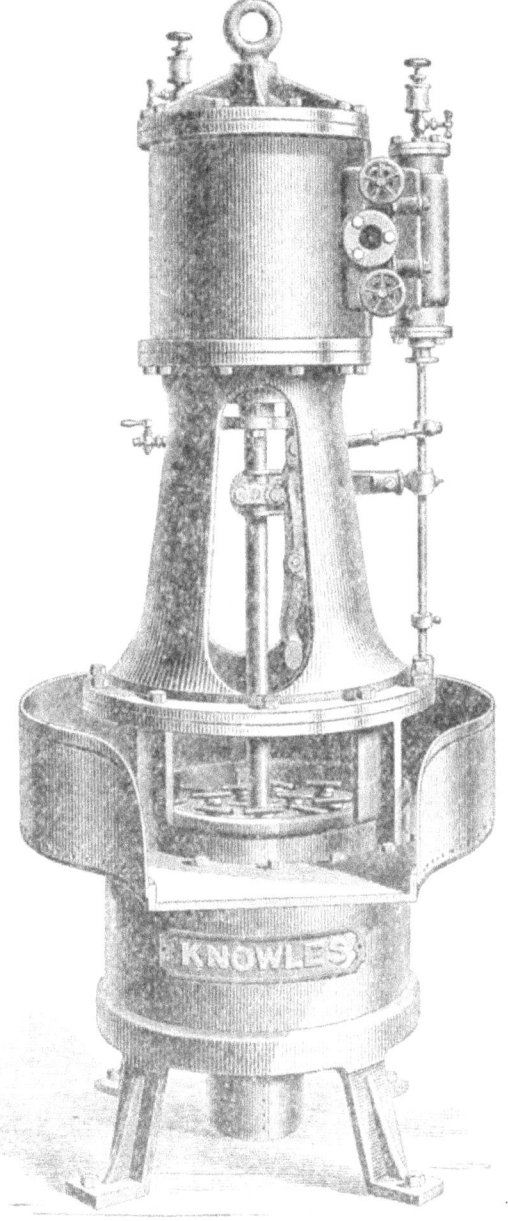

Fig. 162.

important for low lifts than for high ones, because the proportional reduction of total lift is greater. The machines which find application for the purposes mentioned are reciprocating pumps, centrifugal pumps, pulsometers, jet-lifters, air-lift pumps, and siphons. Some of these find application in mining chiefly in furnishing water supply, as for gold-washing, milling purposes, or for boiler use.

6.1.02. *Reciprocating Pumps.* These have been treated in former chapters. For direct-driven, low-lift pumps it is necessary, however, that the steam or compressed-air cylinder be the smaller in proportion to the water cylinder, the less the height is to which the water has to be raised. Some of such pumps are single-acting, and utilize only the suction-lift, being, therefore, only adapted to pump to heights less than that due to the pressure of the atmosphere. The wrecking pump shown in Fig. 162 is of this type.

CHAPTER II.

Centrifugal Pumps.

6.2.01. Where large volumes of water require to be raised to moderate elevations, centrifugal pumps are usually the cheapest, and, under certain conditions, the most efficient machines. They are also well adapted

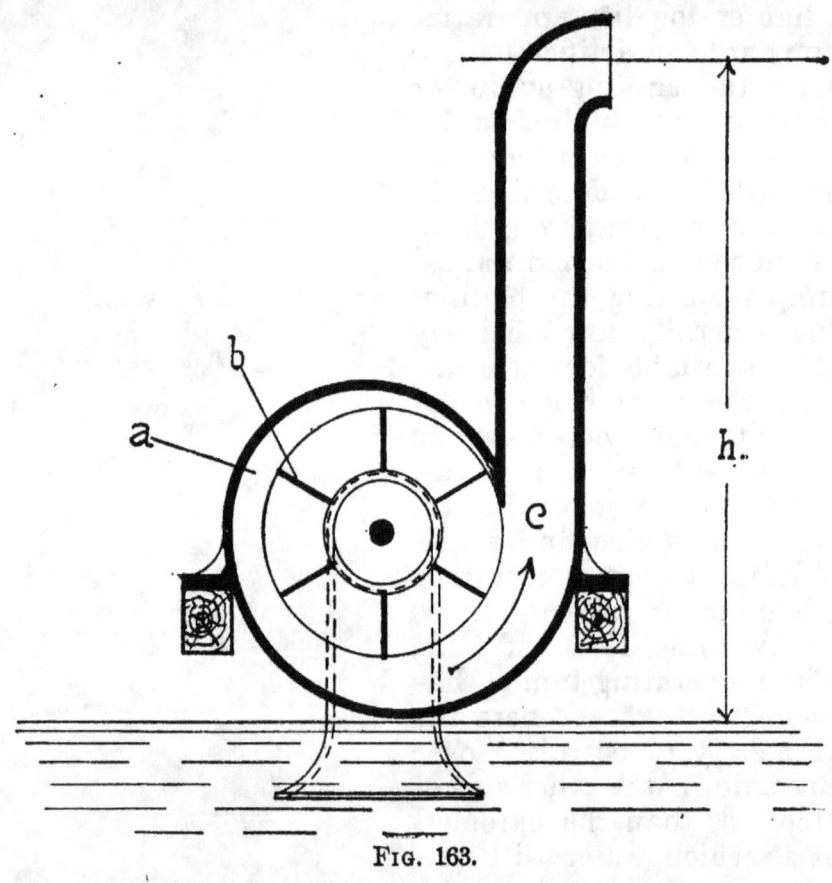

Fig. 163.

to handling muddy and sandy water, and may deliver gravel and even cobbles just large enough to pass through them.

6.2.02. The action of a centrifugal pump may be best described by reference to Fig. 163, in which a is a casing, within which revolve the paddles b. Assuming the pump to be primed and the casing to be filled with water, the latter will have imparted to it the rotary motion of the paddles, by virtue of which centrifugal force or pressure is exerted within the fluid, so that it will escape with that force at any outlet, as at c. If the outlet communicate with an ascending pipe, the liquid will rise in it to a height determined by the amount of centrifugal force. The latter is the greater, the greater the circumferential velocity of the

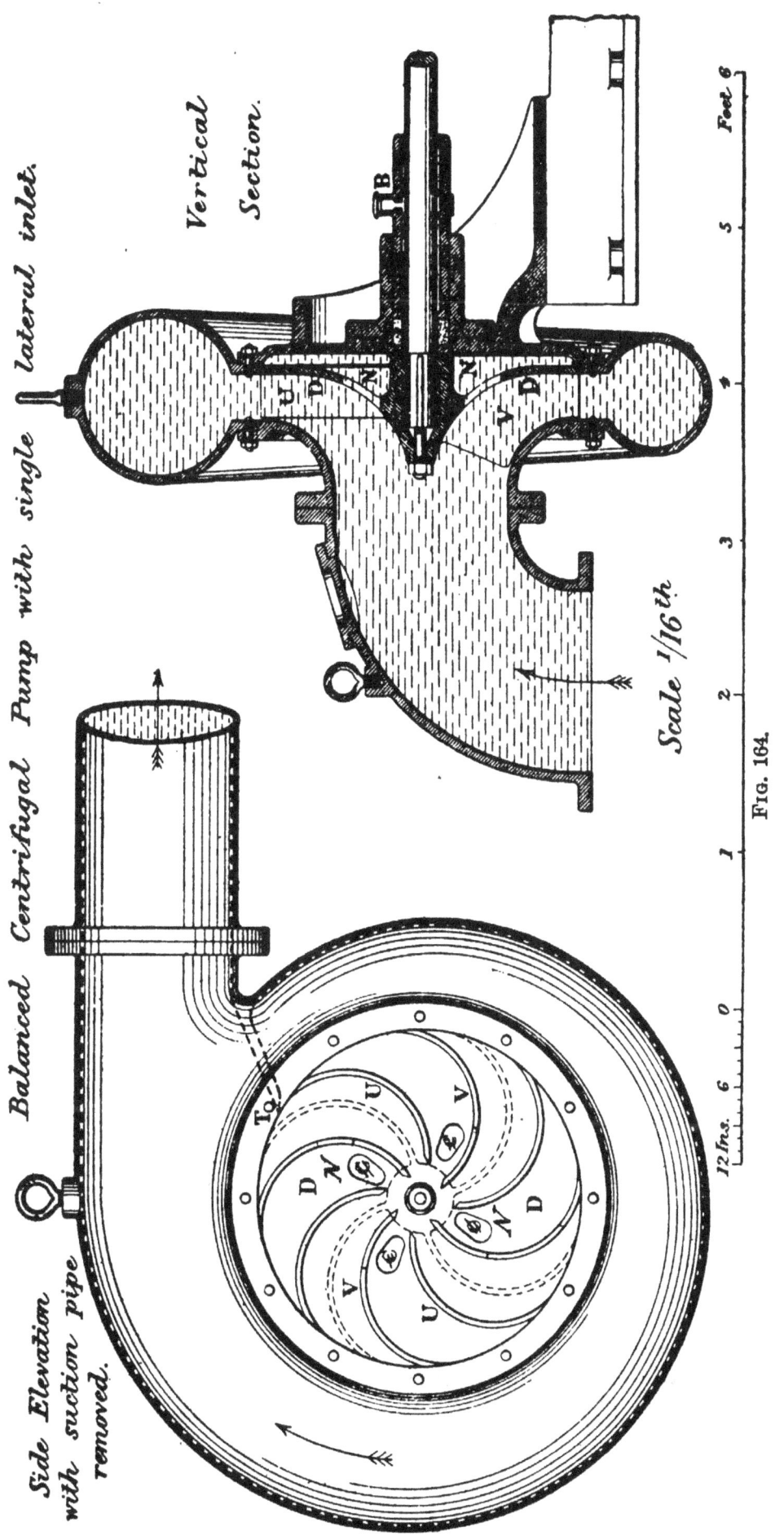

Fig. 164.

paddles. If there were no unavoidable efficiency-losses, the relation of the total lift h to the circumferential velocity V would be expressed by the formula $h = \frac{V^2}{64}$. In practice, h is only about $\frac{4}{5}$ or $\frac{3}{4}$ of $\frac{V^2}{64}$. The lift h is the height to which the fluid will rise in the pipe when there is no discharge from the latter, or when, in case of a discharge, the liquid is caused to lose its energy of motion by being introduced from the space between the paddles into a duct of such width that its velocity will be suddenly much reduced below that of the tips of the paddles. This sudden enlargement of waterway is a feature found in most of the centrifugal pumps manufactured. Figs. 164 and 165 illustrate pumps which possess this feature.

6.2.03. Let us assume, on the other hand, that, by keeping the cross-

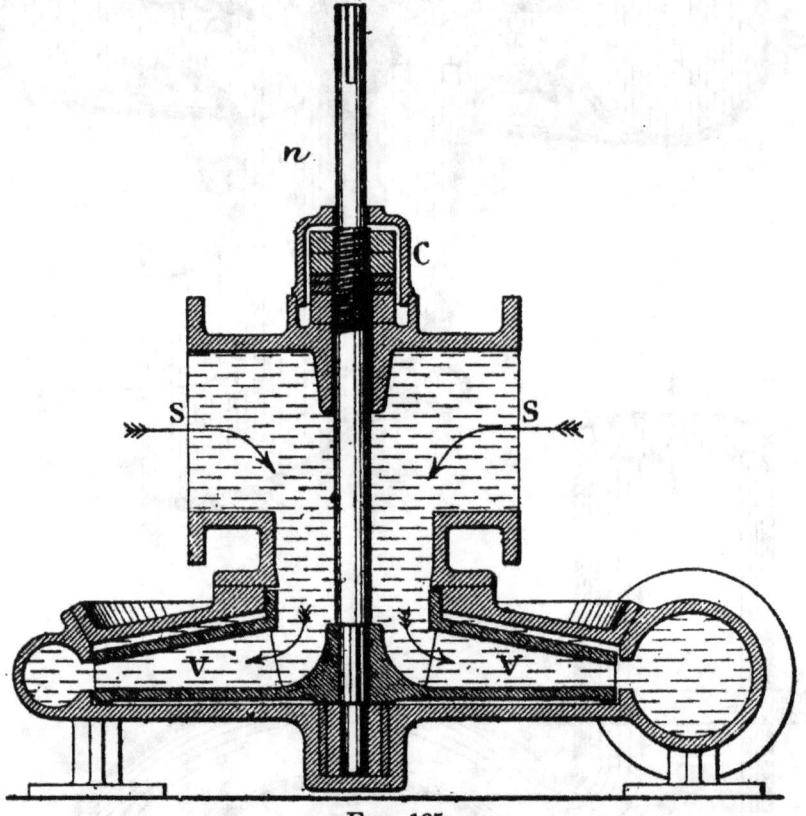

Fig. 165.

section of the spiral duct down to the proper size, the discharge is effected in such a manner that the liquid, which leaves the periphery of the blades with the rotative velocity which it has in common with them, retains this velocity until it reaches the outlet c of the duct, and then has its velocity reduced in a gradual and continuous manner to that admissible in the pipe by a corresponding gradual enlargement of a short part of the discharge-pipe, where it joins the duct, as at d, Fig. 166. In this case most of the energy of motion which has been imparted to the water by the paddles b, is utilized and changed into potential energy or pressure-head additional to that due to centrifugal force alone. This additional height h equals $\frac{V^2}{64}$, if the discharge-pipe is so large that the velocity is insignificant so that $h_1 = h$.* It is apparent, therefore, that

---

* If the velocity u in the discharge-pipe is taken into account the equations become $h = \frac{V^2}{64} - \frac{u^2}{64}$ and $H = \frac{V^2}{32} - \frac{u^2}{64}$. The term $\frac{u^2}{64}$ is the amount of reduction of head.

in an ideal centrifugal pump, the head obtainable with a velocity V of the water at the periphery of the paddles should be $2h = H = \frac{V^2}{32}$. In practice, again this result is not obtainable, but results of $H = 0.86\frac{V^2}{32}$

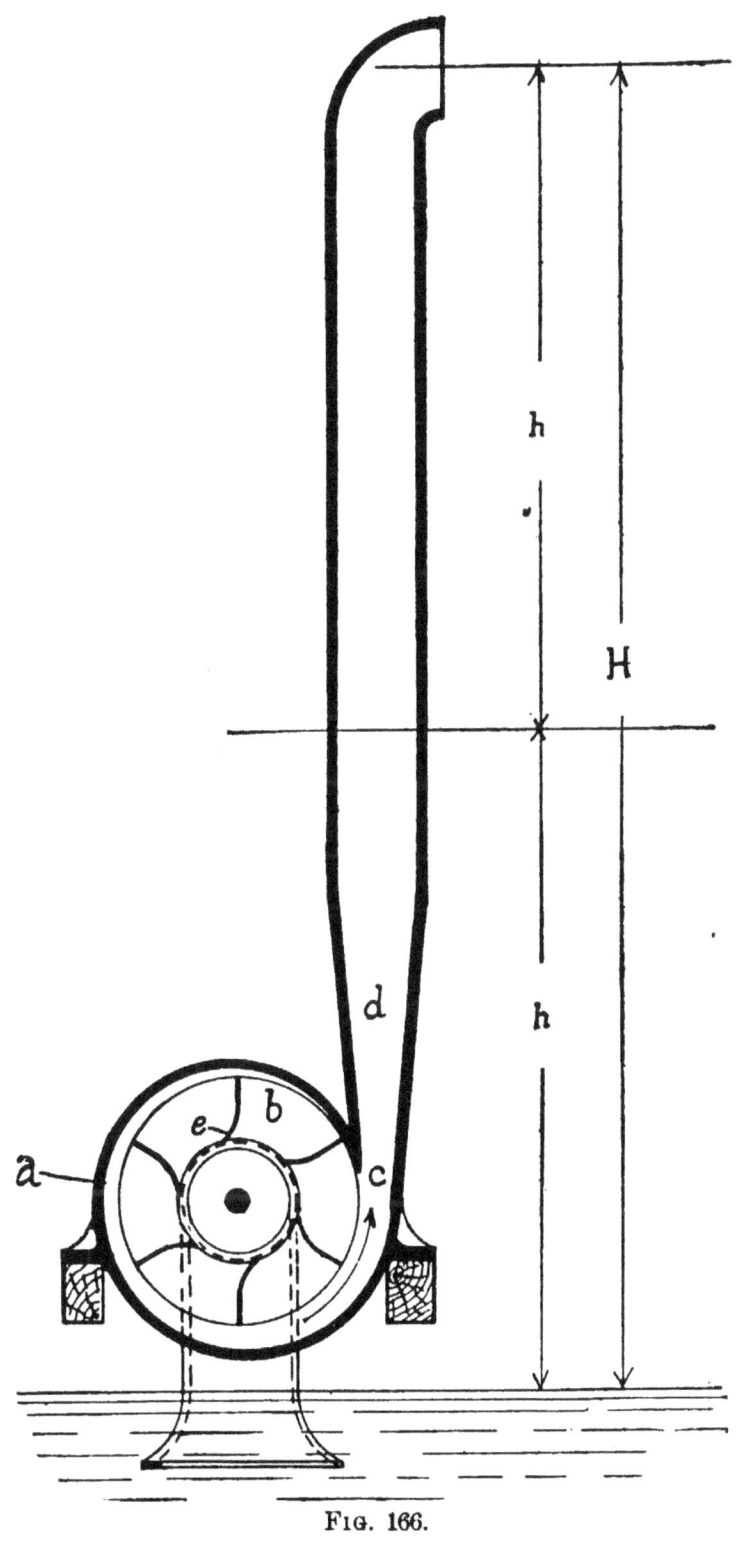

Fig. 166.

have been obtained. It is to be remembered, however, that this height H is not reached when there is no discharge, because then the energy of flow does not exist at c, and cannot develop into additional pressure.

6.2.04. The advantages of utilizing to a greater extent the energy of motion are: first, that more of the work which is spent in imparting to each pound of water that passes through the pump, both centrifugal pressure and an acceleration from rest to the velocity V, is utilized and not allowed to go to waste; second, that water can be raised to a given height H with a less peripheral velocity V of the paddles, which means

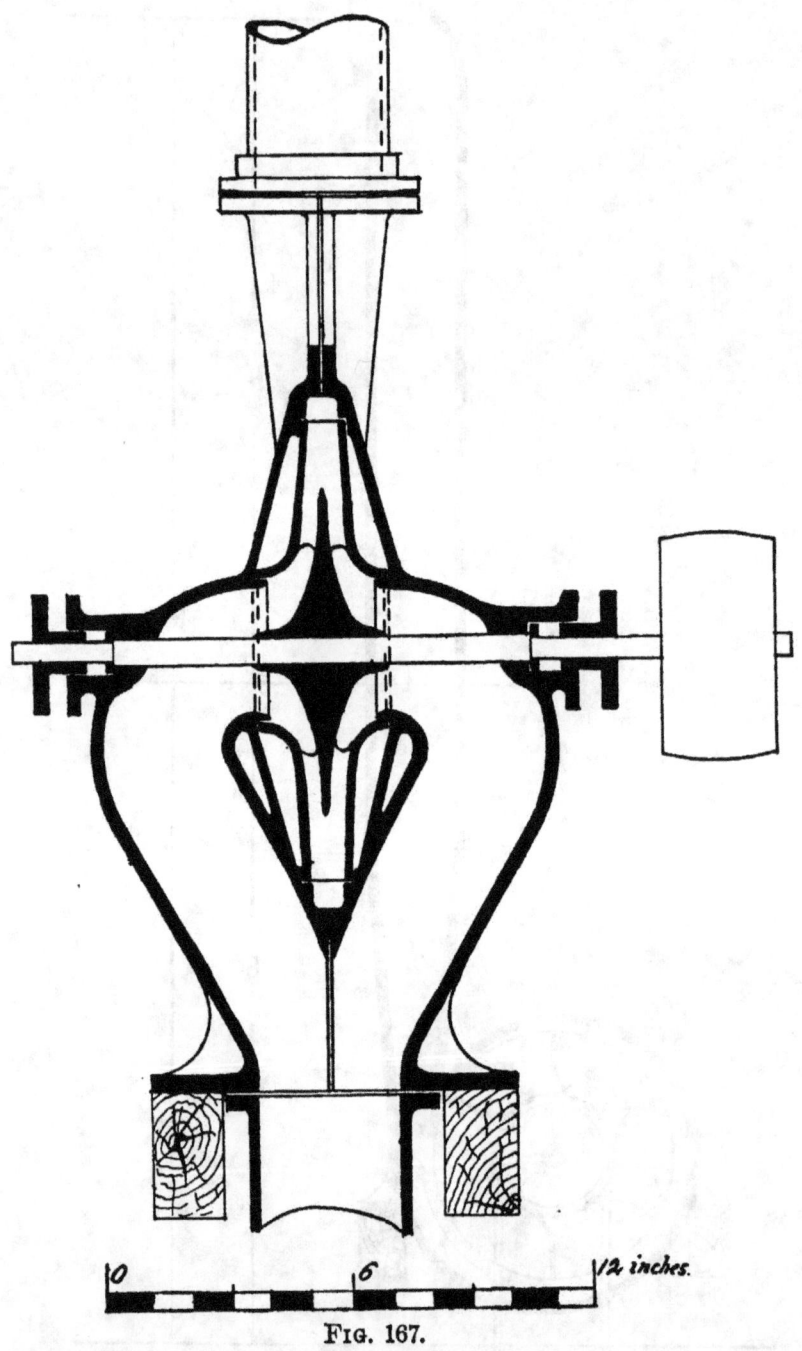

Fig. 167.

less frictional resistance in the pump, and often more convenient connection to motors. The reduced velocity will appear from inversion of the formulæ for h and H.

For the usual imperfect pumps $V = 8\sqrt{h}$ theoretically.*

For a properly constructed pump $v = 5.66\sqrt{h}$ theoretically,* the

---

* Really, if the velocity u in the discharge-pipe is considered, $V = \sqrt{64h + u^2}$ and $v = \sqrt{32h + \frac{u^2}{2}}$

reduction in velocity being nearly 30%. In practice, formulæ for the first case usually make $V = 10\sqrt{h}$ or $11\sqrt{h}$, while the writer has obtained results with properly constructed pumps of $v = 7\sqrt{h}$ and even $6.5\sqrt{h}$.

6.2.05 Another feature ordinarily met with in centrifugal pumps is the excessive backward curvature of the paddles, which causes an unnecessary increase in the velocity required to pump against a given head, thereby increasing fluid friction and helping to account for such relations as $V = 11\sqrt{h}$. Radial blades give a lower value of V or v, but their inner ends should be curved forward, as at e, Fig. 166, so as to scoop up the water with a minimum of shock. The inlet velocity of water to a centrifugal pump should not, if possible, be much over 3' per second.

6.2.06. Where the least amount of fluid friction is desirable in a pump, the runner should be kept small in diameter. A large diameter of paddles or impellers is oftentimes required in order to keep down the number of revolutions to such a limit that the pump can be directly coupled to an engine. For driving by an electric motor the more advantageous small runner is better adapted, as the speed of such motors is usually high. The friction will be less in the case of paddles shrouded at the sides, as in Fig. 167, than where they are open at the side, as in Fig. 164. The open blades, however, permit less leakage past their sides into the suction-pipe than the shrouded blades, as the zone of action of the latter is cut off by the shrouding, while that of the open ones extends by fluid friction somewhat beyond their edges.

6.2.07. If a centrifugal pump is properly constructed so as to utilize that part of the energy imparted to the mass of water as motion, it must nevertheless generally be so arranged with reference to its driving power that it can be run at a somewhat higher speed for a very short time, in order to start the flow in the discharge-pipe necessary afterwards to keep up the extra gain in lift. This increase of speed in starting can be avoided by providing an outlet in the discharge-pipe at one half of the total elevation to which the water is to be pumped, the outlet being opened on starting, so that the water can acquire its speed in the flaring pipe, after which it is closed again, whereupon the water will rise and flow out at the top of the pipe.

6.2.08. Unless centrifugal pumps are submerged, they will not prime themselves like a reciprocating pump with valves, because, in pumping out the air contained in them, they act simply as a fan-blower operating at the lower speed of a centrifugal pump, and are, therefore, capable of raising the water in the suction-pipe only to an insignificant amount. Centrifugal pumps are, therefore, generally provided with means for filling them with water, a foot-valve being generally provided at the lower end of the suction-pipe, in order to prevent its escape until the pump has attained its working condition. The means for filling the pump may be a small hand pump, an ejector, or a pipe from a reservoir into which the pump delivers its waters.

6.2.09. Centrifugal pumps are made in a great variety of forms. Some have single inlets, like the pump in Fig. 164. Others have inlets at each side, by which construction the inlets, and also the diameter of the paddles, may be kept down to a smaller size for the same capacity. Fig. 167, before referred to, illustrates a pump with double inlet, designed by the writer. It exhibits also the features for utilizing the energy of fluid motion before referred to and illustrated in Fig. 166. The flaring discharge and radial paddles contributed to the result

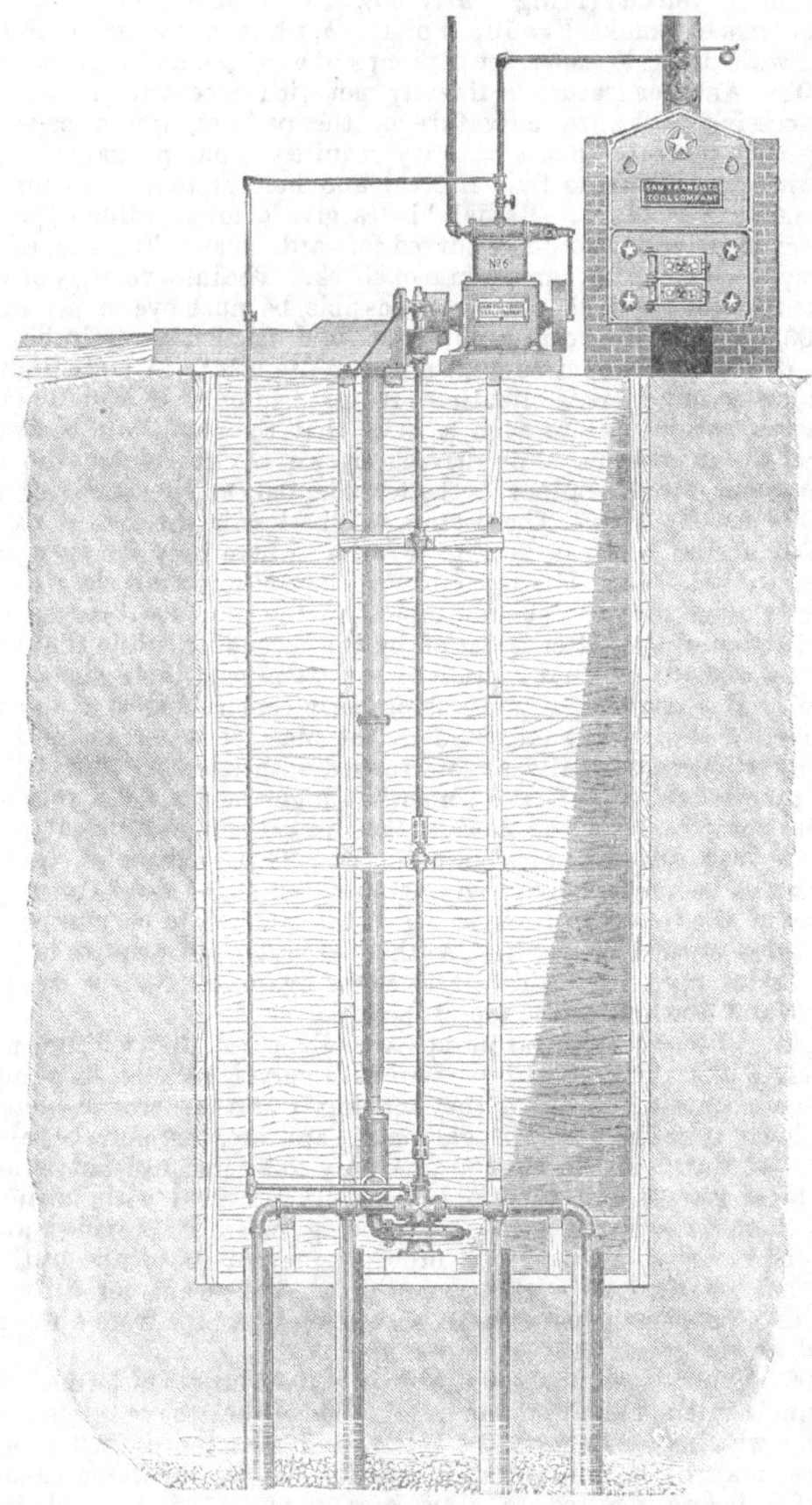

Fig. 168.

anticipated, which was a considerable reduction of the number of revolutions below that of the usual forms, like Figs. 164 and 165, obtainable in the market. The double-inlet pumps have the advantage, in comparison with the ordinary single-inlet, that the lateral pressure against the disk supporting the paddles is balanced. In the Richards single-inlet pump the disk is perforated, as at N, Fig. 164, to allow the pressures on both sides to equalize to a certain extent.

6.2.10. Pumps designed to raise water from wells are frequently arranged with a vertical axis, like Fig. 168, extending to the surface, where the power is applied most conveniently.

6.2.11. Usually centrifugal pumps are only required for low lifts, from 10′ to 20′. They are, however, capable of working efficiently against heads of 100′ and over. For such heads, however, the aim should be to utilize the energy of motion so as to keep the speed and the friction-losses down as low as possible. Heads of 150′ have been overcome by compounding the ordinary centrifugal pumps. Care should always be taken in compounding to convert all the excess of energy of motion acquired by the water in the first pump, into pressure before it is led into the inlet of the second pump with the proper low velocity. Fig. 169 illustrates the proper principle of compounding centrifugal pumps. The same result is obtained by raising the water to half the total height by the first pump,

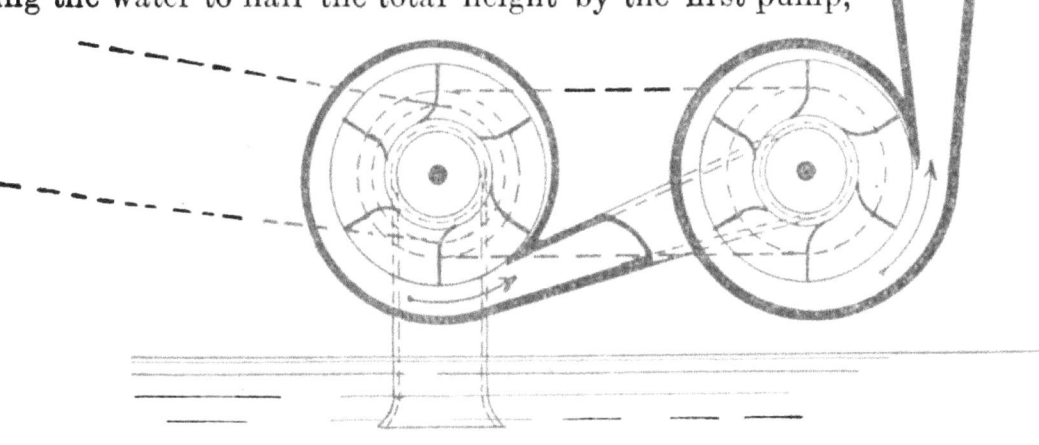

Fig. 169.

and then picking it up and raising it the remaining half by the second pump. In many cases, however, the transmission of power to two separated pumps would be inconvenient. It is the same thing if the water is delivered to the second pump under a pressure equivalent to the head that it would be lifted by the first pump, if the second pump did not pick it up before it was actually lifted.

6.2.12. The capacity of a centrifugal pump is proportional to the speed at which it runs. It therefore also increases with the lift. To increase the discharge for a given head means loss of efficiency, because the speed of the pump and that of the water issuing from the pipe are increased unnecessarily. Reduction of capacity by choking off is also attended with loss by fluid friction, and besides cannot be carried very far without stopping the discharge altogether. This quality of centrifugal pumps makes them less adapted for the ordinary purposes of mining. They are suited for cases where a large quantity of water has to be got rid of in a short time, while the capacity is kept uniform until the supply is exhausted. A variable quantity of water can only be handled with best mechanical efficiency by providing reservoirs, which are alternately filled and then drained by the pump. In mines the smaller sizes only would find application, chiefly in drift mines and open workings, or perhaps in levels of mines operated through shafts.

6.2.13. Centrifugal pumps may be driven by steam or compressed-air engines, gas engines, electric motors, or waterwheels. Horse-powers are also sometimes used for very low lifts. The driving power is either directly coupled to the pump axis, or it is transmitted to a pulley on the axis by means of belting. Where they are driven by horses the average speed should be maintained considerably above the amount required to raise the water, because the horses will not keep up a uniform speed, and will frequently slow down, so that the discharge of the pumps will cease altogether.

6.2.14. The fact that centrifugal pumps do not admit of a wide range of variation of their capacity requires that a great number of patterns should be kept on hand by the manufacturers. For this reason the design and construction of the pump should be as simple and inexpensive as is compatible with the other features to be attained. Where efficiency is desired, it is generally necessary to design a centrifugal pump just to suit the conditions under which it is to operate.

6.2.15. Centrifugal pumps intended for raising coarse gravel, mud, etc., like in dredging, should be designed with a view of adaptation to the material to be handled, wearing qualities, and safety against breakdowns. High efficiency in consumption of power is generally not obtainable under these conditions, and is also of secondary importance here.

CHAPTER III.

Jet-Lifters; Ejectors.

6.3.01. These machines are operated either by steam or a head of water. Compressed air might also find application instead of steam, but no special apparatus for lifting water by means of its jet action can be obtained in the market. The common steam-ejectors could be thus used; with what efficiency, is not known to the writer. It is, however, probable that the efficiency will be extremely low.

6.3.02. Steam-ejectors or water-lifters are simple and cheap devices. Notwithstanding their low efficiency they can be made useful where efficiency does not cut much of a figure, where the work is only temporary, or where the time necessary for installation of apparatus is limited, and the suitable operating power is ready at hand for application. An advantage of machines of this class is also that they occupy little space, and are very light.

6.3.03. Steam-ejectors, if used in the bottom of a shaft or pit, should not be connected to the steam-supply by steam-hose, as the latter is liable to give out at any time. When it is desirable to have the ejector arranged so that it can be conveniently and rapidly let down as the work of sinking progresses, it is necessary to put in a telescope-pipe like those described for direct-acting sinking-pumps in 3.2.08. There should be a stop-valve close to the ejector, to control its operation, and one close to the boiler, so that the steam-pipe can be repaired or extended when the machine is to be lowered There should also be a check-valve in the suction-pipe, so that there is no possibility of steam being blown out through the suction when the sump has been drained. Steam-

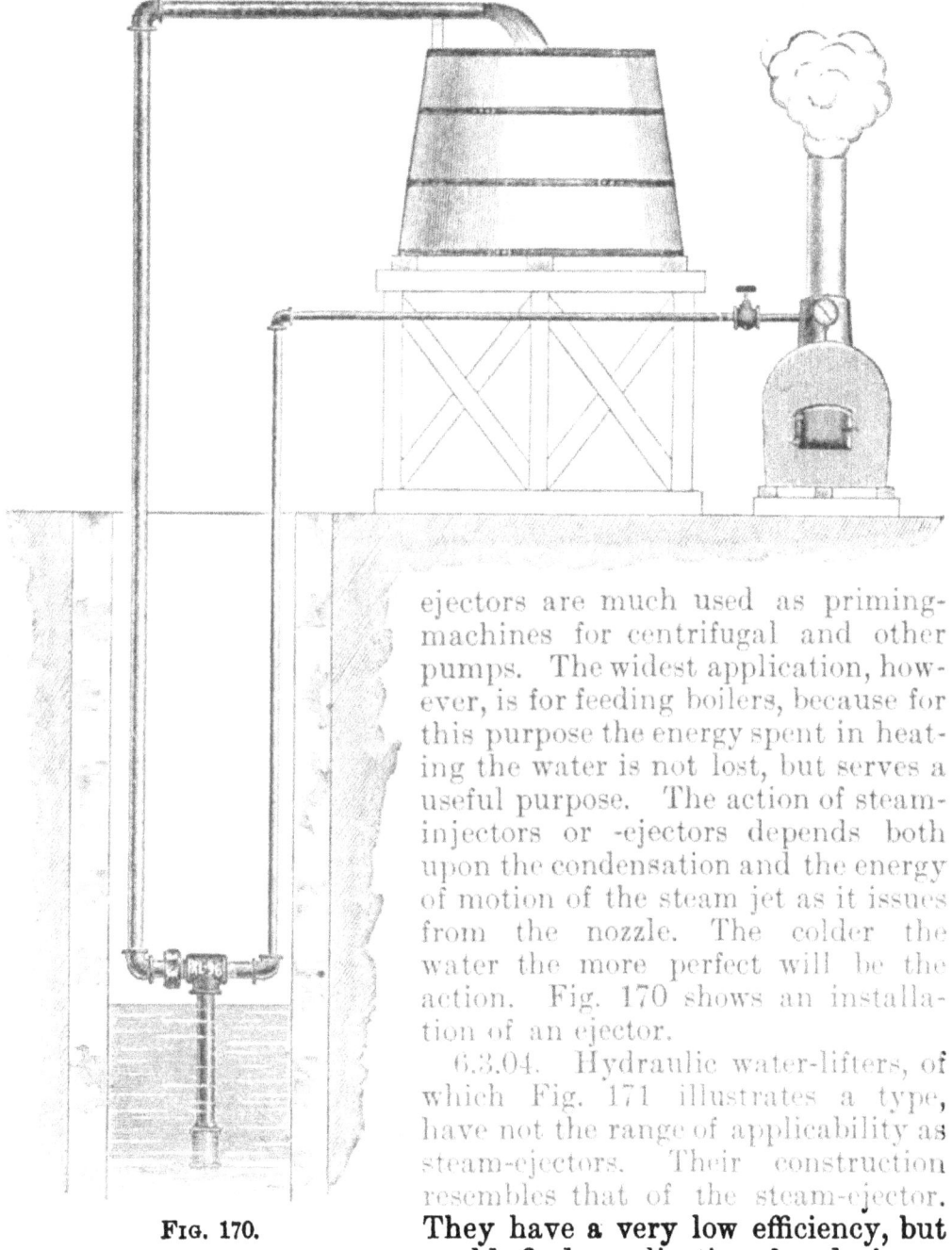

FIG. 170.

ejectors are much used as priming-machines for centrifugal and other pumps. The widest application, however, is for feeding boilers, because for this purpose the energy spent in heating the water is not lost, but serves a useful purpose. The action of steam-injectors or -ejectors depends both upon the condensation and the energy of motion of the steam jet as it issues from the nozzle. The colder the water the more perfect will be the action. Fig. 170 shows an installation of an ejector.

6.3.04. Hydraulic water-lifters, of which Fig. 171 illustrates a type, have not the range of applicability as steam-ejectors. Their construction resembles that of the steam-ejector. They have a very low efficiency, but could find application for drainage purposes under the conditions mentioned in 6.3.02. Where heat is objectionable, as in the bottom of a shaft, and where water-power is at disposal, they would be preferable to steam-ejectors. Where the lift is low, the required driving-head will be correspondingly so. Unless there is a very heavy head, hose may be used for connection to the supply-pipes, thereby affording a more flexible connection than telescope-pipes, which may be an advantage in many cases.

6.3.05. Gritty water soon wears out ordinary ejectors. The nozzles should therefore be made of very hard material where such water is to be lifted by them. Very acid water has the same effect, for which reason ejectors for corrosive liquids are made with hard lead linings, with porcelain nozzles, or entirely of porcelain.

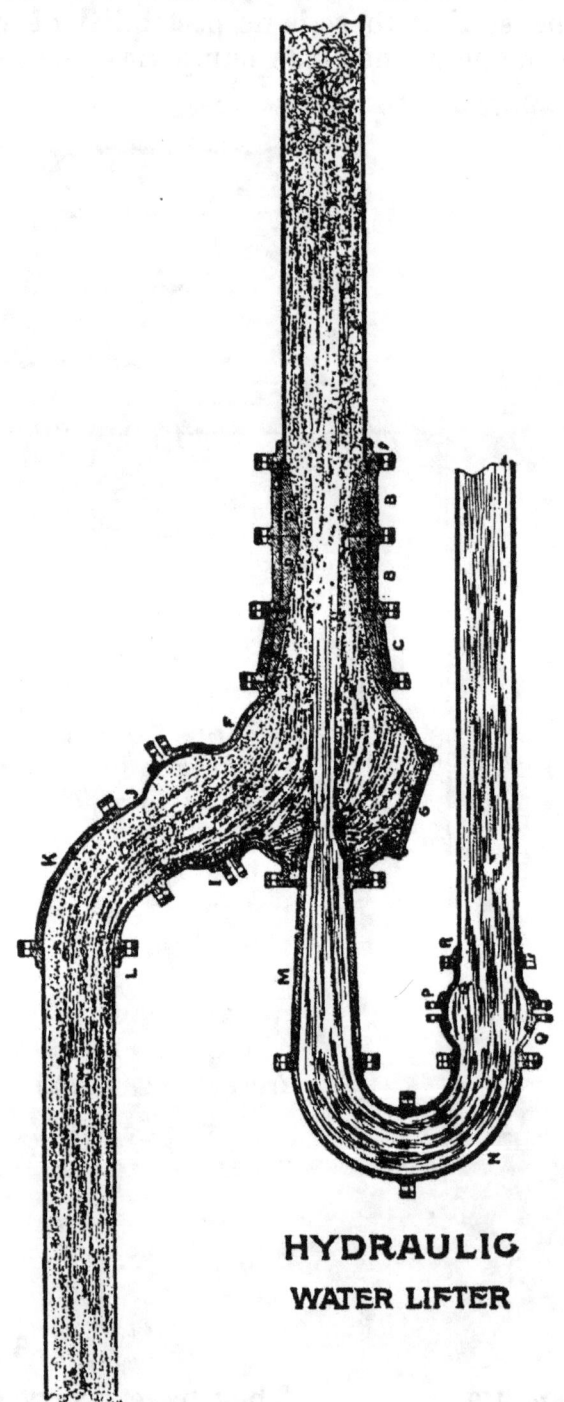

HYDRAULIC WATER LIFTER

Fig. 171.

CHAPTER IV.

Pulsometers.

6.4.01. The invention of C. H. Hall, the pulsometer, is one of the most useful pieces of apparatus for raising sandy or acid water to heights not exceeding 100′. where economy of fuel is less important than quickness of installation, and freedom from risk of breakdowns. When used for sinking operations in shafts, the steam- and water-pipes must be arranged the same as for other sinking apparatus with telescope connections, so that the machine can be lowered quickly or hoisted out of the way when blasting.

6.4.02. By reference to Figs. 172 and 173, it will be seen that to the tapering necks of chambers $A A$ there is attached, by means of a flange joint B, a continuous passage from each chamber, leading to one com-

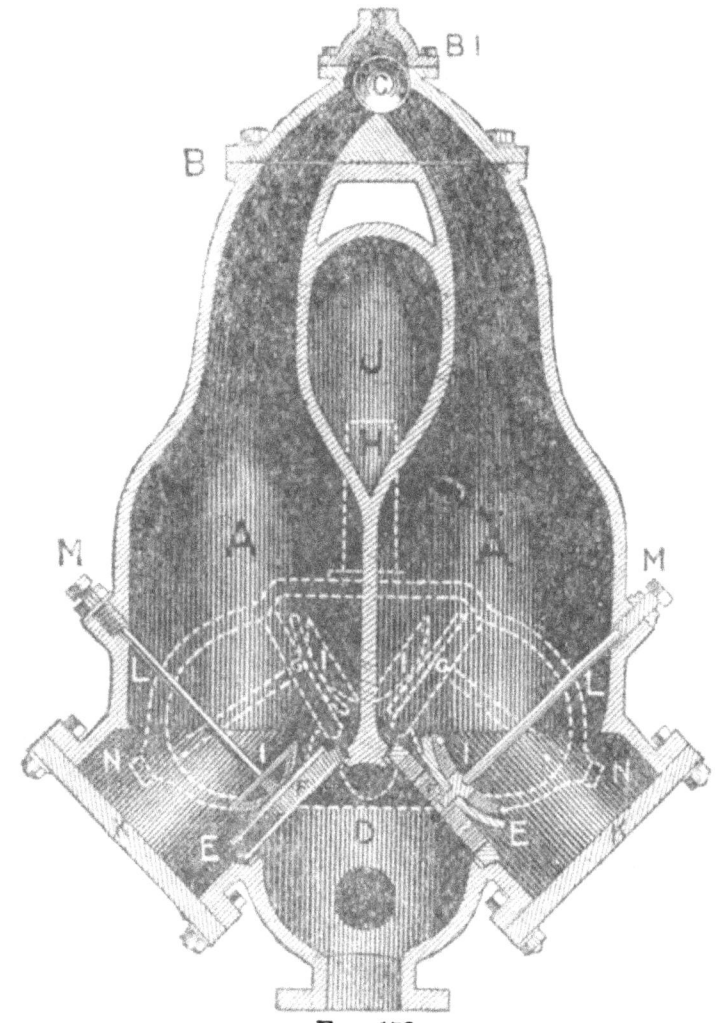

Fig. 172.

mon upright passage, into which a small ball C is fitted so as to oscillate with a slight rolling motion between seats formed in the junction. The chambers $A A$ also connect by means of openings with the vertical induction passage D, which openings are fitted with the valves $E E$ and their seats $F F$.

6.4.03. The delivery passage H communicates with each chamber through openings fitted with the valves and valve-seats $G G$, of the same style as in the induction passage. $I I$ are valve-guards. The vacuum-chamber J between the necks of chambers $A A$ connects only with the induction passage below the valves $E E$. $K K$ are doors covering the openings, affording access to the valves and seats when necessary. Vent plugs are inserted into these flanges for the purpose of drawing off the water to prevent freezing. $L L$ are struts by which the suction seats, valves, and guards are tightly pressed into place. $N N$

are bolts by which the discharge seats, valves, and guards are held in place. A small air check-valve is screwed into the neck of each chamber $A\,A$, and one into the vacuum-chamber J, so that their stems hang downward. The check-valve in the neck of each chamber $A\,A$ allows a small quantity of air to enter above the water, to prevent the steam from agitating it on its first entrance, and thus forms an air-piston, tending to

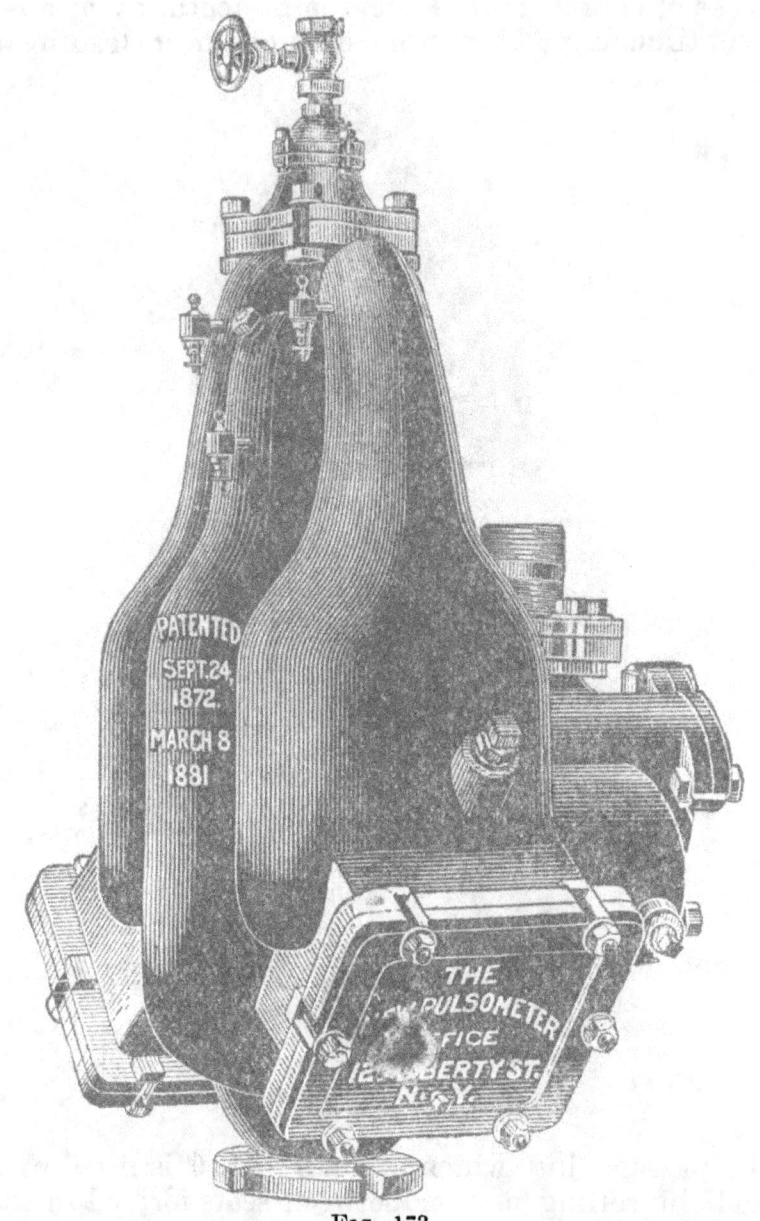

Fig. 173.

prevent condensation. The check-valve in the vacuum-chamber J also admits sufficient air to cushion the ramming action of the water consequent upon the alternate filling of each chamber.

6.4.04. The two working-chambers fill and discharge alternately, like in a steam pump. The steam enters at the top, or neck, and passes into whichever chamber is uncovered by the steam ball-valve, and pressing upon the surface of the water forces it down and out through the discharge-valves, and into the discharge-pipe. As soon as the water-line has been forced downward to the discharge outlet, the steam above it instantly condenses, a partial vacuum is formed, and the chamber in consequence suddenly fills again. Now, while the steam was entering this chamber, which we will designate as the "left-hand" one, the steam ball-valve was seated over the entrance to the "right-hand" chamber, preventing the entrance of steam thereto; but as soon as the sudden collapse of steam occurs, the valve is instantly drawn over to its other seat at the entrance to the "left-hand" chamber. This cuts off the admission of steam thereto, and allows it to enter the other chamber and expel the water therefrom in the same manner as described for the "left-hand" chamber. Steam and water occupy the same chamber successively, and will thus alternate, keeping up a continuous outflow as long as steam and water are supplied.

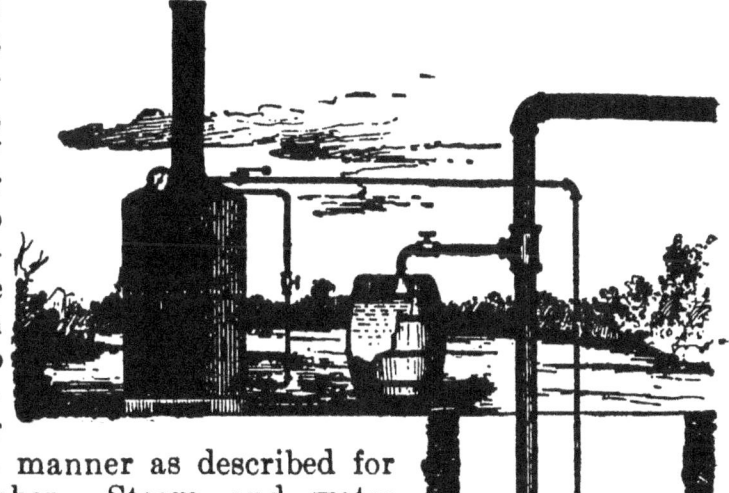

6.4.05. Priming is performed by pouring water

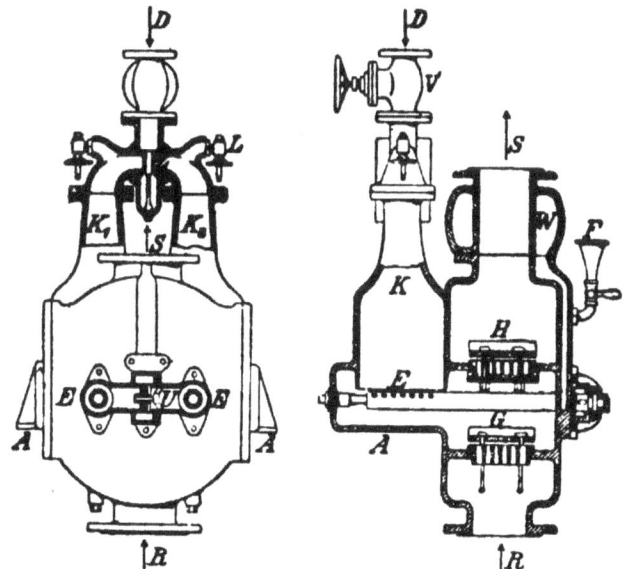

D Steam inlet.
V " valve.
Z Vibrating tongue
L Air-valve.
R Suction-pipe
S Column-pipe.
Π Discharge valve.

G Suction-valve
$K_1 K_2$ Working chambers
E E Injection-pipe.
U " regulator
W Air chamber
F Priming funnel
A Bracket feet

Fig. 175.

Fig. 174.

through the plugged opening in the middle chamber, or through the plugged opening on the discharge outlet side. Care should be taken to replace the plug quickly after priming. Fig. 174 shows a pulsometer in place to pump out a shaft or well.

166   MINE DRAINAGE, PUMPS, ETC.

6.4.06. While the pulsometer is capable of lifting water 100', the most general application is for lower lifts of 25' to 50'. The steam pressure naturally has to be increased with the increase of the forcing part of the lift.

6.4.07. Where the water has to be kept down close in order to enable the men to work in the bottom of the shaft, two pulsometers can be used, one of which will operate while the lengthening of the pipe of the other is proceeded with.

6.4.08. Pulsometers are made in sizes to meet any capacity, the tables of the chief makers running up to 2,000 gals. per minute. The steam consumption is high. Experiments made in Germany show consumption of 200 lbs. and over of steam per horse-power of water lifted per hour.

6.4.09. Pulsometers should be improved, if possible, by preventing condensation of the steam during its entrance into the working-chambers. It should, after filling a chamber, be condensed rapidly by some form of spray-injection like that used in the Korting pulsometer shown in Fig. 175. The steam should again be prevented from condensing during the forcing pulsation. By reducing these two condensation losses to a minimum, Korting claims to have obtained results of less than 100 lbs. of steam per water horse-power per hour, which is a better result than obtained with ordinary steam pumps.

CHAPTER V.

Air-Lift Pumps.

6.5.01. The operation of this apparatus depends upon the buoyancy of air introduced into the column-pipe in bodies alternating with liquid, the air forming virtually a piston, more or less complete, and pushing the water ahead of it. Fig. 176 illustrates the principle involved. The column-pipe is an open pipe, the lower part of which requires to be submerged for such a depth that the hydraulic pressure due to immersion will not quite equal the pressure of the compressed air entering the bottom of the column-pipe by means of the small air-pipe. The less the lift H compared with the submersion h, the greater is the efficiency obtained. This is also greatest when the air pressure exceeds but slightly the pressure due to hydraulic submersion of the air outlet.

6.5.02. The air-lift pump just described is said to have been invented in the last century at Freiberg, Saxony. One of the Siemens brothers made experiments with it more recently. Later still, in 1889, Mr. Ross E. Browne, in conjunction with the writer, made a series of experiments on this apparatus, which had been again invented and patented by Dr. J. G. Pohlé. As there has been considerable inquiry concerning this pump, the writer reproduces a paper prepared and read before the Technical Society of the Pacific Coast by Mr. Browne, giving an account of the experiments and the results obtained.

Dr. Pohlé's Air-Lift Pump.

By Ross E. Browne and H. C. Behr, Members Technical Society.

[Read February 14, 1890.]

During the month of August last, the writers, jointly with Mr. P. M. Randall, conducted a series of tests with Dr. J. G. Pohlé's air-lift pumping apparatus.

Figs. 176 and 177 will show the simplicity of the pump.

A good efficiency being found, and the apparatus having, for many purposes, very apparent advantages over the forms of pump in common use, it is thought that a record of the tests may be of interest.

The pump-column is an open pipe partly submerged in the water to be pumped. A small pipe leading from an air-receiver to the foot of and a short distance into the pump-column, delivers compressed air, which forms in piston-like layers, and rising rapidly in the column, does the work of pumping. The water is discharged in alternate layers with the air.

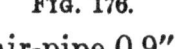

The apparatus tested was erected without due regard to best dimensions, and we deem it proper to state that the efficiencies found could have been increased by a few simple alterations. Pipes of different diameters were not provided, and we were able to change only the length of the pump-column, the amounts of submersion and lift, and the pressure in the receiver, hence the quantity of air supplied.

FIG. 176.

The diameter of the pump-column was 3", of the air-pipe 0.9", and of the air-discharge nozzle ⅝". The air-pipe had four sharp bends and a length of 35' plus the extent of the submersion.

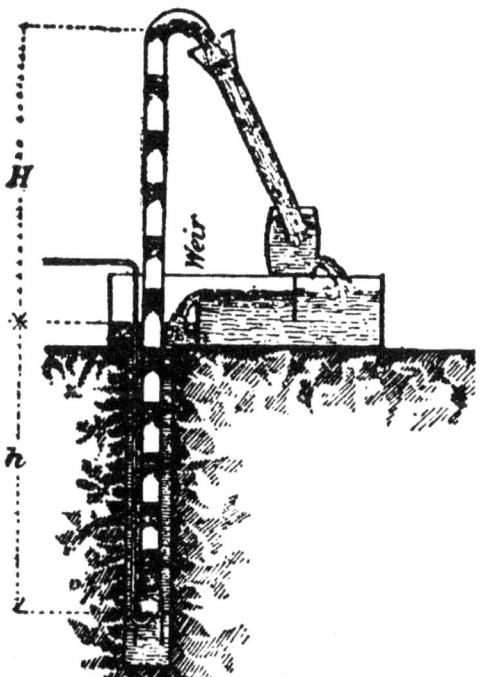

FIG. 177.

The water was pumped from a closed pipe well (55' deep and 10" in diameter), and was discharged into a tank and delivered—over a quadrantal weir—back to the well.

A long mercurial column was connected with the receiver for the purpose of obtaining accurate measurement of pressure.

The quantity of air delivered to the pump was obtained by two methods, as follows:

First Method.—The cubic contents of the receiver were measured. The escape-cocks from the receiver were closed and the compressor was started. Beginning with atmospheric pressure, the increase of pressure was noted for each thirty strokes of the compressor-piston, until a pressure was reached beyond that required in the pump tests. The contents of the receiver were 117 cu. ft.

The following are the results of two separate tests:

The compressor made uniformly one stroke per second. The atmospheric pressure was 2.51' of mercury. The air was unusually dry.

TABLE I.

No. of Strokes of Compressor-Piston.	Temperatures.				Pressures in Receiver Above Atmosphere. Feet of Mercury.	
	Receiver.		Atmosphere.			
	Test No. 1.	Test No. 2.	Test No. 1.	Test No. 2.	Test No. 1.	Test No. 2.
0	78° F.	80° F.	75° F.	77° F.	0	0.01
30	------	------	------	------	(0.76)?	0.94
60	------	------	------	------	1.72	1.77
90	------	------	------	------	2.48	2.56
120	------	------	------	------	3.24	3.31
150	------	------	------	------	3.95	4.08
180	------	------	------	------	4.67	4.81
210	------	------	------	------	5.34	5.54
240	------	------	------	------	6.00	6.29
270	86°	88°	75°	77°	6.66	7.00

These data formed the basis for calculating the number of pounds of air delivered per piston stroke of the compressor, to the receiver at any

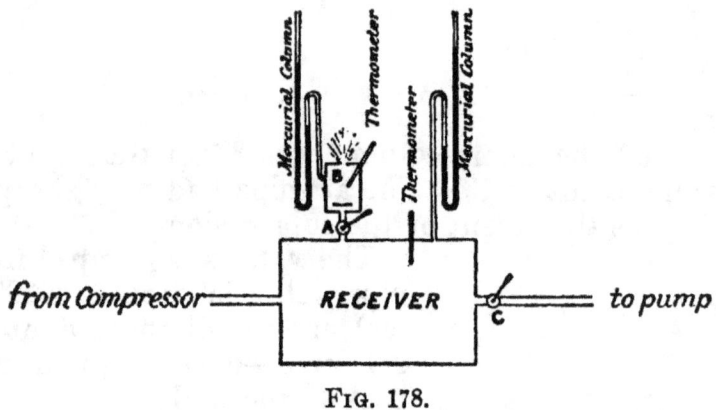

FIG. 178.

required pressure. An average of the results of the two tests was adopted. The following table gives the values obtained:

TABLE II.

Pressure in receiver above atmosphere. Lbs. per square inch.	0	5	10	15	20	25	30	35	40
Lbs. of air delivered per stroke of compressor	.104	.098	.093	.088	.083	.081	.079	.077	.076

Second Method.—A small auxiliary chamber B was attached to the receiver. (See Fig. 178.) Compressed air entering this chamber escaped into the atmosphere through a carefully measured circular orifice in thin plate. After a pump test had been completed, the compressor was kept running, cock C was closed, and cock A opened and adjusted until the conditions in the pump test, regarding number of strokes of compressor per minute and the pressure in the receiver, were repeated and maintained.

The pressures and temperatures of the compressed air in chamber B and of the atmosphere, furnished the data upon which to base a calculation of the quantity of air escaping through the circular orifice.

MINE DRAINAGE, PUMPS, ETC. 169

This quantity was evidently the same as that supplied in the pump test. Such tests were made from time to time, and served to check the values taken from Table II. A few of these are given below. Diameter of orifice was 0.391". Atmospheric pressure, 14.7 lbs. per square inch. Weisbach's and Zeuner's coefficients of efflux were used.

TABLE III.

No. of Pump Test.	No. of Strokes of Compressor per Minute.	Pressures above Atmosphere, lbs. per sq. in.		Temperature Fahr.			Lbs. of Air Delivered per Second.	
		Receiver.	Chamber B.	Receiver.	Chamber B.	Atmosphere.	Table II.	Orifice Test.
1	60	31.1	20.2	77°	77°	68°	.078	.075
5	60	30.6	20.3	74	73	73	.078	.075
10	60	24.1	21.7	78	75	74	.081	.077

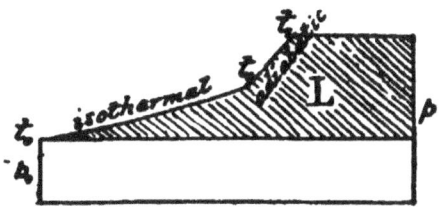

FIG. 179.

The engine used to drive the compressor was built for ten times the power actually applied to the compressor, hence a test of the efficiency of the entire plant was not made.

Table IV gives the results of the pump tests. The "efficiency of the pump" is based upon the least work L theoretically required to compress the air and deliver it to the receiver. (See Fig. 179.)

Atmospheric conditions = p_0, t_0.
Receiver conditions = p_1, t_1.

The values given in the table take no cognizance of the losses of power in the engine and compressor.

If we assume the efficiency of a suitable compressor to be 70%, the efficiency of the pump and compressor together would be 70% of that given in the table for the pump alone.

TABLE IV.

No. of Test.	No. of Strokes of Compressor per minute.	Pressure in Receiver above Atmosphere, lbs. per sq. in.	Temperatures Fahr.			Water Lift "H," in feet.	Submersion "h," in feet.	Pressure corresponding to "h," lbs. per sq. in.	Weight of Air Supplied, lbs. per sec.	Water Pumped, cub. ft. per sec.	Work of Isotherm Compression L, ft. lbs. per sec.	Work of Water-Lift W, ft. lbs. per sec.	Ratio $\frac{H}{h}$	Efficiency of Pump $\frac{W}{L}$ in per ct.
			Receiver	Atmosphere	Water									
1	60	31.1	77	68	68	75.2	53.0	23.0	.078	.1755	2408	824	1.4	34
2	60	30.8	77	72	68	75.4	52.8	22.9	.078	.1799	2445	846	"	34
3	45	27.6	78	71	68	75.3	52.9	22.9	.059	.1488	1716	700	"	41
4	31	25.4	77	74	68	75.3	52.9	22.9	.041	.0757	1156	356	"	31
5	60	30.6	75	72	67	35.1	53.2	23.1	.078	.3136	2459	687	0.6	28
6	46	26.8	78	74	67	35.2	53.1	23.0	.061	.3014	1770	662	"	37
7	30	24.9	78	76	67	35.0	53.3	23.1	.041	.2425	1150	530	"	46
8	22	24.0	78	72	67	35.0	53.3	23.1	.030	.1941	802	424	"	53
9	60	23.8	78	72	70	54.7	33.6	14.6	.081	.1538	2151	525	1.6	24
10	34	17.4	77	72	69	54.7	33.6	14.6	.049	.0824	1056	281	"	27
11	23	16.1	76	73	69	54.5	33.8	14.6	.033	.0576	681	196	"	29
12	60	18.8	76	71	69	69.9	18.4	10.0	.084	.0338	1904	147	3.8	8
13	33	11.9	76	75	69	69.6	18.7	10.0	.050	.0067	837	29	"	3
14	60	20.6	80	77	69	62.1	26.2	11.4	.083	.0931	2041	361	2.4	18
15	38	15.2	80	74	70	62.4	25.9	11.2	.056	.0663	1090	258	"	24
16	19	12.3	79	75	71	62.4	25.9	11.2	.029	.0185	489	72	"	15
17	60	18.9	79	74	67	31.5	20.1	8.7	.084	.1488	1922	292	1.6	15
18	34	12.3	78	72	68	31.5	20.1	8.7	.052	.1126	860	221	"	26
19	20	10.0	76	70	68	31.3	20.3	8.8	.031	.0633	432	124	"	29
20	60	20.3	69	68	67	26.3	25.3	11.0	.083	.2296	2013	377	1.0	19
21	41	15.8	70	66	67	26.3	25.3	11.0	.059	.2050	1178	336	"	29
22	22	12.5	70	67	67	26.3	25.3	11.0	.033	.1420	558	233	"	42
23	60	21.9	72	67	69	20.3	31.3	13.6	.082	.2954	2050	374	0.7	18
24	27	15.1	72	67	69	20.3	31.3	13.6	.040	.2398	769	304	"	39
25	22	14.4	72	67	69	20.3	31.3	13.6	.032	.2086	594	264	"	44
26	60	23.1	74	67	69	15.3	36.3	15.7	.082	.3540	2105	338	0.4	16
27	30	17.4	73	68	69	15.3	36.3	15.7	.043	.3182	918	304	"	33
28	19	16.2	73	69	69	15.3	36.3	15.7	.028	.2558	572	244	"	43
29	60	17.1	74	69	69	36.0	15.6	6.8	.086	.0693	1818	156	2.3	9
30	34	10.1	73	70	70	36.0	15.6	6.8	.052	.0424	749	95	"	13
31	18	7.4	73	70	70	36.0	15.6	6.8	.029	.0093	323	21	"	7
32	60	15.8	76	72	70	41.0	10.6	4.6	.087	.0146	1757	37	3.9	2
33	22	7.1	74	72	70	41.0	10.6	4.6	.035	0	382	0	"	0

An inspection of the above table shows:

First—That, for a given submersion h and lift H, the best efficiency was obtained when the pressure in the receiver did not greatly exceed the pressure due to the submersion.*

Second—That the smaller the ratio $\frac{H}{h}$, the better was the efficiency.

---

*NOTE.—This was only true when the ratio $\frac{H}{h}$ was kept within reasonable limits—i. e., where H was not much greater than h.

We may say in a general way that under the better adapted pressures in the receiver, the pump, as erected, showed the following efficiencies:

For $\frac{H}{h}=0.5$ 50%
" " 1.0 40
" " 1.5 30
" " 2.0 25

It is apparent that the air-pipe should not have been reduced at the discharge end, as such reduction necessitated a greater pressure in the receiver for the delivery of the air to the pump.

Unfortunately, the data is wanting for a reliable estimate of the loss due to the frictional resistance in the small air-pipe. A rough estimate shows that such loss must have been large. The substitution of a $1\frac{1}{2}''$ air-pipe in place of the $1''$ would have appreciably augmented the efficiencies given in the table. In justice to the pump, a considerable allowance should be made for this easily avoidable loss.

The last test (No. 33) shows a limit of lift for a given submersion, beyond which a large excess of pressure is required to pump even an insignificant quantity of water.

For good efficiency, it becomes necessary that the lift should not be very great as compared with the submersion.

Where a shallow sump only is available to pump from, and a considerable lift is to be made, Dr. Pohlé introduces an auxiliary pipe to receive the water, after being pumped to a small height, and act as pump-well for a higher lift. (See Fig. 180.)

We have not attempted an analytic treatment of the action of this pump. Such treatment would have little value without coefficients, derived from a more comprehensive set of tests.

The simplicity of this pump commends it for many uses.

Among the numerous applications which Dr. Pohlé proposes for this air-lift may be mentioned: the drainage of mines; the supply of water from deep wells; the lifting of liquids which damage the working parts of the pumps ordinarily used; the increase of the lift and capacity of other pumps by introducing an air-jet into the pump-column.

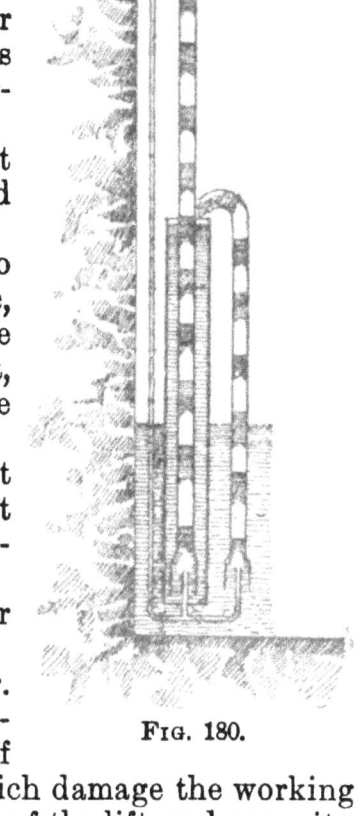

Fig. 180.

6.5.03. Although the air-lift pump can, under certain conditions, be made a comparatively efficient machine, particularly if the compressing-plant is such, its application to mine drainage is generally inconvenient, and its use for such purposes will certainly never be extensive. It can, however, be made a useful auxiliary to a sinking-pump, where a flooded mine, affording a chance for the necessary submersion, is to be pumped out, and where air-compressing machinery is at hand. Special provisions of air-compressing plant for operating the air-lift pump would hardly pay. The capacity of air-lift pumps cannot be varied in very wide limits, without greatly reducing the efficiency.

6.5.04. Compressed air is also sometimes introduced into the column-pipe of a pump, in order to increase the lift beyond that otherwise admissible on the pump. This is probably the most useful application of the air-lift pump in mine drainage. To start such an apparatus the air pressure must be sufficient in the beginning to balance the column full of water; it can then be reduced to that necessary to support the mixed column of water and air when the flow has been started. A better plan is, though, to drain the column-pipe to such a level that the air pressure will overbalance the hydrostatic head and start the flow. The application of air-lift pump just described is very conveniently made, as it requires no special submersion column.

6.5.05. Air-lift pumps must be placed in vertical shafts. No data on the working of inclined air-lift pumps are known to the writer. It may be doubted if they would work at all if the inclination were made appreciable, as the air bubbles would hug the high side of the pipe and afford more chance for back-flow of water on the lower side.

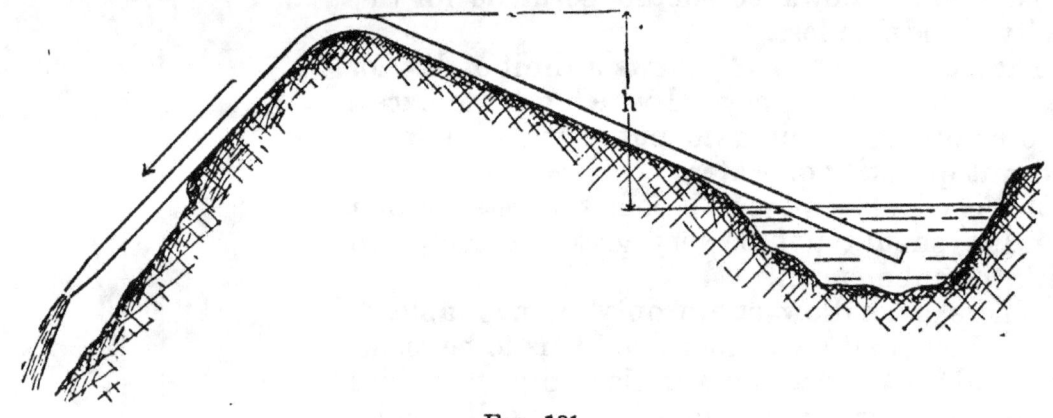

Fig. 181.

CHAPTER VI.

Siphons.

6.6.01. Though not a water-*raising* appliance in the proper sense of the term, since the water by this apparatus can only be lifted or transported over an eminence limited in height and discharged on the other side of it at a level lower than that of the supply, siphons can find application in many instances for forming a water communication over an elevation between two distant points; also, for draining levels or open workings, thereby doing away with the necessity of installing a pump.

6.6.02. In construction, the siphon is a very simple piece of apparatus, but the conditions that govern its working are many, and it is therefore proper to consider the principles involved, and the means employed in securing or aiding its proper action.

6.6.03. A siphon consists essentially of a pipe curved downward at each end, as shown in Fig. 181, with one end dipping into the supply-reservoir, and the other discharging at a level lower than that of the supply.

6.6.04. The height h, over which a siphon may automatically transport a liquid, depends, firstly, upon the specific gravity of the liquid to be lifted. Thus, for example, a heavy liquid like quicksilver cannot be raised by a siphon over the same height as water, and water containing heavy substances in solution or suspension cannot be raised over the same height as pure water. Secondly, it depends upon the barometric pressure, the possible lift being therefore less at high altitudes than at sea-level, and varying also with the state of the weather. In designing a siphon which is to operate at all times, it is therefore necessary to base the calculations upon the lowest observed barometric pressure. The working of a siphon depends also very materially upon the temperature of the liquid. If this temperature be higher than that of the boiling-point which the liquid has at the low absolute pressure existing at the highest part of the siphon, the liquid will begin to boil at that point and give off vapor, which will fill the siphon and cause it to stop flowing. It is well, therefore, to shed-over the rising and high parts of long siphons, so that the heat of the sun will not raise the temperature of the liquid. A similar effect will be produced by air or other gases held in solution. Such gases are liberated when the pressure is reduced, and cause stoppage of flow in the same way as the vapors.

6.6.05. These conditions limit, as with all water-suction apparatus, the height of the column of the liquid which can be supported by the overpressure of the atmosphere. This height will be further reduced by an amount necessary to cause the required velocity of flow and to overcome the frictional resistance.

6.6.06. When a charged siphon is not in operation, the air accumulates at the highest point; but when there is a flow, the latter presses the bubble of air ahead so that it occupies a position ahead of the highest point, the position depending upon the energy of the current and the grade of the descending leg of the siphon.

6.6.07. The air and gases held in water are liberated even at a moderate reduction of pressure, and, as nearly all water is charged more or less with gases, these will be liberated in any siphon, and will cause thereby a gradual increase of pressure, a little beyond the highest point, which thereby gradually reduces the available flow-producing-head, so that the flow becomes less and less, and when the pressure equals the acceleration-head, ceases altogether. Long siphons will run less time in this manner than short ones, because, in the former, there is a greater volume of water containing air in the pipe, and more time and surface afforded for the liberation of gases in the longer passage of the water.

6.6.08. When the siphon is not too long, and when the acceleration-head is sufficient to give the water a considerable velocity, the air and gases may be entrained by the rapid current and carried out at the end of the discharge branch, if the latter is not too steep. In most cases, however, it is necessary to provide artificial means to remove the accumulated gases, either periodically or continuously. The means employed for this purpose is usually a hand air-pump connected with its suction to the highest point of the siphon. In order to collect the gases at one point, the siphon should have the shape shown by Fig. 182, where the descending branch falls more abruptly, in order to prevent entraining any of the air accumulated in the chamber a. The pump b may also serve to prime the siphon, if it has sufficient capacity. There should be no level pipe in the siphon, but it should ascend all the way toward the accumulator-chamber, which latter should present a large opening to the pipe, so that the air will readily find its way into it and not rush past into the descending leg, where it might be retained by the force of the current. Where there is fall available for the water discharged from the siphon, it may be utilized to run a small waterwheel for driving, by means of suitable transmission, the air-pump at the highest point of the siphon, so as to continuously remove the air.

6.6.09. A siphon, arranged as in Fig. 181, should either have a reduced discharge opening or a regulating valve at the lower end, or

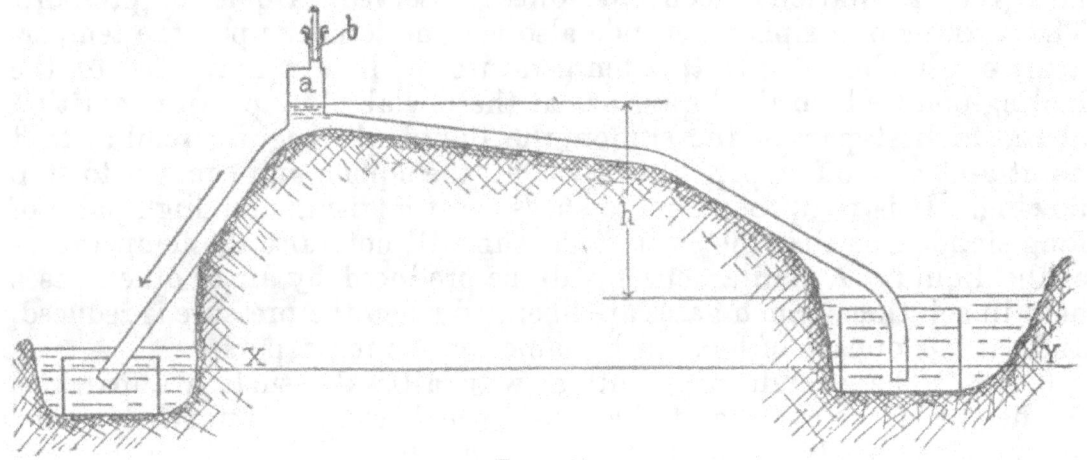

Fig. 182.

the descending leg should be smaller in size than the ascending one. If such precautions are not taken, the water may run out of the descending leg faster than it can flow into the ascending one, with the result that air will enter by way of the descending leg and stop the operation of the siphon. This result may be avoided in any siphon by always having both ends submerged, or by turning them upwards, as in Fig. 183. If a level line $x\,y$ intersects the two upturned branches, the water

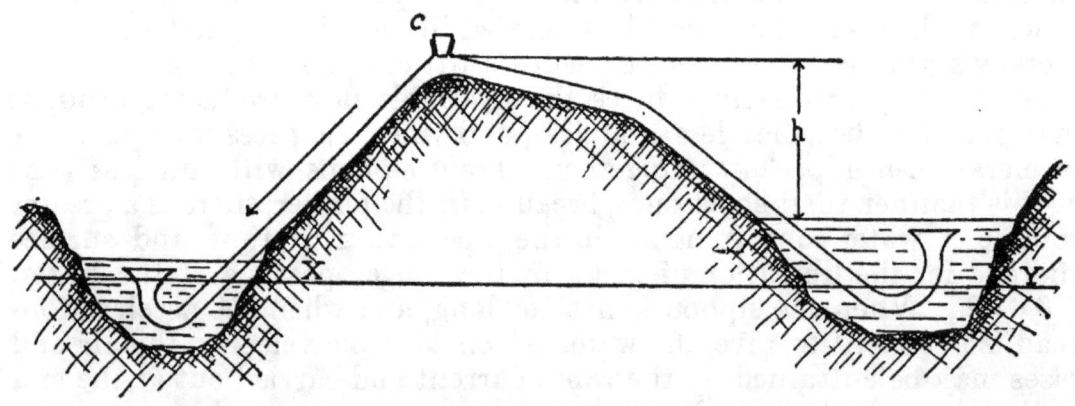

Fig. 183.

will not run out of the siphon, and air cannot enter it, when, from any cause, the supply level sinks below the top of the upturned entrance branch. This arrangement also secures the self-starting of the siphon thus stopped, when the supply level again rises above the edge of the branch. The submerged ends are advantageously made flaring or bell-mouthed, so that the water will be gradually accelerated as it enters the siphon, and will leave it with an easy flow. In this manner a few inches of lift or a somewhat increased flow may be gained.

6.6.10. There should be valves at each end of the siphon, which can be closed, when it becomes necessary to prime it, by filling it with water either through the plug-hole c, Fig. 183, at the highest point, or by means of the pump, Fig. 182.

6.6.11. It is important that siphons, particularly long ones, should be absolutely tight, so that no air can enter them; otherwise, this also will have to be removed with that liberated from the water.

6.6.12. Advantage should be taken of the action of siphons wherever possible, not merely by themselves, but generally more frequently to aid pumping-plants, the pipes from which, in order to reach the point of discharge, have to pass over intervening elevations. The lower the total lift in such a case the greater will be the proportional gain by properly arranging the siphon part of the plant. Occasionally, also, the reverse proceeding may be advisable, and a large siphon may be aided by installing low-lift pumping machinery.

CHAPTER VII.

Water-Raising Appliances of Small Capacity Operated by Men or Animals.

6.7.01. These are more frequently for temporary uses, as in prospect work, or draining shallow holes in drifts or in river channels. Much of this work is done by men and horses or mules, because its short and uncertain duration does not warrant the outlay for mechanical power apparatus.

6.7.02. The power of men can be applied in various ways to pumping. It is generally by hand-power that the small water-raising machines used in our mines are operated. Hand-power may be exerted in a reciprocating manner, or rotatively by means of a crank.

6.7.03. Often the hand pumps are constructed at the mine to suit the conditions required. Fig. 184 illustrates a hand pump of this kind. The barrel is an ordinary piece of gas-pipe or tubing. The foot-valve a is simply a piece of leather, or sheet rubber cut out of the sheet, as shown by the figure, and clamped between two washers, so that the part b serves as a hinge. The seat c is made of a circular piece of sheet-iron with a central hole somewhat smaller than the valve. The valve, with its seat and a lower gasket, is clamped between the flanges for connecting the suction-pipe to the pump-barrel. An ordinary vertical check-valve may also be used instead of this construction. The bucket d is made of a piece of leather rolled together in a conical form, the smaller end being nailed to the

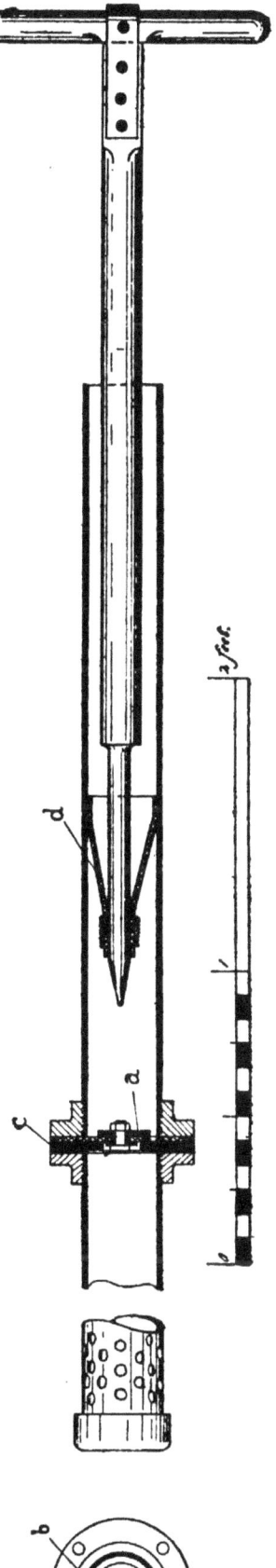

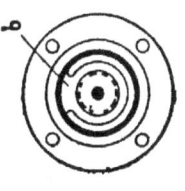

FIG. 184.

wooden pumprod, and further secured by marline or copper wire. The edges of the leather, where they overlap, are beveled so that they can slip past each other and allow the cone to collapse on the downstroke, so that it can pass down through the water. On the working-stroke the cone again spreads out, and the upper edge is pressed against the side of the pump-barrel by the water, thus serving as valve and piston-packing at the same time. By making the cross-section of the rod equal to half the area of the pipe, the pump will be double-acting, and will discharge at each stroke half the amount of water which it draws in during the suction-stroke. Sometimes pumps of this kind are made with wooden barrels of square cross-section. The pump, Fig. 185, is only suitable for suction lift, but can handle a large amount of water.

Fig. 185.

6.7.04. The work of men is most advantageously utilized in reciprocating motion of the hands; that is, men can do more and work longer than in any other manner, if they perform the work with a horizontal, rowing motion, at which they are seated, and can brace their feet.

6.7.05. In raising water, the work of men, when transmitted by a crank, can be applied either to bailing by means of a winch, or to operating pumps by secondary or driven cranks. Where there is one double-acting pump, or two single-acting pumps with opposite cranks, the crank-angle should be such with reference to the hand-crank that the greatest resistance will occur at such a point when the hand-crank is in the most favorable position to utilize the effort of the operator with the least fatigue to himself. Where there are a number of pumps with the cranks so disposed relatively that the resistance at the hand-crank will be almost uniform in the direction of rotation, a flywheel of sufficient weight should be mounted on the hand-crank shaft, in order to distribute the less fatiguing variable effort of the operation.

6.7.06. The crank can also be employed to operate a Chinese pump or water-elevator. The Chinese pump, Fig. 186, may be constructed in various ways. One of the most usual forms consists essentially of an endless canvas or rubber belt passing over two pulleys, one close to the point of discharge, and the other submerged in the water to be raised. On the outside of the belt are fastened a series of blocks about 18" to 24" apart. The upper pulley is rotated by means of a hand-crank, or by a belt on a pulley, if by animal or mechanical power. The ascending side of the belt is encased in a rectangular wooden pipe, into which the blocks on the belt fit as closely as possible without risk of jamming fast. The blocks in ascending carry up the water between them, minus the leakage, and push it out at the top of the wooden pipe. Similar pumps are also made with chain belts instead of canvas or rubber, as in Fig. 187.

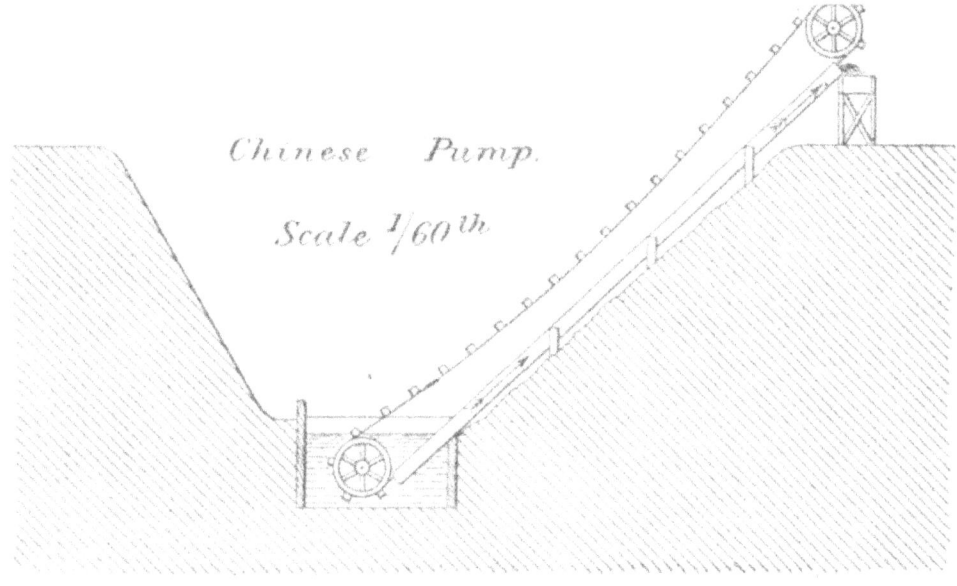

FIG. 186.

6.7.07. The work of horses, mules, and cattle in raising water is, like in most other employment of such animals, nearly always utilized in the form of traction. Occasionally they are found operating machines of the treadmill character, in which the animal raises its own weight, as when it is walking up hill, except that the "hill" slips down as much as the animal raises itself, so that the latter remains in a fixed position, only moving its legs in a climbing motion and pushing back the surface beneath it. Such apparatus, however, requires special training of the animals, and traction animals are therefore rarely used in that way for water-raising. Working animals trained for traction purposes can always be readily obtained. For this reason, it is best to use such power machines for which the training of the animals already fits them.

6.7.08. Animals may exert tractive force either in a straight or in a circular path. In the former they work more efficiently than in the latter, because of the constant change in direction of effort; but in the former they generally require an attendant to direct the reversal of their motion at the ends of the path.

6.7.09. The work in the straight path can only be used in bailing. This application requires no apparatus except a sheave and rope, but is attended with some inconvenience, as the rope and bucket have to be lowered by the attendant. The work of animals is, therefore, most usually employed by causing them to exert their tractive force in a circular path by means of horse-powers or horse-winches.

6.7.10. The horse-winch, as its name implies, is a hoisting machine, and is frequently used for bailing small quantities of water. The animal must reverse its direction of travel when the bucket reaches the

Fig. 187.

top, and again at the bottom. For this reason, an attendant is usually required to direct the operations of the animal. Fig. 188 illustrates a common form of horse-winch.

MINE DRAINAGE, PUMPS, ETC. 179

Fig. 188.

180   MINE DRAINAGE, PUMPS, ETC.

6.7.11. In the horse-power the animal maintains the same direction of travel. When applied to bailing it is arranged with a geared hoisting-drum fitted with a brake and clutch, or device for disengaging the gears. When the bucket reaches the top, the brake is applied and the animal stops. The clutch or gear is disengaged after emptying the bucket, and the latter is then lowered by means of the brake. When filled, the clutch or gear is again thrown in, and the animal started up. The horse-power is, however, better adapted to operate pumps by means of cranks. A usual form of horse-power adapted to this purpose is shown in Fig. 189. The driving arm or radius pole, at the end of which the animal exerts its pull, should not be less than 16′ long, so as to modify as much as possible the curvature of the path. As the speed of the animal is limited in doing work, the number of revolutions per minute made by the arm is very few. In order to secure a more advantageous speed, the horse-powers, like the one in Fig. 189, are geared, by means of suitable toothed wheels, to a horizontal shaft, which makes a higher number of revolutions in accordance with the proportion of the gearing A flywheel is generally mounted on the end of the shaft to distribute the resistance and cause it to be more uniform at the point where the animals apply their work. A crankpin in the side of the flywheel drives the pump by means of a connecting-rod, sometimes coupled to an intermediate working-beam. Pumps which are fitted with cranks can be operated by means of a pulley and belt from the flywheel shaft of the horse-power.

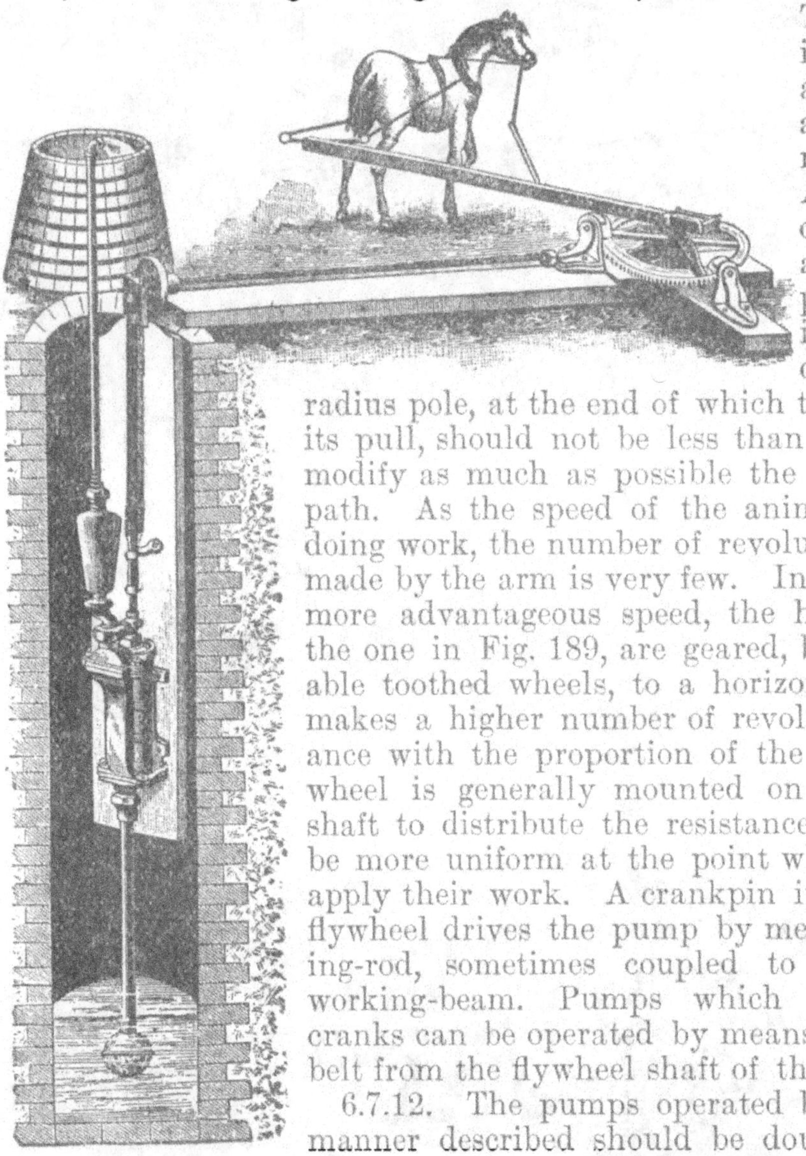

Fig. 189.

6.7.12. The pumps operated by animals in the manner described should be double-acting, or, if single-acting, two pumps worked from opposite ends of a working-beam should be used, or, instead of one of them, a balance weight. The difficulties of efficiently operating centrifugal pumps by means of horses were pointed out in 6.2.13. Where mechanical efficiency is not required, however, they may, on account of the uniform resistance and their simplicity, find application for raising water to moderate heights by means of animals. Chinese pumps and water-elevators can also be readily operated by horse-powers, and cause a uniform resistance.

6.7.13. Where the work of men or animals is required for water-lifting in mines, it is generally needed at once, so that there is no time afforded for the design and construction of special machinery. The plant should, for this reason, as well as on account of cheapness, be composed as much as possible of apparatus which can be obtained ready from a stock in the market.

6.7.14. The power of men or animals depends upon individual qualities of strength, weight, and endurance, as well as upon the time occupied in work. It also varies with the manner of application of the power, the existing temperature, and the amount and quality of food. The power or rate of work of an individual is greater with frequent intervals of rest, and increases also with the period of rest. The average daily capacity of a man may be taken at about one twelfth of a horse-power, while exertions of the short duration of a few seconds have been noted where the power exerted for the time being exceeded one horse-power. The average power of a horse or mule is about one half of a mechanical horse-power. The average power of cattle is less, and that of donkeys much less. As remarked before, however, these data are subject to considerable variation, due to difference in individual qualities and conditions.

SECTION VII.

CONCLUDING REMARKS ON MINE-DRAINAGE PLANTS.

CHAPTER I.

7.1.01. In providing for the drainage of a mine there are, after fixing upon the capacity, two things chiefly to be borne in mind. The first is the commercial efficiency of the installation, considered with due reference to the mining risk and the length of time that the plant will probably be in use. The second is the degree of safety against drowning out which the plant affords. Drainage tunnels can sometimes be used, where the conditions are favorable, to partially relieve an existing pumping-plant which has to handle a large quantity of water, by reducing the pumping height, and thereby either saving expense of operation, or enabling an existing pumping-plant to handle a larger quantity of water.

7.1.02. The capacity of a pumping-plant should be liberally measured, as upon it depends the welfare of the whole mine.

7.1.03. The best safeguards against the flooding of a mine are either large and rapid bailing-capacity, or an ample pump-compartment and a pumping-plant admitting of rapidly introducing and attaching to the piping movable reserve pumps kept in readiness at the surface for such emergencies.

7.1.04. Where sinking is abandoned for a long period, it is a good plan, where the conditions admit, to increase the capacity of the sump by running drifts, which, in case of short stoppages of the pumping-plant, retard the rise of water in the shaft, by the amount of time required to fill them.

7.1.05. In determining upon the general type of plant, the kind of power available or already at hand may be of importance. Where steam is the power adopted, and several kinds of fuel are available, the boiler-plant should be arranged with a view of using either of the fuels, as thereby the competition of the different dealers could be taken advantage of to secure fuel at more reasonable rates than otherwise. The price of fuel is generally higher the greater its evaporative effect; transportation, however, is generally the same per ton over the same route, so that the more high-priced may be the cheapest to get. But as prices vary the conditions may change, and it is, therefore, well to be prepared to take advantage of the conditions of the market. Boiler-plants should be of ample capacity, so that one or more of the boilers can be laid off for cleaning during regular operation, while all the boilers can be used when an extra flow of water is struck.

7.1.06. In case water-power is to be considered, its safety against stoppage from breaks in ditches, flumes, or pipe-lines is of vital importance. The possible occurrence of such accidents may necessitate a relay of steam-power to be kept in readiness, so as to prevent stoppage of pumps or bailing appliances.

7.1.07. Where electric transmission is available for operating pumps, it is still to be considered that the burning out of an armature would hang up the pumps connected therewith and expose the mine to the danger of flooding. Spare armatures, with shaft and all attachments complete, should be on hand for immediate replacement of the one burnt out. The motors should be of such construction as to admit of rapidly making repairs or changes.

7.1.08. In starting a shaft it is generally advisable to provide an ample pump-compartment, to afford space for pumps as well as for lowering these or parts of them. It is generally not possible to determine beforehand which is commercially the more advantageous: to permit all the water from different levels of the mine to collect at the sump, and bring it to the surface in as few lifts as the kind of pumps used admit of, or to collect the water at the different levels where it issues, and from there deliver it to the surface.

7.1.09. A preliminary plant is usually required before the plan for the final installation is decided upon. The appliances for preliminary use should be of a type and size most readily and quickly obtainable in the market.

7.1.10. Where the quantity of water that may be encountered is beyond conjecture, as is often the case, the Cornish system is not advisable, as it does not generally lend itself to considerable increase of capacity without discarding the entire machinery. It is also inefficient where variable quantities of water are taken in at different levels, because the pumps have to be adjusted to their proper relative capacity by permitting back-flow of water already pumped. Where a Cornish plant has been installed at great cost for large capacity, its degree of efficiency, commercially considered, will decrease considerably when the quantity of water that it is called upon to handle decreases, much more so than that of direct-driven pumping-plants.

7.1.11. In by far the greatest number of cases the use of direct-driven pumps will be the most advisable; it is impracticable, however, to consider, in a general treatise, the conditions that might influence the choice of the most suitable plant. Each special case generally develops so many characteristic features, that only a careful study of the conditions, by experienced and trained engineers, can lead to satisfactory results.

7.1.12. The statements of efficiency of pumps, engines, compressors, etc., contained in the many trade catalogues floating around through the mining settlements, must be taken very cautiously. The same applies to many of the so-called practical tests of pumping-plants. Such data should only be accepted when a full account of the test and a complete description of the methods and appliances used in observation, together with detailed data of observations, is given by parties who are known to be competent and disinterested experimenters.

7.1.13. Generally, plans and complete specifications having reference to the quality of the work are required, in order to obtain good workmanship under the conditions of keen competition so prevalent now.

SECTION VIII.

APPENDIX.

CHAPTER I.

Water-Raising Machinery for Irrigation or Land Drainage.

8.1.01. *General Remarks.* As stated in the Preface, it has been considered that this Bulletin would be incomplete without some reference to water-raising machinery for other than mine-drainage purposes. There are useful machines for raising water which can find no application in draining mines, but which may be of interest because they can serve, under certain conditions, to furnish a supply of water, when required, for other useful purposes, in mining as well as for irrigation or land drainage.

8.1.02. A feature which usually attaches to such water-raising machinery is that, except perhaps quite often in land drainage, it is not required to operate at capacities varying widely from those at which it is designed to give the best mechanical efficiency. Generally, also, the conditions admit of varying the capacity by varying the time of operation of the machinery.

8.1.03. The two sources of water supply to be considered for the purposes of irrigation or mining are natural or artificial watercourses and bored wells. The chief sources for mining supply are watercourses. In the mountainous regions wells are generally of smaller capacity than in the great valleys, and, therefore, are only rarely utilized.

8.1.04. Frequently it is not possible to bring the water by means of canals, ditches, flumes, or inverted siphons to the places where it is required, and then pumping must be resorted to, and the water conveyed under pressure in pipes to its destination. Sometimes, also, the first cost, with interest and maintenance expenses of an artificial watercourse, exceeds the corresponding items plus the operating expenses of a pumping-plant. For land drainage, canals and ditches are often impracticable, and the water has to be raised by machinery.

8.1.05. The sources of power for operating such water-raising machinery may be water-power, steam and gas engines, wind, or animals. Windmills have a wide application for the familiar small irrigation plants. Horses, mules, and cattle are used only to a limited extent, while gas engines have recently been applied quite extensively for small operations. For larger plants only water- or steam-power can be considered. These may be applied to drive the water-lifting machinery, either directly, or, as in the case of reciprocating or centrifugal pumps, by means of transmission, such as wire ropes, compressed air, or electricity. It is often most advantageous to subdivide the transmission, so as to operate several smaller favorably located pumping units from one large central power-plant.

8.1.06. The kind of water-raising machinery employed may consist of reciprocating pumps, including deep-well and direct-driven pumps, centrifugal pumps, water-elevators, Chinese pumps, air-lift pumps, bucket-wheels, paddle-wheels, or rams. The bulk of these, constituting

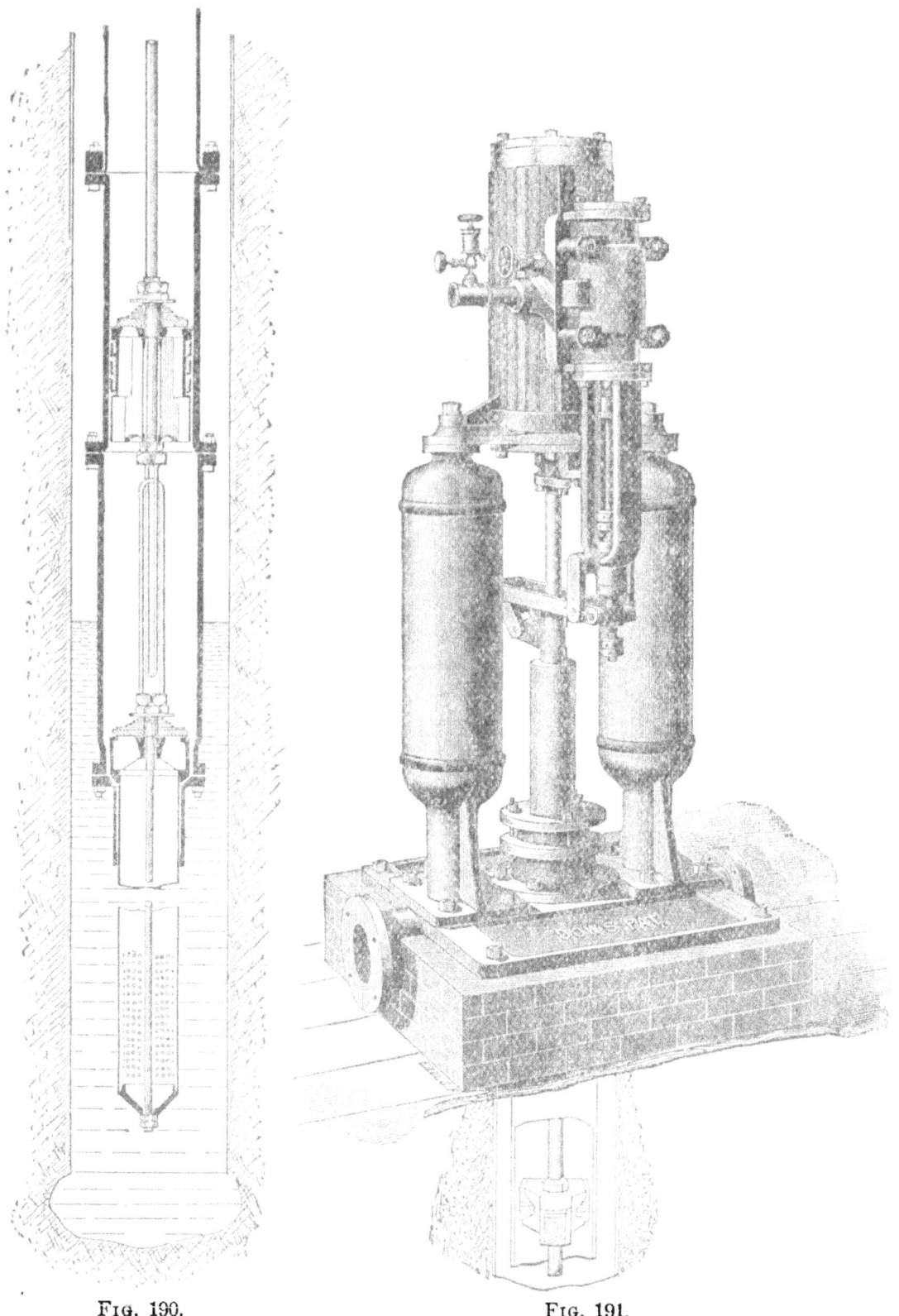

Fig. 190.   Fig. 191.

those which find application in mine drainage, have been described in the body of this Bulletin. It remains yet to describe more in detail those machines not treated before, viz.: certain kinds of reciprocating pumps, bucket-wheels, paddle-wheels, and rams, together with such power appliances as may be particularly suited to operate them, and also to speak of the machines already described in connection with their application to irrigation and drainage, and of the methods and means for driving them.

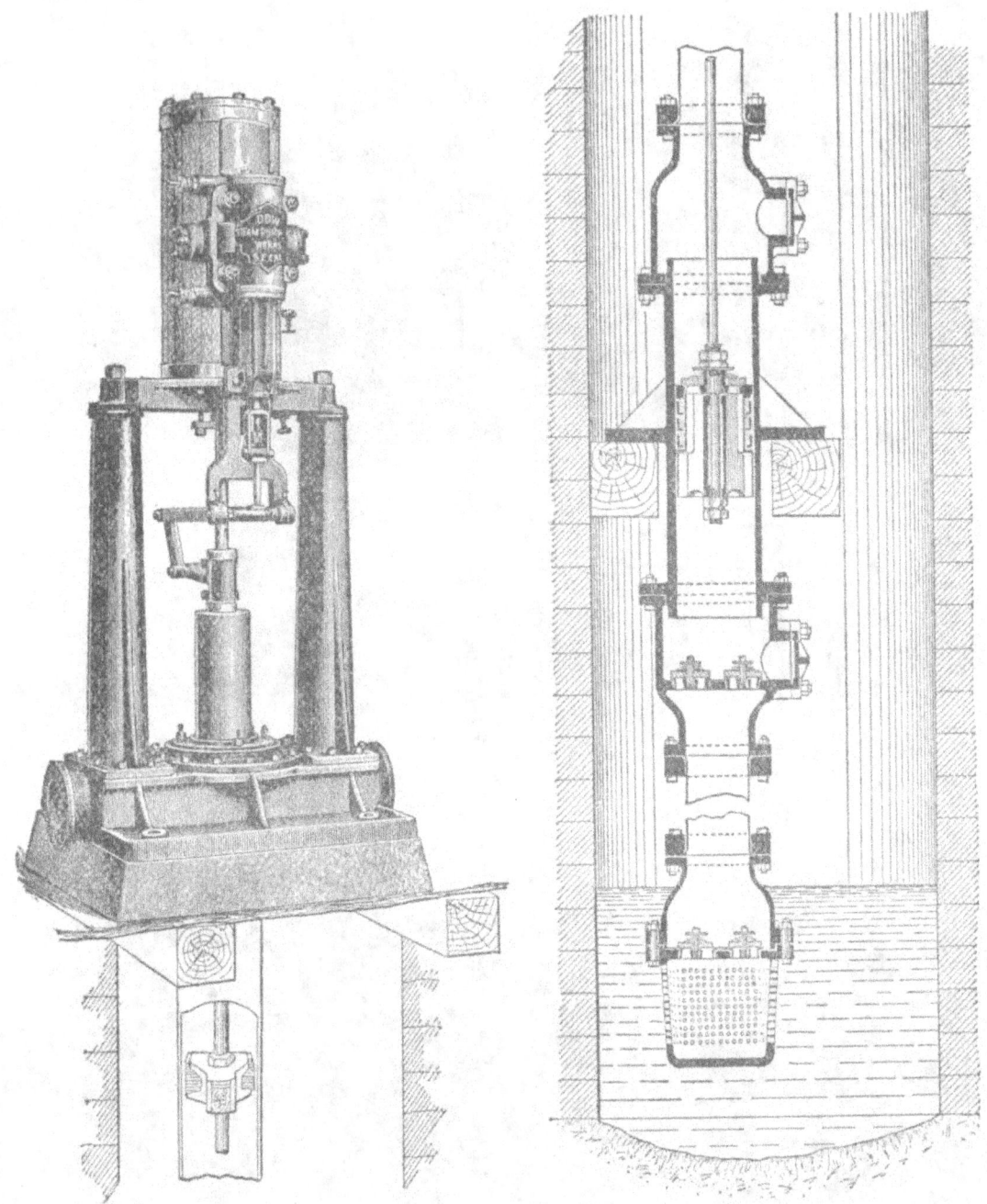

Fig. 192.

8.1.07. *Reciprocating Pumps.* These may be plunger or bucket-lift pumps operated through pumprods, either for pumping against higher heads from tube- or shaft-wells, or for low heads in draining land, as used in Holland; or, they may be direct-driven pumps, similar to the types used in mine drainage, but permitted to be made lighter and with proportionally smaller steam cylinders to suit the generally lower pumping-head.

8.1.08. Rod-actuated well-pumps find a wide application on this coast. They exist in various forms, and of all capacities, from the small windmill pump to the large deep-well pump driven by a compound engine, and similar in many respects to a mine-drainage pump.

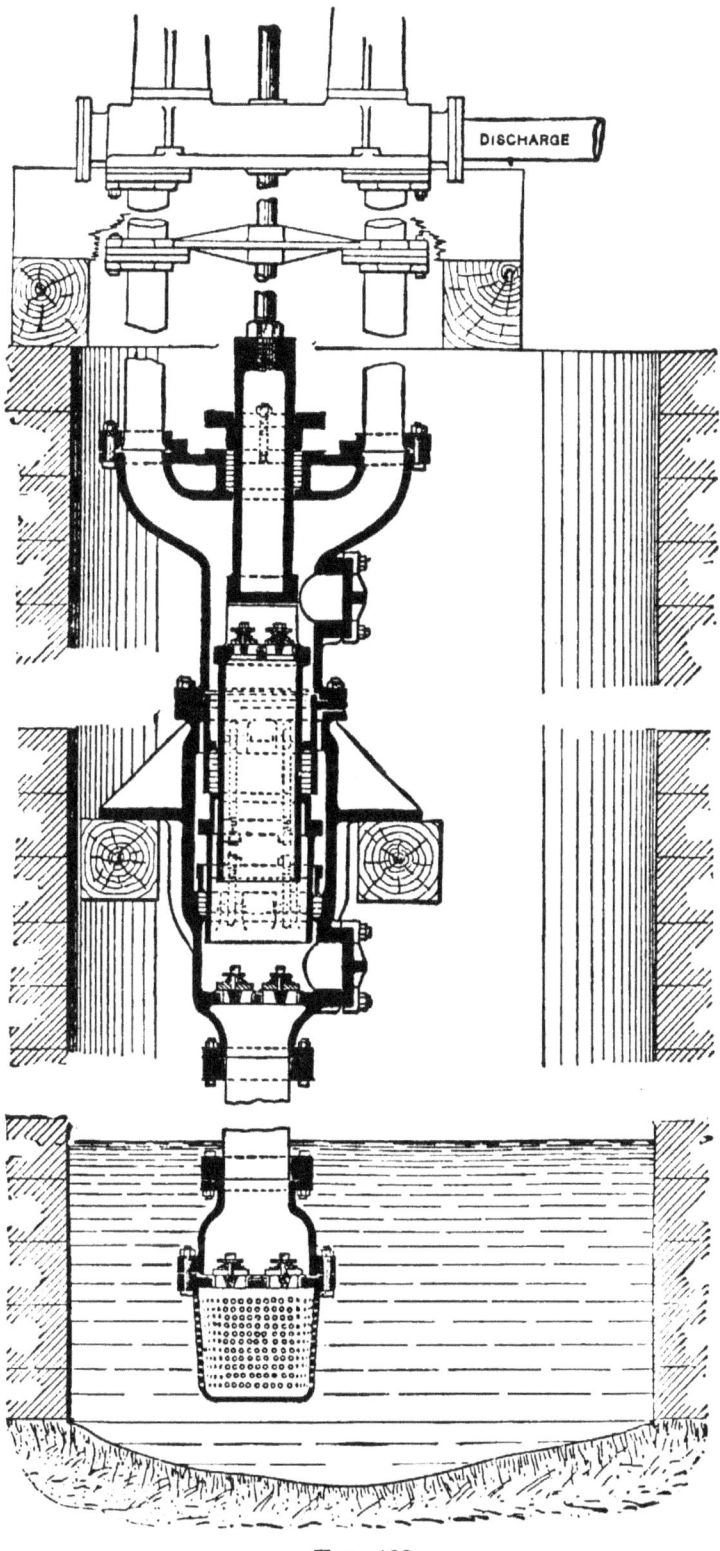

FIG. 193.

8.1.09. Figs. 190 and 191 illustrate a large-tube well-pump and the steam engine for operating it from the surface. The pump-column carries at the upper end a head, with outlet at side and stuffing-box at the top, through which passes a plunger, the area of which is equal to half that of the pump-bucket. By this construction the pump is made double-acting, so that it discharges equal amounts of water during both strokes. It is to be noted, however, that the pump resistance itself is not equal for the two strokes, because the plunger, being high above the bucket, is not subjected on its lower face to the same pressure per square inch as the top of the bucket. This defect will be less if the water be raised to a considerable height above the top of the well, and it can be entirely overcome by making the plunger larger in diameter to compensate for the lower pressure, in which case, however, the discharge will

Fig. 194.

be more unequal. Continuing the plunger of half the area of the bucket down to the latter will generally be no advantage, on account of the weight of the plunger counterbalancing the gain in water pressure. The suction-valve-seat and suction-pipe hang simply by their own weight in the bottom of the sleeve bolted below the suction-valve-chamber. The suction-valve is provided with a long, upwardly projecting, rigid link, which hooks into a similar link depending from the bottom of the pump-bucket. The length and width of the links are such that they are out of contact while the pump is in running adjustment. When repairs become necessary and the bucket is hauled up, the suction-valve and suction-pipe follow it. Where there is much sand in the water the bucket leathers naturally wear out in a very short time. Fig. 192 shows a bucket pump, and Fig. 193 a differential plunger pump for raising water from a shaft-well. The descriptions of similar operating devices in preceding sections of this Bulletin make further explanations of the illustrations superfluous.

8.1.10. Where electric motors or gasoline engines are used to operate deep-well pumps, the top of the tube is arranged with suitable gearing, as in Fig. 194, the pulley of which is driven by belting from the motor. When driven by electric motors, the resistance should be more uniform than is afforded by simple equalization of the two strokes. If two wells

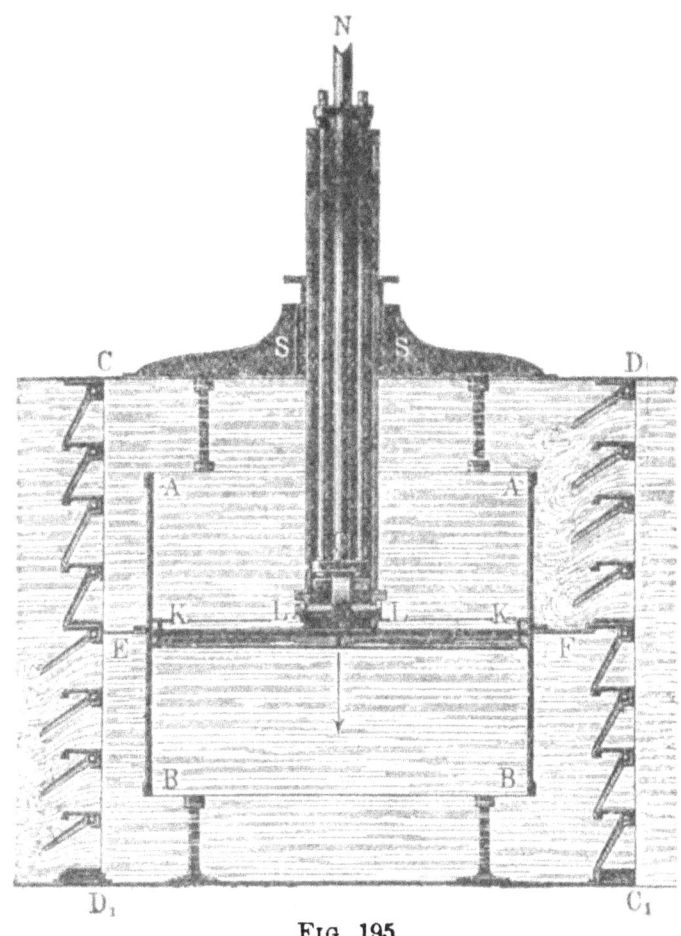

Fig. 195.

with pumps at right angles cannot be used, some such arrangement as described in 2.5.21 could be applied.

8.1.11. A well-known but interesting reciprocating pump designed for land-drainage purposes by the Dutch engineer Fynje, is the so-called box pump, a vertical section of which is shown in Fig. 195. This pump is always vertical and double-acting, and its characteristic feature is the arrangement of the suction- and discharge-valves, which are disposed in opposite sides of a box, surrounding the pump-barrel, and divided horizontally at the middle by the partition F into an upper and a lower chamber. Pumps of this kind are inserted in or built against a bulkhead separating the supply-water from the discharge-water. They are made of very large capacity.

8.1.12. *Centrifugal Pumps.* These have been described in 6.2.01 to 6.2.14, but their application to irrigation and land drainage, as well as to analogous purposes in mining, admits of so much more favorable arrangement and connection with power machinery than for mine drainage, that further reference to them is of interest. When used to pump from open watercourses they are often of very large size and capacity; for example, the five pumps at Khatetbeh, Egypt, built by Farcot, of Paris, for irrigating the province of Behera. Each of these five pumps has a capacity of 140,000,000 gals. in twenty-four hours, at forty revolutions per minute, and at a lift of 10'. These are the largest centrifugal pumps ever built, the runners being over 12' in diameter.

8.1.13. Centrifugal pumps can generally be operated when submerged, except direct-driven centrifugal pumps with horizontal axes, such as are

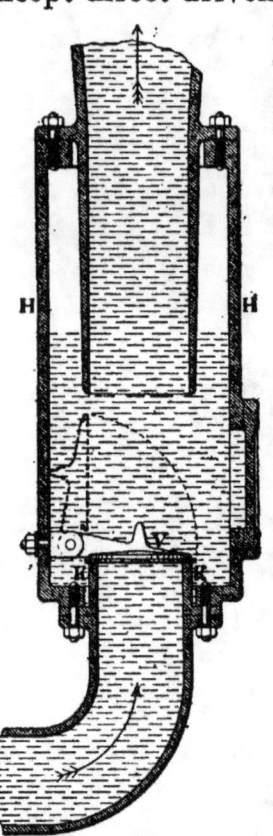

FIG. 196.

generally employed in pumping from open watercourses, which must, on account of the connected power-plant, be located above the highest level that the suction-water may be expected to reach. Where the runner-axis is vertical, it may be made so long that the driving-power connected to its upper end shall always be above reach of the highest suction-water-level.

8.1.14. It is a good plan, particularly for high lifts, to use a check-valve in the discharge-pipe, and to arrange an air-chamber above it, as in Fig. 196, so as to cushion the column of water which falls back and closes the check-valve, as soon as the pump slows down below a certain speed. The check-valve swings clear of the current when open, so that the sand and gravel in the water do not wear its face out too soon.

8.1.15. Centrifugal pumps, when employed to draw water from bored wells, necessarily do so by suction. When the suction-distance to the water-level in the well exceeds the power of the pump to lift water in this way, which is generally at about 20', the centrifugal pump is placed lower down by preparing a dug well or shaft for its reception, several tube wells being sunk 50' to 150' below the bottom of the pit. In most cases such pumps are arranged with a vertical axis, as in Fig. 168, the power being applied at the surface.

8.1.16. Foot-valves cannot be used in the tube wells, on account of lack of space there. Therefore, steam-ejectors or other appliances are required to prime the pumps. The check-valve in the discharge-pipe, if tight, will hold the water in the pump for a time.

8.1.17. Well-water frequently contains a large amount of carbonic acid, which becomes liberated at the upper end of the suction-pipe, and interferes with the action of the pump, if it is not carried along by the force of the current or removed automatically by special appliances, such as a small air-pump driven from the axis of the pump. Pockets or valves, where the gas may lodge, should therefore be avoided in suction-pipes.

8.1.18. One difficulty with centrifugal pumps having long vertical shafts attached to them, is the friction due to the weight of these shafts, and the unbalanced pressure on the pump disk. Where electric motors are used to drive the pumps, the long vertically extended shafts can be avoided, if a motor with vertical axis can be obtained for connection close to the pump.

8.1.19. *Power, and Its Transmission to Reciprocating and Centrifugal Pumps.* Steam-, and sometimes low-head water-power operating by means of turbines, can be applied for driving directly reciprocating and centrifugal pumps.

8.1.20. A power-plant can often be located more advantageously to its operating expense at a distance from the pumping-plant or -plants, either by reason of cheaper fuel, due to saving in freight, or on account of the availability of a water-power. In such cases the power must be transmitted over a distance to the pumping-plant.

8.1.21. It is generally advisable, then, if the power be adequate to drive a number of pumping-plants, to operate them all from one large power-plant, for the reason that a larger plant can be equipped with machinery of higher mechanical efficiency, while the cost of attendance will be less in proportion to the number of pumping subdivisions.

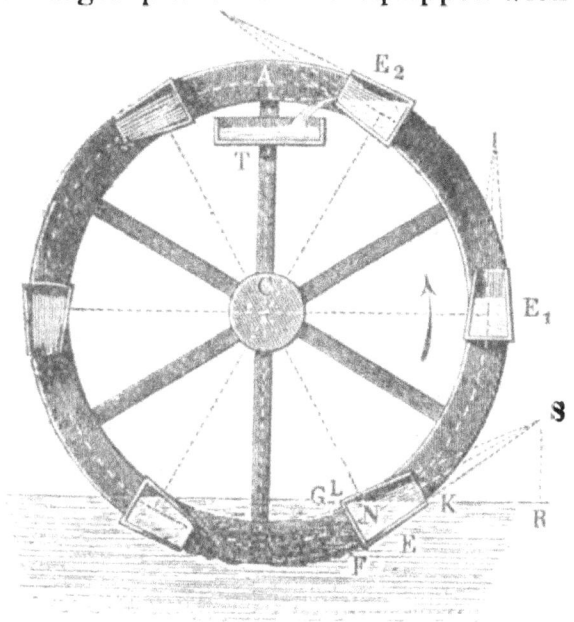

Fig. 197.

8.1.22. Wire rope transmissions would have but a rare application, the distances to which they are suited being limited.

8.1.23. Compressed air, reheated at the pump engines, might be the best method in cases where the pumping is variable, where a good efficiency is desirable, and where steam is the prime motive power; also, where the distance of transmission is not too great, and where reciprocating pumps, which lend themselves best to operation by steam or compressed air, constitute the water-raising machinery.

8.1.24. For great distances, and for operating high-speed, rotary, water-raising machines, like centrifugal pumps, electricity is in general the best mode of transmission. It is not well adapted to cases where great variation of speed is required.

8.1.25. Occasionally, small portable plants, consisting of a centrifugal pump with steam engine and boiler on wheels, find application. These are moved about by means of horses from place to place along a line of ditches or canals. Sometimes a pumping-plant is mounted on a barge floated on the canal.

8.1.26. *Bucket-Wheels.* One of the oldest water-raising appliances for moderate lifts, is the bucket-wheel. Fig. 197 shows a common form. The wheel is rotated either by animal- or engine-power, or, as is most usually the case, by the current of a stream from which it lifts the water, being fitted in this case with paddles. The paddles of the wheel are best made curved, or bent at an angle for the sake of simplicity of construction, as in Fig. 198, so that they leave the water in a vertical direction. The efficiency of paddle-wheels, particularly where running in an unconfined current, is very low. The wheels require to be of great width to obtain even small amounts of power. Where the water-level of a rapid stream does not vary much, stream wheels, though inefficient mechanically, afford a very cheap source of power for raising moderate quantities of water. Where the streams can be confined and dammed so as to raise the water in front and produce a head which acts on the wheel by its weight, the efficiency is much better, and the power much greater, if the wheel is properly constructed. Wheels so situated make a greater number of revolutions, because the paddles then travel with the same velocity as the water which leaves the wheel.

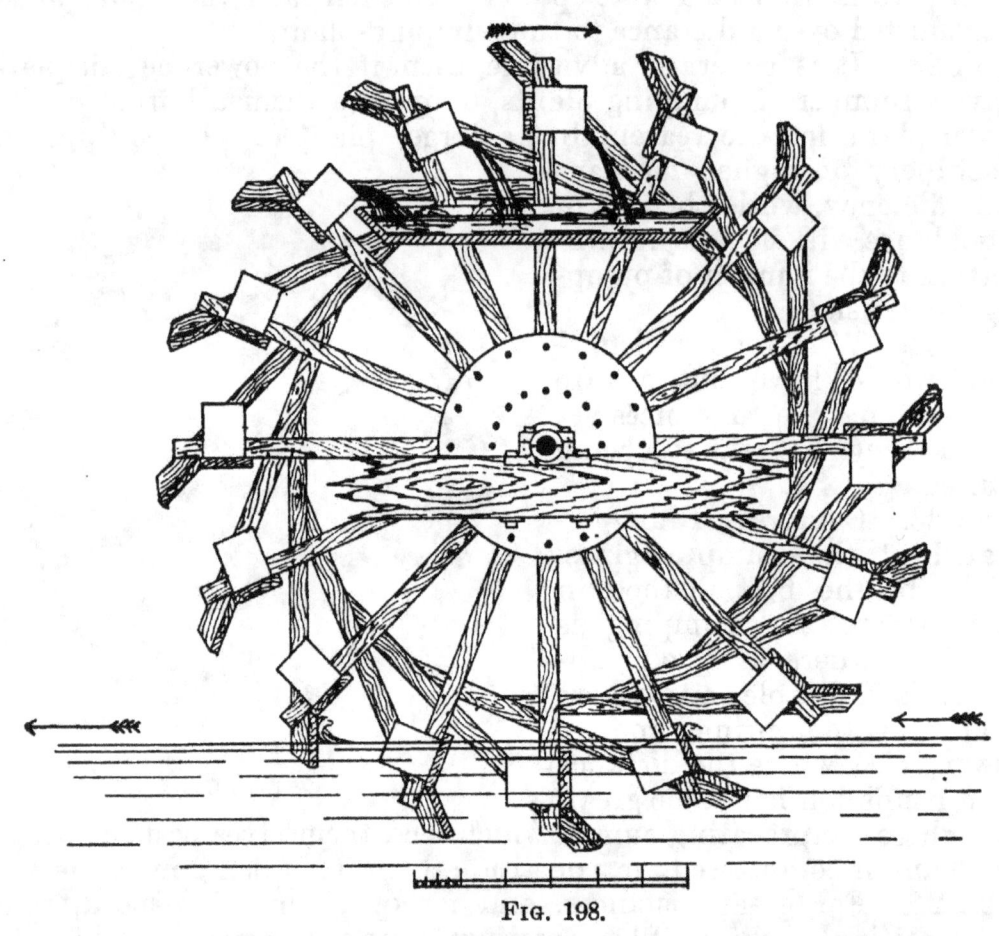

Fig. 198.

8.1.27. Bucket-wheels are suitable only for moderate capacities. The buckets are often made only of common tin cans nailed to the wooden arms or rim of the paddle-wheel.

8.1.28. In fixing the diameter of bucket-wheels, it must be remembered that the level of the discharge-trough is considerably below the top of the wheel. Also, that the distributing-troughs must have sufficient grade to deliver the required quantity of water at more or less distance in a given time.

8.1.29. *Paddle-Wheels.* Where large quantities of water are to be lifted only to a small height, like in some of the drainage undertakings in Holland, paddle-wheels revolved by engine-power in a curb, as in Fig. 199, give very good results. The curb should fit the wheel as close as possible without touching it. The paddles should be inclined so that the water will flow from their surface rapidly, and not be thrown higher than is necessary. Such wheels should be made of iron; otherwise, they will swell or shrink and either jam in the curb or leave too much clearance for back leakage. The back-flow of water is prevented, when the wheel is stopped, by a check-gate at a.

8.1.30. *Hydraulic Rams.* The hydraulic ram is a machine in which a body of water in a pipe under a generally low drive-head intermittently acquires velocity and energy of motion, by virtue of which a part of the water is raised to a height generally greater than the drive-head, while a larger part is permitted to escape to a lower level during the time that the water acquires its velocity. Like in any other utilization of water-

FIG. 199.

power, the conditions for the operation of a ram require an available fall for the discharge of power-water below the level of the supply-reservoir.

8.1.31. The essential arrangement of a hydraulic ram is as shown schematically by Fig. 200, in which A is the supply-reservoir, from which water is to be raised to the elevation H. The drive-pipe B enters the air-chamber C of the ram at the bottom, where the opening is provided with a check-valve D, to arrest the back-flow of the water discharged into the air-chamber. The discharge-pipe E leads from a low point of the air-chamber, in the manner as described in connection with pumps. Close to where the drive-pipe enters the air-chamber, there is located a valve-chamber F, its lower end open to the pipe and its cover on top fitted with an inwardly and downwardly opening check-valve G, called the overflow-valve.

8.1.32. To explain the operation of the machine, suppose, first, that the valves D and G are both closed, and that the discharge-pipe E, as well as the drive-pipe D, are filled with water, while the air-chamber C contains water and air. If, now, the overflow-valve G is opened by forcing it down from its seat, it will remain open by its own weight for a short time, while water will start to flow from the opening, and the water in the drive-pipe will acquire velocity until the pressure below the valve will close it suddenly, so that at the lower end of the drive-pipe there occurs a rise of pressure, which, if sufficient, will open the discharge-valve D against the pressure on its upper surface and force water into the air-chamber, compressing the air therein, which in turn drives an equivalent amount of water out through the discharge-pipe and delivers it at the level L. The rise in pressure at the lower end of the drive-pipe and below the discharge-valve D increases with the length of the drive-pipe B, and with the velocity acquired by the water contained in it. When the energy of the water flowing in the drive-pipe has spent itself in compressing the air in the chamber C, the pressure of the air on re-expanding, and while forcing water up through the discharge-pipe, also forces part of the water in the chamber back into the drive-pipe, before the discharge-valve D has time

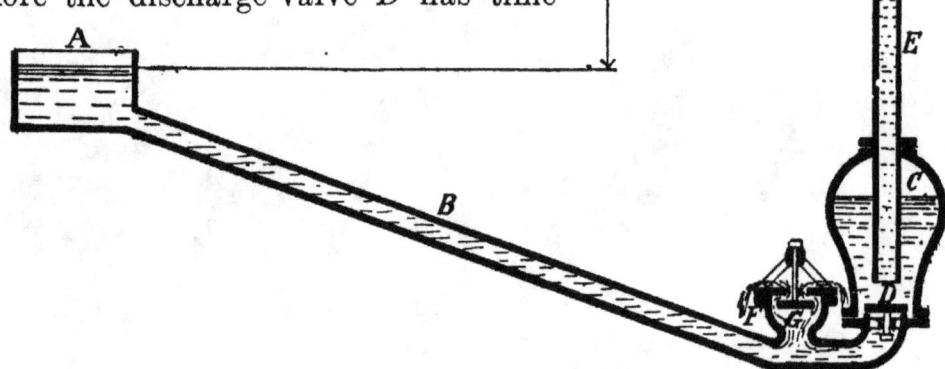

Fig. 200.

to close, thus starting a backward flow in the drive-pipe. The discharge-valve D, however, suddenly closes, and the acquired return motion of the water in the drive-pipe reduces the pressure at its lower end and below the overflow-valve G sufficiently, so that it will open by the pressure of the atmosphere combined with its own weight, and thus permit water to escape from the drive-pipe, thereby again starting a flow toward the ram, when the operations as before described are repeated continuously.

8.1.33. The length of the drive-pipe has an important bearing on the action of a ram. It should be the longer the higher the water is to be raised by a given fall of power-water. It can be shorter, if this fall constitutes a large proportion of the lift, or equals or exceeds it. There should be as few bends and obstructions as possible in the drive-pipe.

8.1.34. The weight of the overflow-valve should generally be small, but should be capable of adjustment, so that the duration of overflow and the velocity acquired by the water in the drive-pipe can be regulated to suit the lift. The overflow-valve and discharge-valve should be located as close together as possible.

8.1.35. The ordinary rams obtainable in the market are only suitable

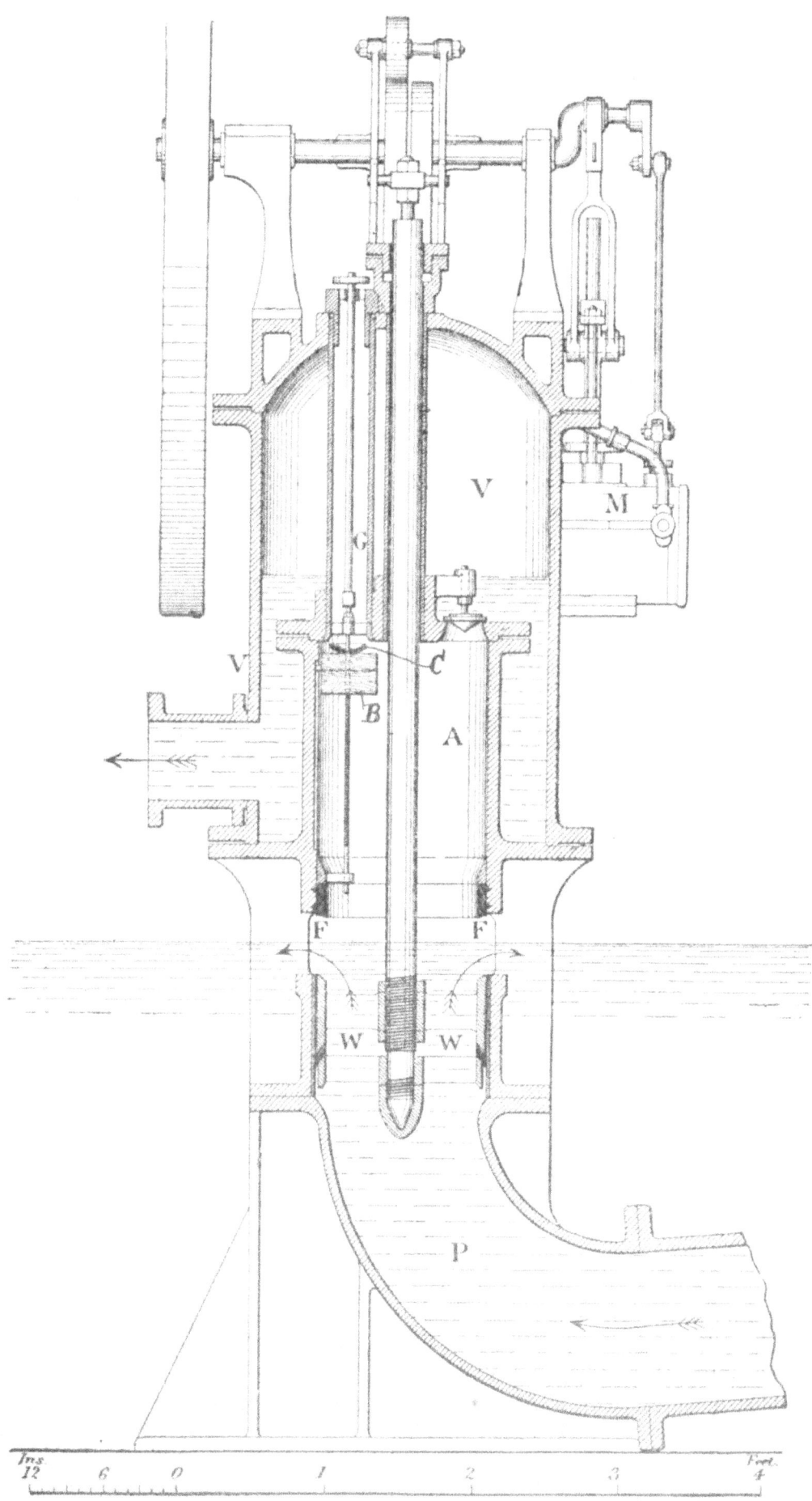

Fig. 201.

for small capacities and moderate lifts. Special rams have been constructed to meet such conditions as a supply of 300,000 gals. per day.

8.1.36. The common forms of rams are very inefficient appliances mechanically, particularly when used for higher lifts. The blow occurring on the closing of the overflow-valve causes a great loss of

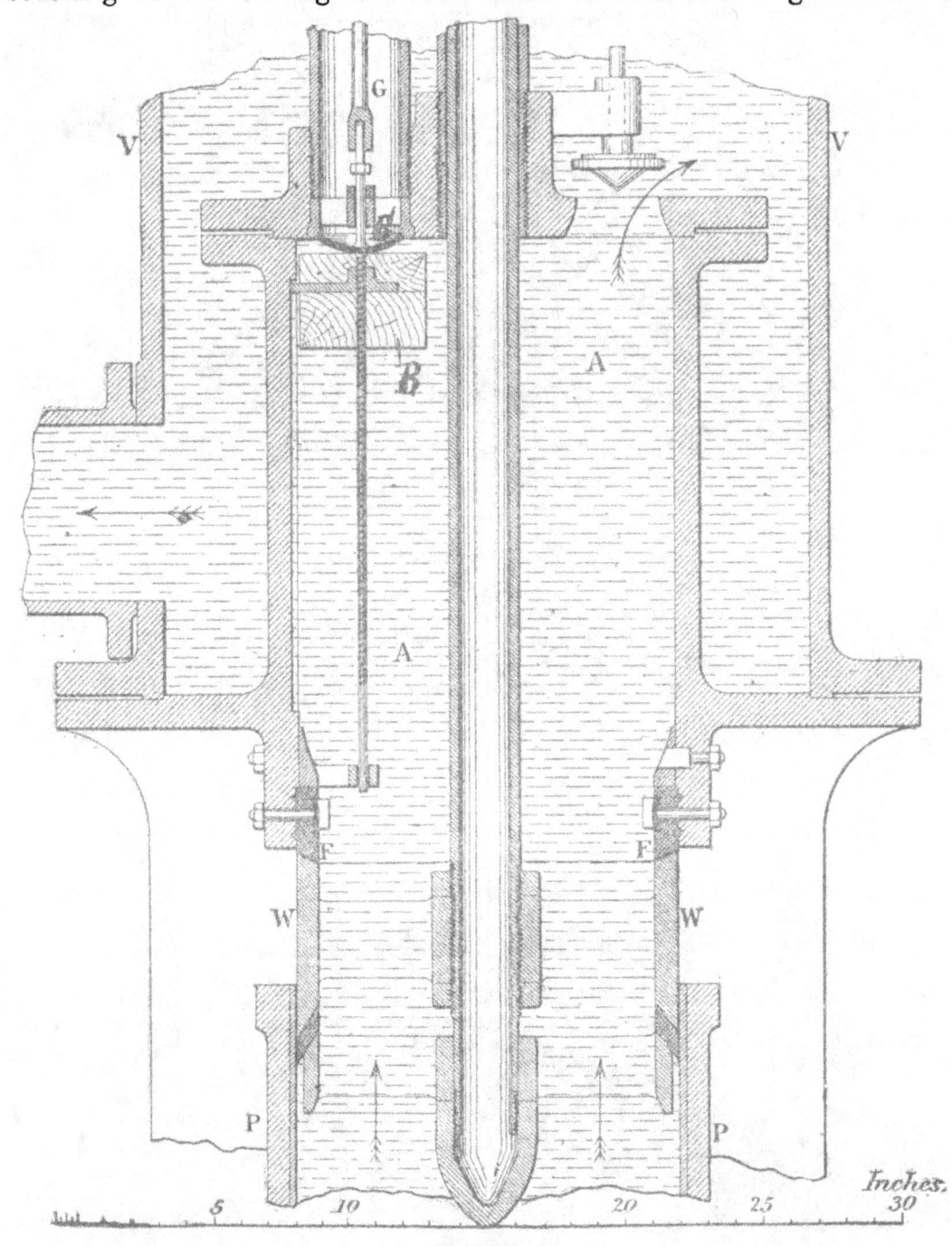

Fig. 202.

energy, and limits the size and capacity of the ordinary machines. The pretty general impression prevails that the blow or shock on the closing of the overflow-valve is a necessary function without which a ram could not operate. That this is erroneous is shown by some very large high-lift rams of recent construction. The most important of these is the ram designed and patented by H. D. Pearsall, which is illustrated in Figs. 201 and 202. In this machine the functions of the balanced overflow-valve W are made directly independent of the action of the water in the drive-pipe. The opening and closing of W is controlled by a small compressed-air engine M mounted on the air-chamber V and operated by the compressed air contained therein, the amount used being replenished at each pulsation by air trapped in the chamber A. This chamber, which is called the ante-chamber, is kept filled with air at atmospheric pressure coming in through the tube G up to the time that the water on rising in A reaches the wooden float B with the valve C and closes the lower end of G, thus cutting off communication with the outer air. The air remaining in A is compressed until it overbalances the pressure on top of the discharge-valves, when it rises into the air-chamber in advance of the water. The discharge of the water from

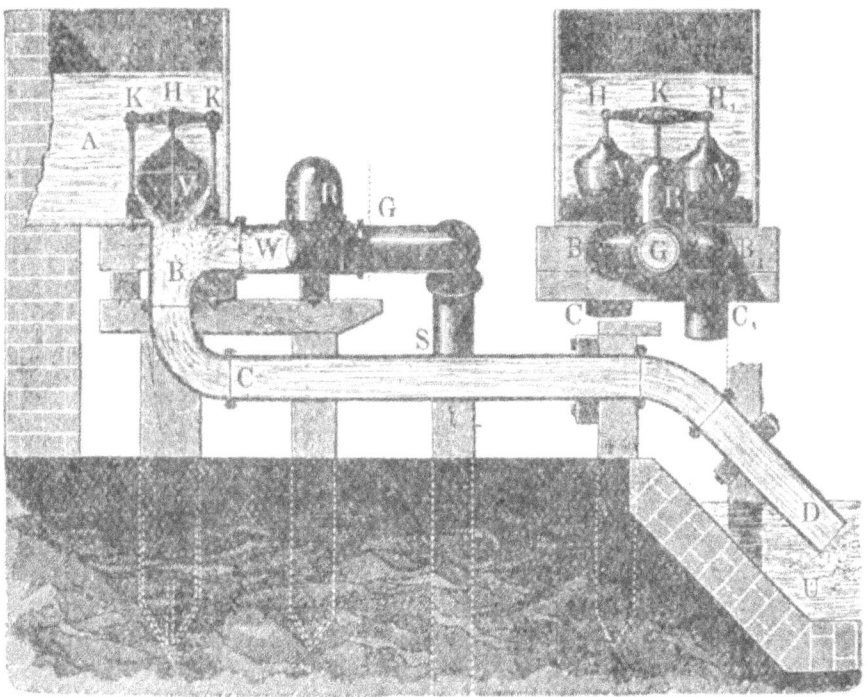

Fig. 203.

V is effected by the compressed air in the same manner as in the ordinary rams. As the column of water in A and in the drive-pipe P falls back the float B and valve C sink, uncovering the lower end of G and admitting air into A. At the same time the compressed-air engine M again opens the overflow-valve W, and the water in P begins to acquire velocity for a new impulse. Instead of operating the valve W by a compressed-air engine, it could also be worked by a small waterwheel driven by a nozzle from the discharge-pipe.

8.1.37. None of the rams heretofore described are suitable for raising water by suction, and they require to be placed at the level of the overflow from the drive-pipe. It is, however, often convenient, particularly for irrigation and land-drainage purposes, to locate the ram at a higher level, in which case it must necessarily draw the water up from the supply-reservoir.

8.1.38. A double-acting ram of this description was designed by the Belgian engineer Leblanc, for raising water from a level below that of the discharge, by means of an independent supply of power-water situated at a higher level. It is illustrated in Fig. 203. A is the supply-reservoir for the power-water; *B C D* the discharge-pipe leading therefrom and corresponding to the drive-pipe of the ordinary ram. It might in this arrangement be properly termed the draft-pipe. *S G* is the upper part of the suction-pipe leading up from the excavation to be drained. The power-water here flows past the open valve *V* into the

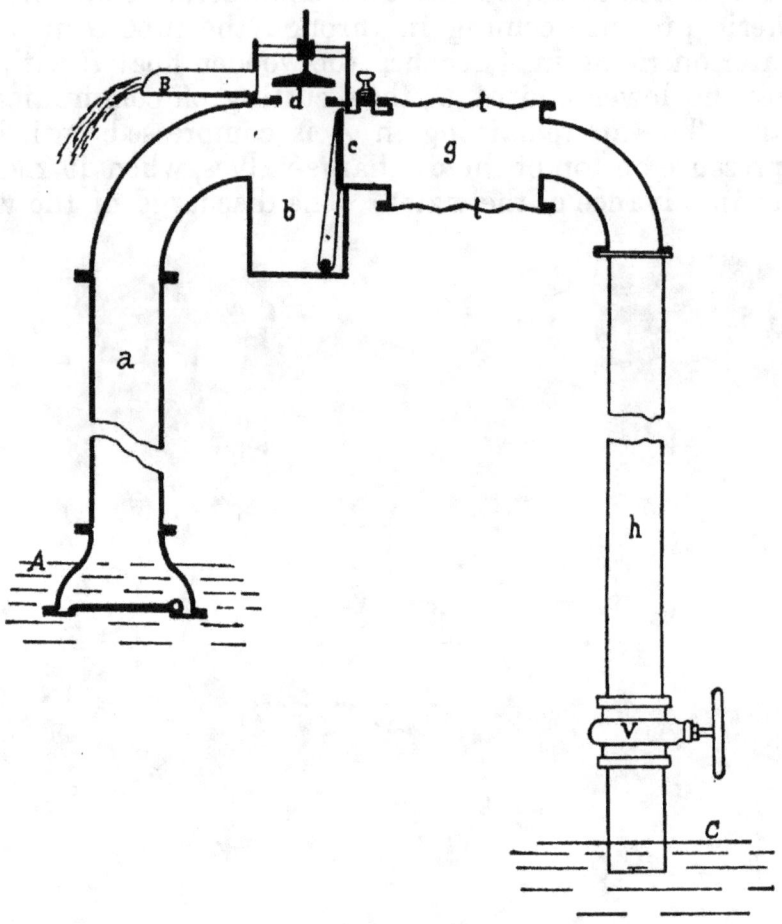

Fig. 204.

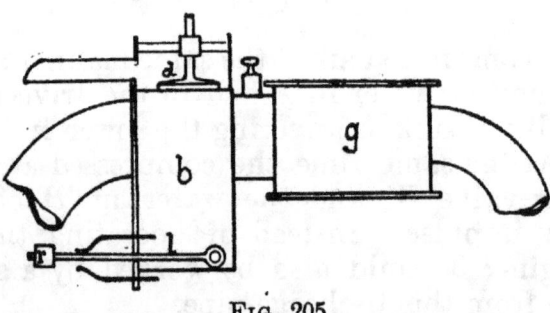

Fig. 205.

discharge-pipe *B C D*, therein acquiring velocity, increasing until the pressure beneath the nearly buoyant valve *V* is reduced so that the pressure above it forces it to its seat. As the water in *B C D* continues its motion it will create a suction effect in front of the suction-valve *W*, which then opens and allows water from the suction-pipe *S G* to follow the water in *B C D*. As soon as the energy of flow has spent itself, the column in both pipes begins a retrograde motion, and the valve *W* is suddenly closed, thereby cutting off the exit of the water in *B C D*, which then spends its remaining energy of return flow in forcing open the valve V, and thus prepares the conditions for the next discharge flow in $B\ C\ D$.

8.1.39. The machine illustrated is made double-acting by the use of two discharge-pipes, two suction-valves W, and two arresting-valves V and V', the latter being balanced by suspension from the opposite ends of a double-armed lever K. By this arrangement a more perfect functioning of the apparatus is secured, because the discharge flow in one discharge-pipe occurs at the same time as the more feeble return flow in the other, so that the more powerful suction-action of the discharge, by closing its arresting-valve V, will, at the same time, aid in lifting the other valve V' over the returning column in the other discharge-pipe.

8.1.40. Another machine of the ram type, which also raises water by suction, but which is designed to work under different conditions than the Leblanc ram, is the so-called siphon water-elevator of Lemichel & Co. of Paris, which was exhibited at the Midwinter Fair in San Francisco. This machine, which is illustrated in Fig. 204, is intended to raise water from a supply at the level A to a higher level B, by means of a discharge to a lower level C; these conditions being similar to those under which the ordinary ram is called on to operate, but with the difference that in this case the machine raises water by suction, and is located at the highest level B, identical with that of the delivery of the water intended to be raised for a useful purpose.

8.1.41. This machine employs the principles of action both of the ram and of the siphon, and should, therefore, more properly be called the "siphon ram" instead of "siphon elevator."

8.1.42. The operation of the machine is as follows: On opening the valve V in the discharge- or draft-pipe h, Fig. 204, the water in the siphon begins to move in the direction of the discharge level C, falling in h and rising in the suction-pipe a, and acquiring velocity of flow until the force of the latter is sufficient to close the check-valve c in the chamber b. The exit of the moving water in a being thus cut off suddenly, the momentum of the water spends itself in raising the outlet-valve d and discharging a portion of the column over the edge of the valve-seat. During this time the downward momentum of the water in the discharge-column h causes a reduction of pressure within the regulator g (which here performs the functions of the air-chamber in the ordinary ram), so that the elastic corrugated heads $t\ t_1$ are forced inward by the overpressure of the atmosphere until the energy of the water in h is spent, when a return flow takes place toward the check-valve c, which now opens through the combined action of the return flow both in h and in a, assisted by the weight r on the level l, Fig. 205. The functions described take place in a very brief period of time, the number of pulsations being from 150 to 400 per minute. The chamber g, with elastic heads, is here substituted for the air-chamber on the ordinary ram, because under the low pressure the air would not have much cushioning effect and would also cause the pulsations to be too slow. The air, which is liberated at the exceedingly low pressure at the highest point, just like in a siphon, is here expelled with the water through the discharge-valve d. If this were not the case the apparatus would not operate for many minutes.

8.1.43. It is claimed that by this machine the water may be raised at sea-level to a height of about 30' above the supply-reservoir A. For greater lifts a series of superposed siphons may be used, the upper ones decreasing regularly in capacity, as they can raise only a part of the water raised by the siphon below them.

8.1.44. Fig. 206 shows an application of the siphon ram or elevator, in which the water delivered to irrigate land below a main ditch is utilized to raise a less quantity of water to a small ditch at a higher level for the purpose of irrigating land situated above the main ditch.

8.1.45. The siphon ram is offered in capacities of from 250 to 3,000,000 gals. per twenty-four hours.

8.1.46. All rams, in order to operate efficiently, should be specially designed to suit the conditions under which they are expected to work. They will work efficiently only within narrow limits of variation of capacity, because the proper period of the pulsations is fixed by the proportions of lift to fall, and the lengths of the pipes.

8.1.47. In obtaining a supply of water for useful purposes, the ques-

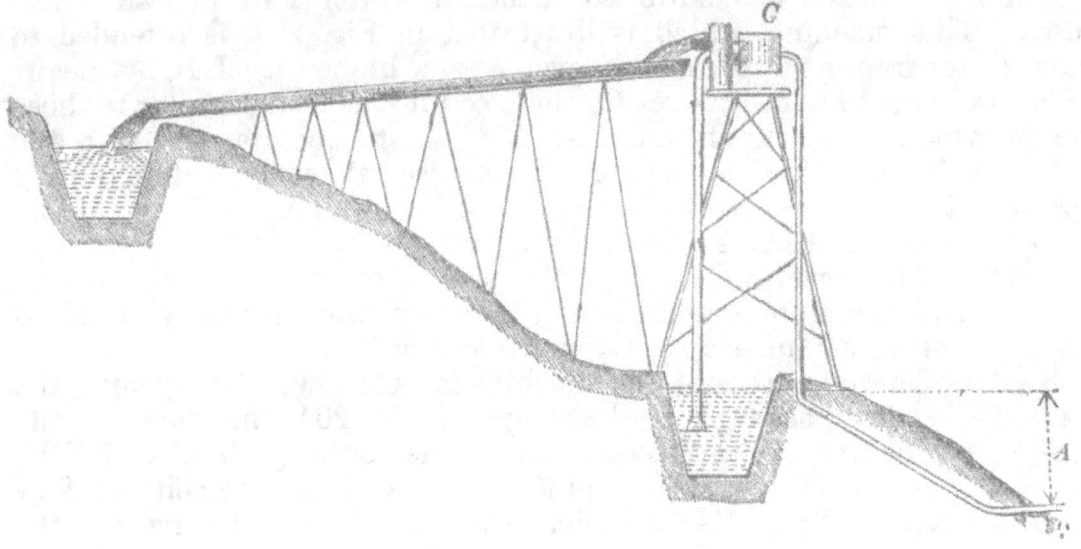

FIG. 206.

tion often arises as to which is the cheapest, both in first cost and in operating expense: a long ditch-line or flume, or a shorter one starting at and delivering the water to a lower level, in combination with a pumping-plant to bring the water up the remaining height to the required level. It must also be considered whether it is necessary to raise all the water to the entire elevation, and if part of it may not be delivered at the lower level accessible by the shorter ditch.

8.1.48. It is to be determined also whether such a pumping-plant is required to operate during the entire year, or only for a part of the time. Like mine-draining by pumps, the probable number of years during which the plant will be needed is also a factor which enters into the choice of arrangement.

8.1.49. It is impossible, in a treatise of this kind, to give more than suggestions as to apparatus and mode of operating it which are best suited to the requirements. Conditions in practice are so varied and present so many unforeseen problems that it would go beyond the scope of this Bulletin to attempt detailed consideration of particular cases. The proper treatment of such can only be carried out in special articles devoted to each case.

www.GoldRushBooks.com

More Books On Mining

Visit: www.goldrushbooks.com to order your copies or ask your favorite book seller to offer them.

Mining Books by Kerby Jackson

Gold Dust: Stories From Oregon's Mining Years

Oregon mining historian and prospector, Kerby Jackson, brings you a treasure trove of seventeen stories on Southern Oregon's rich history of gold prospecting, the prospectors and their discoveries, and the breathtaking areas they settled in and made homes. 5" X 8", 98 ppgs. Retail Price: $11.99

The Golden Trail: More Stories From Oregon's Mining Years

In his follow-up to "Gold Dust: Stories of Oregon's Mining Years", this time around, Jackson brings us twelve tales from Oregon's Gold Rush, including the story about the first gold strike on Canyon Creek in Grant County, about the old timers who found gold by the pail full at the Victor Mine near Galice, how Iradel Bray discovered a rich ledge of gold on the Coquille River during the height of the Rogue River War, a tale of two elderly miners on the hunt for a lost mine in the Cascade Mountains, details about the discovery of the famous Armstrong Nugget and others. 5" X 8", 70 ppgs. Retail Price: $10.99

Oregon Mining Books

Geology and Mineral Resources of Josephine County, Oregon

Unavailable since the 1970's, this important publication was originally compiled by the Oregon Department of Geology and Mineral Industries and includes important details on the economic geology and mineral resources of this important mining area in South Western Oregon. Included are notes on the history, geology and development of important mines, as well as insights into the mining of gold, copper, nickel, limestone, chromium and other minerals found in large quantities in Josephine County, Oregon. 8.5" X 11", 54 ppgs. Retail Price: $9.99

Mines and Prospects of the Mount Reuben Mining District

Unavailable since 1947, this important publication was originally compiled by geologist Elton Youngberg of the Oregon Department of Geology and Mineral Industries and includes detailed descriptions, histories and the geology of the Mount Reuben Mining District in Josephine County, Oregon. Included are notes on the history, geology, development and assay statistics, as well as underground maps of all the major mines and prospects in the vicinity of this much neglected mining district. 8.5" X 11", 48 ppgs. Retail Price: $9.99

The Granite Mining District

Notes on the history, geology and development of important mines in the well known Granite Mining District which is located in Grant County, Oregon. Some of the mines discussed include the Ajax, Blue Ribbon, Buffalo, Continental, Cougar-Independence, Magnolia, New York, Standard and the Tillicum. Also included are many rare maps pertaining to the mines in the area. 8.5" X 11", 48 ppgs. Retail Price: $9.99

Ore Deposits of the Takilma and Waldo Mining Districts of Josephine County, Oregon

The Waldo and Takilma mining districts are most notable for the fact that the earliest large scale mining of placer gold and copper in Oregon took place in these two areas. Included are details about some of the earliest large gold mines in the state such as the Llano de Oro, High Gravel, Cameron, Platerica, Deep Gravel and others, as well as copper mines such as the famous Queen of Bronze mine, the Waldo, Lily and Cowboy mines. This volume also includes six maps and 20 original illustrations. 8.5" X 11", 74 ppgs. Retail Price: $9.99

Metal Mines of Douglas, Coos and Curry Counties, Oregon

Oregon mining historian Kerby Jackson introduces us to a classic work on Oregon's mining history in this important re-issue of Bulletin 14C Volume 1, otherwise known as the Douglas, Coos & Curry Counties, Oregon Metal Mines Handbook. Unavailable since 1940, this important publication was originally compiled by the Oregon Department of Geology and Mineral Industries includes detailed descriptions, histories and the geology of over 250 metallic mineral mines and prospects in this rugged area of South West Oregon. 8.5" X 11", 158 ppgs. Retail Price: $19.99

Metal Mines of Jackson County, Oregon

Unavailable since 1943, this important publication was originally compiled by the Oregon Department of Geology and Mineral Industries includes detailed descriptions, histories and the geology of over 450 metallic mineral mines and prospects in Jackson County, Oregon. Included are such famous gold mining areas as Gold Hill, Jacksonville, Sterling and the Upper Applegate. 8.5" X 11", 220 ppgs. Retail Price: $24.99

Metal Mines of Josephine County, Oregon

Oregon mining historian Kerby Jackson introduces us to a classic work on Oregon's mining history in this important re-issue of Bulletin 14C, otherwise known as the Josephine County, Oregon Metal Mines Handbook. Unavailable since 1952, this important publication was originally compiled by the Oregon Department of Geology and Mineral Industries includes detailed descriptions, histories and the geology of over 500 metallic mineral mines and prospects in Josephine County, Oregon. 8.5" X 11", 250 ppgs. Retail Price: $24.99

Metal Mines of North East Oregon

Oregon mining historian Kerby Jackson introduces us to a classic work on Oregon's mining history in this important re-issue of Bulletin 14A and 14B, otherwise known as the North East Oregon Metal Mines Handbook. Unavailable since 1941, this important publication was originally compiled by the Oregon Department of Geology and Mineral Industries and includes detailed descriptions, histories and the geology of over 750 metallic mineral mines and prospects in North Eastern Oregon. 8.5" X 11", 310 ppgs. Retail Price: $29.99

Metal Mines of North West Oregon

Oregon mining historian Kerby Jackson introduces us to a classic work on Oregon's mining history in this important re-issue of Bulletin 14D, otherwise known as the North West Oregon Metal Mines Handbook. Unavailable since 1951, this important publication was originally compiled by the Oregon Department of Geology and Mineral Industries and includes detailed descriptions, histories and the geology of over 250 metallic mineral mines and prospects in North Western Oregon. 8.5" X 11", 182 ppgs. Retail Price: $19.99

Mines and Prospects of Oregon

Mining historian Kerby Jackson introduces us to a classic mining work by the Oregon Bureau of Mines in this important re-issue of The Handbook of Mines and Prospects of Oregon. Unavailable since 1916, this publication includes important insights into hundreds of gold, silver, copper, coal, limestone and other mines that operated in the State of Oregon around the turn of the 19th Century. Included are not only geological details on early mines throughout Oregon, but also insights into their history, production, locations and in some cases, also included are rare maps of their underground workings. 8.5" X 11", 314 ppgs. Retail Price: $24.99

Lode Gold of the Klamath Mountains of Northern California and South West Oregon

(See California Mining Books)

Mineral Resources of South West Oregon

Unavailable since 1914, this publication includes important insights into dozens of mines that once operated in South West Oregon, including the famous gold fields of Josephine and Jackson Counties, as well as the Coal Mines of Coos County. Included are not only geological details on early mines throughout South West Oregon, but also insights into their history, production and locations. 8.5" X 11", 154 ppgs. Retail Price: $11.99

Chromite Mining in The Klamath Mountains of California and Oregon

(See California Mining Books)

Southern Oregon Mineral Wealth

Unavailable since 1904, this rare publication provides a unique snapshot into the mines that were operating in the area at the time. Included are not only geological details on early mines throughout South West Oregon, but also insights into their history, production and locations. Some of the mining areas include Grave Creek, Greenback, Wolf Creek, Jump Off Joe Creek, Granite Hill, Galice, Mount Reuben, Gold Hill, Galls Creek, Kane Creek, Sardine Creek, Birdseye Creek, Evans Creek, Foots Creek, Jacksonville, Ashland, the Applegate River, Waldo, Kerby and the Illinois River, Althouse and Sucker Creek, as well as insights into local copper mining and other topics. 8.5" X 11", 64 ppgs. Retail Price: $8.99

Geology and Ore Deposits of the Takilma and Waldo Mining Districts

Unavailable since the 1933, this publication was originally compiled by the United States Geological Survey and includes details on gold and copper mining in the Takilma and Waldo Districts of Josephine County, Oregon. The Waldo and Takilma mining districts are most notable for the fact that the earliest large scale mining of placer gold and copper in Oregon took place in these two areas. Included in this report are details about some of the earliest large gold mines in the state such as the Llano de Oro, High Gravel, Cameron, Platerica, Deep Gravel and others, as well as copper mines such as the famous Queen of Bronze mine, the Waldo, Lily and Cowboy mines. In addition to geological examinations, insights are also provided into the production, day to day operations and early histories of these mines, as well as calculations of known mineral reserves in the area. This volume also includes six maps and 20 original illustrations. **8.5" X 11", 74 ppgs. Retail Price: $9.99**

Gold Mines of Oregon

Oregon mining historian Kerby Jackson introduces us to a classic work on Oregon's mining history in this important re-issue of Bulletin 61, otherwise known as "Gold and Silver In Oregon". Unavailable since 1968, this important publication was originally compiled by geologists Howard C. Brooks and Len Ramp of the Oregon Department of Geology and Mineral Industries and includes detailed descriptions, histories and the geology of over 450 gold mines Oregon. Included are notes on the history, geology and gold production statistics of all the major mining areas in Oregon including the Klamath Mountains, the Blue Mountains and the North Cascades. While gold is where you find it, as every miner knows, the path to success is to prospect for gold where it was previously found. **8.5" X 11", 344 ppgs. Retail Price: $24.99**

Mines and Mineral Resources of Curry County Oregon

Originally published in 1916, this important publication on Oregon Mining has not been available for nearly a century. Included are rare insights into the history, production and locations of dozens of gold mines in Curry County, Oregon, as well as detailed information on important Oregon mining districts in that area such as those at Agness, Bald Face Creek, Mule Creek, Boulder Creek, China Diggings, Collier Creek, Elk River, Gold Beach, Rock Creek, Sixes River and elsewhere. Particular attention is especially paid to the famous beach gold deposits of this portion of the Oregon Coast. **8.5" X 11", 140 ppgs. Retail Price: $11.99**

Chromite Mining in South West Oregon

Originally published in 1961, this important publication on Oregon Mining has not been available for nearly a century. Included are rare insights into the history, production and locations of nearly 300 chromite mines in South Western Oregon. **8.5" X 11", 184 ppgs. Retail Price: $14.99**

Mineral Resources of Douglas County Oregon

Originally published in 1972, this important publication on Oregon Mining has not been available for nearly forty years. Included are rare insights into the geology, history, production and locations of numerous gold mines and other mining properties in Douglas County, Oregon. **8.5" X 11", 124 ppgs. Retail Price: $11.99**

Mineral Resources of Coos County Oregon

Originally published in 1972, this important publication on Oregon Mining has not been available for nearly forty years. Included are rare insights into the geology, history, production and locations of numerous gold mines and other mining properties in Coos County, Oregon. **8.5" X 11", 100 ppgs. Retail Price: $11.99**

Mineral Resources of Lane County Oregon

Originally published in 1938, this important publication on Oregon Mining has not been available for nearly seventy five years. Included are extremely rare insights into the geology and mines of Lane County, Oregon, in particular in the Bohemia, Blue River, Oakridge, Black Butte and Winberry Mining Districts. 8.5" X 11", 82 ppgs. Retail Price: $9.99

Mineral Resources of the Upper Chetco River of Oregon: Including the Kalmiopsis Wilderness

Originally published in 1975, this important publication on Oregon Mining has not been available for nearly forty years. Withdrawn under the 1872 Mining Act since 1984, real insight into the minerals resources and mines of the Upper Chetco River has long been unavailable due to the remoteness of the area. Despite this, the decades of battle between property owners and environmental extremists over the last private mining inholding in the area has continued to pique the interest of those interested in mining and other forms of natural resource use. Gold mining began in the area in the 1850's and has a rich history in this geographic area, even if the facts surrounding it are little known. Included are twenty two rare photographs, as well as insights into the Becca and Morning Mine, the Emmly Mine (also known as Emily Camp), the Frazier Mine, the Golden Dream or Higgins Mine, Hustis Mine, Peck Mine and others. 8.5" X 11", 64 ppgs. Retail Price: $8.99

Gold Dredging in Oregon

Originally published in 1939, this important publication on Oregon Mining has not been available for nearly seventy five years. Included are extremely rare insights into the history and day to day operations of the dragline and bucketline gold dredges that once worked the placer gold fields of South West and North East Oregon in decades gone by. Also included are details into the areas that were worked by gold dredges in Josephine, Jackson, Baker and Grant counties, as well as the economic factors that impacted this mining method. This volume also offers a unique look into the values of river bottom land in relation to both farming and mining, in how farm lands were mined, re-soiled and reclamated after the dredges worked them. Featured are hard to find maps of the gold dredge fields, as well as rare photographs from a bygone era. 8.5" X 11", 86 ppgs. Retail Price: $8.99

Quick Silver Mining in Oregon

Originally published in 1963, this important publication on Oregon Mining has not been available for over fifty years. This publication includes details into the history and production of Elemental Mercury or Quicksilver in the State of Oregon. 8.5" X 11", 238 ppgs. Retail Price: $15.99

Mines of the Greenhorn Mining District of Grant County Oregon

Originally published in 1948, this important publication on Oregon Mining has not been available for over sixty five years. In this publication are rare insights into the mines of the famous Greenhorn Mining District of Grant County, Oregon, especially the famous Morning Mine. Also included are details on the Tempest, Tiger, Bi-Metallic, Windsor, Psyche, Big Johnny, Snow Creek, Banzette and Paramount Mines, as well as prospects in the vicinities in the famous mining areas of Mormon Basin, Vinegar Basin and Desolation Creek. Included are hard to find mine maps and dozens of rare photographs from the bygone era of Grant County's rich mining history. 8.5" X 11", 72 ppgs. Retail Price: $9.99

Geology of the Wallowa Mountains of Oregon: Part I (Volume 1)

Originally published in 1938, this important publication on Oregon Mining has not been available for nearly seventy five years. Included are details on the geology of this unique portion of North Eastern Oregon. This is the first part of a two book series on the area. Accompanying the text are rare photographs and historic maps. **8.5" X 11", 92 ppgs. Retail Price: $9.99**

Geology of the Wallowa Mountains of Oregon: Part II (Volume 2)

Originally published in 1938, this important publication on Oregon Mining has not been available for nearly seventy five years. Included are details on the geology of this unique portion of North Eastern Oregon. This is the first part of a two book series on the area. Accompanying the text are rare photographs and historic maps. **8.5" X 11", 94 ppgs. Retail Price: $9.99**

Field Identification of Minerals For Oregon Prospectors

Originally published in 1940, this important publication on Oregon Mining has not been available for nearly seventy five years. Included in this volume is an easy system for testing and identifying a wide range of minerals that might be found by prospectors, geologists and rockhounds in the State of Oregon, as well as in other locales. Topics include how to put together your own field testing kit and how to conduct rudimentary tests in the field. This volume is written in a clear and concise way to make it useful even for beginners. **8.5" X 11", 158 ppgs. Retail Price: $14.99**

Idaho Mining Books

Gold in Idaho

Unavailable since the 1940's, this publication was originally compiled by the Idaho Bureau of Mines and includes details on gold mining in Idaho. Included is not only raw data on gold production in Idaho, but also valuable insight into where gold may be found in Idaho, as well as practical information on the gold bearing rocks and other geological features that will assist those looking for placer and lode gold in the State of Idaho. This volume also includes thirteen gold maps that greatly enhance the practical usability of the information contained in this small book detailing where to find gold in Idaho. **8.5" X 11", 72 ppgs. Retail Price: $9.99**

Geology of the Couer D'Alene Mining District of Idaho

Unavailable since 1961, this publication was originally compiled by the Idaho Bureau of Mines and Geology and includes details on the mining of gold, silver and other minerals in the famous Coeur D'Alene Mining District in Northern Idaho. Included are details on the early history of the Coeur D'Alene Mining District, local tectonic settings, ore deposit features, information on the mineral belts of the Osburn Fault, as well as detailed information on the famous Bunker Hill Mine, the Dayrock Mine, Galena Mine, Lucky Friday Mine and the infamous Sunshine Mine. This volume also includes sixteen hard to find maps. **8.5" X 11", 70 ppgs. Retail Price: $9.99**

Utah Mining Books

Fluorite in Utah

Unavailable since 1954, this publication was originally compiled by the USGS, State of Utah and U.S. Atomic Energy Commission and details the mining of fluorspar, also known as fluorite in the State of Utah. Included are details on the geology and history of fluorspar (fluorite) mining in Utah, including details on where this unique gem mineral may be found in the State of Utah. **8.5" X 11", 60 ppgs. Retail Price: $8.99**

California Mining Books

The Tertiary Gravels of the Sierra Nevada of California

Mining historian Kerby Jackson introduces us to a classic mining work by Waldemar Lindgren in this important re-issue of The Tertiary Gravels of the Sierra Nevada of California. Unavailable since 1911, this publication includes details on the gold bearing ancient river channels of the famous Sierra Nevada region of California. **8.5" X 11", 282 ppgs. Retail Price: $19.99**

The Mother Lode Mining Region of California

Unavailable since 1900, this publication includes details on the gold mines of California's famous Mother Lode gold mining area. Included are details on the geology, history and important gold mines of the region, as well as insights into historic mining methods, mine timbering, mining machinery, mining bell signals and other details on how these mines operated. Also included are insights into the gold mines of the California Mother Lode that were in operation during the first sixty years of California's mining history. **8.5" X 11", 176 ppgs. Retail Price: $14.99**

Lode Gold of the Klamath Mountains of Northern California and South West Oregon

Unavailable since 1971, this publication was originally compiled by Preston E. Hotz and includes details on the lode mining districts of Oregon and California's Klamath Mountains. Included are details on the geology, history and important lode mines of the French Gulch, Deadwood, Whiskeytown, Shasta, Redding, Muletown, South Fork, Old Diggings, Dog Creek (Delta), Bully Choop (Indian Creek), Harrison Gulch, Hayfork, Minersville, Trinity Center, Canyon Creek, East Fork, New River, Denny, Liberty (Black Bear), Cecilville, Callahan, Yreka, Fort Jones and Happy Camp mining districts in California, as well as the Ashland, Rogue River, Applegate, Illinois River, Takilma, Greenback, Galice, Silver Peak, Myrtle Creek and Mule Creek districts of South Western Oregon. Also included are insights into the mineralization and other characteristics of this important mining region. **8.5" X 11", 100 ppgs. Retail Price: $10.99**

Mines and Mineral Resources of Shasta County, Siskiyou County, Trinity County, California

Unavailable since 1915, this publication was originally compiled by the California State Mining Bureau and includes details on the gold mines of this area of Northern California. Also included are insights into the mineralization and other characteristics of this important mining region, as well as the location of historic gold mines. **8.5" X 11", 204 ppgs. Retail Price: $19.99**

Geology of the Yreka Quadrangle, Siskiyou County, California

Unavailable since 1977, this publication was originally compiled by Preston E. Hotz and includes details on the geology of the Yreka Quadrangle of Siskiyou County, California. Also included are insights into the mineralization and other characteristics of this important mining region. **8.5" X 11", 78 ppgs. Retail Price: $7.99**

Mines of San Diego and Imperial Counties, California

Originally published in 1914, this important publication on California Mining has not been available for a century. This publication includes important information on the early gold mines of San Diego and Imperial County, which were some of the first gold fields mined in California by early Spanish and Mexican miners before the 49ers came on the scene. Included are not only details on early mining methods in the area, production statistics and geological information, but also the location of the early gold mines that helped make California "The Golden State". Also included are details on the mining of other minerals such as silver, lead, zinc, manganese, tungsten, vanadium, asbestos, barite, borax, cement, clay, dolomite, fluospar, gem stones, graphite, marble, salines, petroleum, stronium, talc and others. **8.5" X 11", 116 ppgs. Retail Price: $12.99**

Mines of Sierra County, California

Unavailable since 1920, this publication was originally compiled by the California State Mining Bureau and includes details on the gold mines of Sierra County, California. Also included are insights into the mineralization and other characteristics of this important mining region, as well as the location of historic gold mines. **8.5" X 11", 156 ppgs. Retail Price: $19.99**

Mines of Plumas County, California

Unavailable since 1918, this publication was originally compiled by the California State Mining Bureau and includes details on the gold mines of Plumas County, California. Also included are insights into the mineralization and other characteristics of this important mining region, as well as the location of historic gold mines. **8.5" X 11", 200 ppgs. Retail Price: $19.99**

Mines of El Dorado, Placer, Sacramento and Yuba Counties, California

Originally published in 1917, this important publication on California Mining has not been available for nearly a century. This publication includes important information on the early gold mines of El Dorado County, Placer County, Sacramento County and Yuba County, which were some of the first gold fields mined by the Forty-Niners during the California Gold Rush. Included are not only details on early mining methods in the area, production statistics and geological information, but also the location of the early gold mines that helped make California "The Golden State". Also included are insights into the early mining of chrome, copper and other minerals in this important mining area. **8.5" X 11", 204 ppgs. Retail Price: $19.99**

Mines of Los Angeles, Orange and Riverside Counties, California

Originally published in 1917, this important publication on California Mining has not been available for nearly a century. This publication includes important information on the early gold mines of Los Angeles County, Orange County and Riverside County, which were some of the first gold fields mined in California by early Spanish and Mexican miners before the 49ers came on the scene. Included are not only details on early mining methods in the area, production statistics and geological information, but also the location of the early gold mines that helped make California "The Golden State". **8.5" X 11", 146 ppgs. Retail Price: $12.99**

Mines of San Bernadino and Tulare Counties, California

Originally published in 1917, this important publication on California Mining has not been available for nearly a century. This publication includes important information on the early gold mines of San Bernadino and Tulare County, which were some of the first gold fields mined in California by early Spanish and Mexican miners before the 49ers came on the scene. Included are not only details on early mining methods in the area, production statistics and geological information, but also the location of the early gold mines that helped make California "The Golden State". Also included are details on the mining of other minerals such as copper, iron, lead, zinc, manganese, tungsten, vanadium, asbestos, barite, borax, cement, clay, dolomite, fluospar, gem stones, graphite, marble, salines, petroleum, strontium, talc and others. **8.5" X 11", 200 ppgs. Retail Price: $19.99**

Chromite Mining in The Klamath Mountains of California and Oregon

Unavailable since 1919, this publication was originally compiled by J.S. Diller of the United States Department of Geological Survey and includes details on the chromite mines of this area of Northern California and Southern Oregon. Also included are insights into the mineralization and other characteristics of this important mining region, as well as the location of historic mines. Also included are insights into chromite mining in Eastern Oregon and Montana. **8.5" X 11", 98 ppgs. Retail Price: $9.99**

Mines and Mining in Amador, Calaveras and Tuolumne Counties, California

Unavailable since 1915, this publication was originally compiled by William Tucker and includes details on the mines and mineral resources of this important California mining area. Included are details on the geology, history and important gold mines of the region, as well as insights into other local mineral resources such as asbestos, clay, copper, talc, limestone and others. Also included are insights into the mineralization and other characteristics of this important portion of California's Mother Lode mining region. **8.5" X 11", 198 ppgs. Retail Price: $14.99**

Alaska Mining Books

Ore Deposits of the Willow Creek Mining District, Alaska

Unavailable since 1954, this hard to find publication includes valuable insights into the Willow Creek Mining District near Hatcher Pass in Alaska. The publication includes insights into the history, geology and locations of the well known mines in the area, including the Gold Cord, Independence, Fern, Mabel, Lonesome, Snowbird, Schroff-O'Neil, High Grade, Marion Twin, Thorpe, Webfoot, Kelly-Willow, Lane, Holland and others. **8.5" X 11", 96 ppgs. Retail Price: $9.99**

Arizona Mining Books

Mines and Mining in Northern Yuma County Arizona

Originally published in 1911, this important publication on Arizona Mining has not been available for over a hundred years. Included are rare insights into the gold, silver, copper and quicksilver mines of Yuma County, Arizona together with hard to find maps and photographs. Some of the mines and mining districts featured include the Planet Copper Mine, Mineral Hill, the Clara Consolidated Mine, Viati Mine, Copper Basin prospect, Bowman Mine, Quartz King, Billy Mack, Carnation, the Wardwell and Osbourne, Valensuella Copper, the Mariquita, Colonial Mine, the French American, the New York-Plomosa, Guadalupe, Lead Camp, Mudersbach Copper Camp, Yellow Bird, the Arizona Northern (Salome Strike), Bonanza (Harqua Hala), Golden Eagle, Hercules, Socorro and others. **8.5" X 11", 144 ppgs. Retail Price: $11.99**

The Aravaipa and Stanley Mining Districts of Graham County Arizona

Originally published in 1925, this important publication on Arizona Mining has not been available for nearly ninety years. Included are rare insights into the gold and silver mines of these two important mining districts, together with hard to find maps. **8.5" X 11", 140 ppgs. Retail Price: $11.99**

More Mining Books

Prospecting and Developing A Small Mine

Topics covered include the classification of varying ores, how to take a proper ore sample, the proper reduction of ore samples, alluvial sampling, how to understand geology as it is applied to prospecting and mining, prospecting procedures, methods of ore treatment, the application of drilling and blasting in a small mine and other topics that the small scale miner will find of benefit. **8.5" X 11", 112 ppgs, Retail Price: $11.99**

Timbering For Small Underground Mines

Topics covered include the selection of caps and posts, the treatment of mine timbers, how to install mine timbers, repairing damaged timbers, use of drift supports, headboards, squeeze sets, ore chute construction, mine cribbing, square set timbering methods, the use of steel and concrete sets and other topics that the small underground miner will find of benefit. This volume also includes twenty eight illustrations depicting the proper construction of mine timbering and support systems that greatly enhance the practical usability of the information contained in this small book. **8.5" X 11", 88 ppgs. Retail Price: $10.99**

Timbering and Mining

A classic mining publication on Hard Rock Mining by W.H. Storms. Unavailable since 1909, this rare publication provides an in depth look at American methods of underground mine timbering and mining methods. Topics include the selection and preservation of mine timbers, drifting and drift sets, driving in running ground, structural steel in mine workings, timbering drifts in gravel mines, timbering methods for driving shafts, positioning drill holes in shafts, timbering stations at shafts, drainage, mining large ore bodies by means of open cuts or by the "Glory Hole" system, stoping out ore in flat or low lying veins, use of the "Caving System", stoping in swelling ground, how to stope out large ore bodies, Square Set timbering on the Comstock and its modifications by California miners, the construction of ore chutes, stoping ore bodies by use of the "Block System", how to work dangerous ground, information on the "Delprat System" of stoping without mine timbers, construction and use of headframes and much more. This volume provides a reference into not only practical methods of mining and timbering that may be employed in narrow vein mining by small miners today, but also rare insights into how mines were being worked at the turn of the 19th Century. **8.5" X 11", 288 ppgs. Retail Price: $24.99**

A Study of Ore Deposits For The Practical Miner

Mining historian Kerby Jackson introduces us to a classic mining publication on ore deposits by J.P. Wallace. First published in 1908, it has been unavailable for over a century. Included are important insights into the properties of minerals and their identification, on the occurrence and origin of gold, on gold alloys, insights into gold bearing sulfides such as pyrites and arsenopyrites, on gold bearing vanadium, gold and silver tellurides, lead and mercury tellurides, on silver ores, platinum and iridium, mercury ores, copper ores, lead ores, zinc ores, iron ores, chromium ores, manganese ores, nickel ores, tin ores, tungsten ores and others. Also included are facts regarding rock forming minerals, their composition and occurrences, on igneous, sedimentary, metamorphic and intrusive rocks, as well as how they are geologically disturbed by dikes, flows and faults, as well as the effects of these geologic actions and why they are important to the miner. Written specifically with the common miner and prospector in mind, the book will help to unlock the earth's hidden wealth for you and is written in a simple and concise language that anyone can understand. **8.5" X 11", 366 ppgs. Retail Price: $24.99**

www.ingramcontent.com/pod-product-compliance
Lightning Source LLC
Chambersburg PA
CBHW081115170526
45165CB00008B/2455